Lecture Notes in Computer Science 15609

Founding Editors

Gerhard Goos
Juris Hartmanis

Editorial Board Members

Elisa Bertino, *Purdue University, West Lafayette, IN, USA*
Wen Gao, *Peking University, Beijing, China*
Bernhard Steffen ®, *TU Dortmund University, Dortmund, Germany*
Moti Yung ®, *Columbia University, New York, NY, USA*

The series Lecture Notes in Computer Science (LNCS), including its subseries Lecture Notes in Artificial Intelligence (LNAI) and Lecture Notes in Bioinformatics (LNBI), has established itself as a medium for the publication of new developments in computer science and information technology research, teaching, and education.

LNCS enjoys close cooperation with the computer science R & D community, the series counts many renowned academics among its volume editors and paper authors, and collaborates with prestigious societies. Its mission is to serve this international community by providing an invaluable service, mainly focused on the publication of conference and workshop proceedings and postproceedings. LNCS commenced publication in 1973.

Bing Xue · Luca Manzoni · Illya Bakurov
Editors

Genetic Programming

28th European Conference, EuroGP 2025
Held as Part of EvoStar 2025
Trieste, Italy, April 23–25, 2025
Proceedings

Editors
Bing Xue
Victoria University of Wellington
Wellington, New Zealand

Luca Manzoni
University of Trieste
Trieste, Italy

Illya Bakurov
Michigan State University
East Lansing, MI, USA

ISSN 0302-9743 ISSN 1611-3349 (electronic)
Lecture Notes in Computer Science
ISBN 978-3-031-89990-4 ISBN 978-3-031-89991-1 (eBook)
https://doi.org/10.1007/978-3-031-89991-1

© The Editor(s) (if applicable) and The Author(s), under exclusive license
to Springer Nature Switzerland AG 2025, corrected publication 2025

This work is subject to copyright. All rights are solely and exclusively licensed by the Publisher, whether the whole or part of the material is concerned, specifically the rights of translation, reprinting, reuse of illustrations, recitation, broadcasting, reproduction on microfilms or in any other physical way, and transmission or information storage and retrieval, electronic adaptation, computer software, or by similar or dissimilar methodology now known or hereafter developed.
The use of general descriptive names, registered names, trademarks, service marks, etc. in this publication does not imply, even in the absence of a specific statement, that such names are exempt from the relevant protective laws and regulations and therefore free for general use.
The publisher, the authors and the editors are safe to assume that the advice and information in this book are believed to be true and accurate at the date of publication. Neither the publisher nor the authors or the editors give a warranty, expressed or implied, with respect to the material contained herein or for any errors or omissions that may have been made. The publisher remains neutral with regard to jurisdictional claims in published maps and institutional affiliations.

This Springer imprint is published by the registered company Springer Nature Switzerland AG
The registered company address is: Gewerbestrasse 11, 6330 Cham, Switzerland

If disposing of this product, please recycle the paper.

Preface

This volume contains the proceedings of EuroGP 2025, the 28th European Conference on Genetic Programming. The conference was part of Evo*, the leading event on bio-inspired computation in Europe, and was held in Trieste, Italy, as a hybrid event, between Wednesday, April 23, and Friday, April 25, 2025.

EuroGP is the premier annual conference on Genetic Programming (GP), the oldest and the only meeting worldwide devoted specifically to this branch of Evolutionary Computation. At the same time, under the Evo* umbrella, EvoAPPS focused on the applications of Evolutionary Computation, EvoCOP targeted Evolutionary Computation in combinatorial optimization, and EvoMUSART was dedicated to evolved and bio-inspired music, sound, art, and design. The proceedings for these co-located events are available in the LNCS series.

Genetic Programming (GP) is a unique branch of Evolutionary Computation that has to automatically solve design problems, in particular computer program design, without requiring the user to know or specify the form or structure of the solution in advance. It uses the principles of Darwinian evolution to approach problems in the synthesis, improvement, and repair of computer programs. The universality of computer programs and their importance in so many areas of our lives means that automating these tasks is an exceptionally ambitious challenge with far-reaching implications. It has attracted a very large number of researchers, and a vast amount of theoretical and practical contributions are available by consulting the GP bibliography[1].

Since the first EuroGP event in Paris in 1998, EuroGP has been the only conference exclusively devoted to the evolutionary design of computer programs and other computational structures. In fact, EuroGP represents one of the largest venues at which GP researchers meet. It plays an important role in the success of the field by serving as a forum for expressing new ideas, meeting fellow researchers, and initiating collaborations. It attracts scholars from all over the world. In a friendly and welcoming atmosphere authors present the latest advances in the field, also presenting GP-based solutions to complex real-world problems.

EuroGP 2025 received 27 submissions from around the world. The articles underwent a rigorous double-blind peer review process, each reviewed by at least three members of the Program Committee and a senior member of the Program Committee.

We selected 10 of these papers for full oral presentation, while 5 works were presented in short oral presentations and as posters. In 2025, papers submitted to EuroGP could also be assigned to the "Evolutionary Machine Learning Track", with one of them accepted for a short oral presentation. The authors of both categories of papers also had the opportunity to present their work in poster sessions to promote the exchange of ideas in a carefree manner. All accepted contributions, regardless of the presentation format, appear as full papers in this volume.

---

[1] http://gpbib.cs.ucl.ac.uk.

An event of this kind would not be possible without the contribution of a large number of people:

- We express our gratitude to the authors for submitting their works and to the members of the Program Committee for devoting selfless effort to the review process.
- We also thank Nuno Lourenço (University of Coimbra, Portugal) for his dedicated work as Submission System Coordinator.
- We thank the Evo* Graphic Identity Team, Sérgio Rebelo and Jéssica Parente (University of Coimbra, Portugal), for their dedication and excellence in graphic design.
- We are grateful to Francisco Chicano (University of Málaga, Spain) and João Correia (University of Coimbra, Portugal) for their impressive work managing and maintaining the Evo* website and handling the publicity, respectively.
- We credit the invited keynote speakers, Tea Tušar (Jožef Stefan Institute, Slovenia) and Daniela Besozzi (University of Milano-Bicocca, Italy), for their fascinating and inspiring presentations.
- We would like to express our gratitude to the Steering Committee of EuroGP for helping to organize the conference.
- Special thanks to Luca Manzoni, Eric Medvet, Giorgia Nadizar, and Gloria Pietropolli (University of Trieste, Italy) as local organizers and to the University of Trieste, Italy, for organizing and providing an enriching conference venue.
- We are grateful for the support provided by SPECIES, the Society for the Promotion of Evolutionary Computation in Europe and its Surroundings, for the coordination and financial administration.

Finally, we express our continued appreciation to Anna I. Esparcia-Alcázar, from SPECIES, Europe, whose considerable efforts in managing and coordinating Evo* helped build a unique, vibrant, and friendly atmosphere.

April 2025

Bing Xue
Luca Manzoni
Illya Bakurov

Organization

Program Chairs

Bing Xue	Victoria University of Wellington, New Zealand
Luca Manzoni	University of Trieste, Italy

Publication Chair

Illya Bakurov — Michigan State University, USA

Local Chairs

Luca Manzoni	University of Trieste, Italy
Eric Medvet	University of Trieste, Italy
Giorgia Nadizar	University of Trieste, Italy
Gloria Pietropolli	University of Trieste, Italy

Publicity Chair

João Correia — University of Coimbra, Portugal

Conference Administration

Anna Esparcia-Alcázar — Universitat Politècnica de València, Spain

Program Committee

Illya Bakurov	Michigan State University, USA
Ying Bi	Victoria University of Wellington, New Zealand
Stefano Cagnoni	University of Parma, Italy
Mauro Castelli	Universidade NOVA de Lisboa, Portugal
Qi Chen	Victoria University of Wellington, New Zealand
Ernesto Costa	University of Coimbra, Portugal

Sylvain Cussat-Blanc	University of Toulouse, France
Andrea De Lorenzo	University of Trieste, Italy
Mario Giacobini	University of Turin, Italy
Steven Gustafson	Noonum Inc., USA
Jin-Kao Hao	University of Angers, France
Malcolm Heywood	Dalhousie University, Canada
Ting Hu	Queen's University, Canada
Zhixing Huang	Victoria University of Wellington, New Zealand
Giovanni Iacca	University of Trento, Italy
Domagoj Jakobović	University of Zagreb, Croatia
W. B. Langdon	University College London, UK
Andrew Lensen	Victoria University of Wellington, New Zealand
Nuno Lourenço	University of Coimbra, Portugal
Evelyne Lutton	INRAE, France
Penousal Machado	University of Coimbra, Portugal
Luca Manzoni	University of Trieste, Italy
Luca Mariot	University of Twente, Netherlands
James McDermott	University of Galway, Ireland
Eric Medvet	University of Trieste, Italy
Yi Mei	Victoria University of Wellington, New Zealand
Alberto Moraglio	University of Birmingham, UK
Aidan Murphy	University College Dublin, Ireland
Stjepan Picek	Radboud University, Netherlands
Gloria Pietropolli	University of Trieste, Italy
Peter Rockett	The University of Sheffield, UK
Lukas Sekanina	Brno University of Technology, Czech Republic
Sara Silva	Universidade de Lisboa, Portugal
Moshe Sipper	Ben-Gurion University of the Negev, Israel
Ernesto Tarantino	ICAR-CNR, Italy
Andrea Tettamanzi	Université Côte d'Azur, France
Leonardo Vanneschi	Universidade NOVA de Lisboa, Portugal
Man Leung Wong	Lingnan University, China
Bing Xue	Victoria University of Wellington, New Zealand
Mengjie Zhang	Victoria University of Wellington, New Zealand

Contents

Long Presentations

Ghost Swarms: Learning Swarm Rules from Environmental Changes Alone 1
   Khulud Alharthi, Zahraa S. Abdallah, and Sabine Hauert

A Systematic Evaluation of Evolving Highly Nonlinear Boolean Functions
in Odd Sizes 18
   *Claude Carlet, Marko Đurasević, Domagoj Jakobović, Stjepan Picek,
and Luca Mariot*

Exploring the Impact of Data Scale on Mutation Step Size in SLIM-GSGP 35
   Davide Farinati, Gloria Pietropolli, and Leonardo Vanneschi

Multi-objective Evolutionary Design of Explainable EEG Classifier 52
   Martin Hurta, Anna Ovesna, Vojtech Mrazek, and Lukas Sekanina

On the Effectiveness of Crossover Operators in Cartesian Genetic
Programming 68
   *Mark Kocherovsky, Marzieh Kianinejad, Illya Bakurov,
and Wolfgang Banzhaf*

Population Diversity, Information Theory and Genetic Improvement 85
   William B. Langdon and David Clark

Introducing Crossover in SLIM-GSGP 103
   *Gloria Pietropolli, Davide Farinati, Luca Manzoni, Mauro Castelli,
Sara Silva, and Leonardo Vanneschi*

Exploring the Integration of Cellular Structures in Genetic
Programming-Based Methods 120
   *Luigi Rovito, Lorenzo Bonin, Davide Farinati, Leonardo Vanneschi,
Luca Manzoni, Andrea De Lorenzo, and Gloria Pietropolli*

Ant-Based Metaheuristics Struggle to Solve the Cartesian Genetic
Programming Learning Task 139
   Julian Trautwein, Michael Heider, Henning Cui, and Jörg Hähner

Designing Lookahead Relocation Rules for the Container Relocation
Problem with Genetic Programming 156
   Marko Đurasević, Mateja Đumić, Francisco Javier Gil-Gala,
   and Domagoj Jakobović

Short Presentations

Evolved and Transparent Pipelines for Biomedical Image Classification 173
   Camilo De La Torre, Giorgia Nadizar, Yuri Lavinas, Robin Schwob,
   Camille Franchet, Hervé Luga, Dennis Wilson, and Sylvain Cussat-Blanc

Unified Piecewise Symbolic Regression 190
   Guillaume Doquet

Was Tournament Selection All We Ever Needed? A Critical Reflection
on Lexicase Selection 207
   Alina Geiger, Martin Briesch, Dominik Sobania, and Franz Rothlauf

The Role of Stepping Stones in MAP-Elites: Insights from Search
Trajectory Networks 224
   Giorgia Nadizar, Francesco Rusin, Eric Medvet, and Gabriela Ochoa

Micro-step Time-Series Regression: Insights from System Identification
Using Symbolic Regression 240
   Hengzhe Zhang, Alberto Tonda, Qi Chen, Bing Xue, Evelyne Lutton,
   and Mengjie Zhang

Correction to: Introducing Crossover in SLIM-GSGP C1
   Gloria Pietropolli, Davide Farinati, Luca Manzoni, Mauro Castelli,
   Sara Silva, and Leonardo Vanneschi

Author Index 257

Ghost Swarms: Learning Swarm Rules from Environmental Changes Alone

Khulud Alharthi[1,3]($\boxtimes$)[iD], Zahraa S. Abdallah[2][iD], and Sabine Hauert[1,2][iD]

[1] Bristol Robotics Laboratory, University of Bristol, Bristol, UK
khulud.alharthi@bristol.ac.uk
[2] School of Engineering Mathematics and Technology, University of Bristol, Bristol, UK
[3] Department of Computer Science, College of Computers and Information Technology, Taif University, Taif, Saudi Arabia

Abstract. Swarm behaviours emerge from agents interacting with their local environment following simple rules. While directly observing each agent can be challenging, their collective behaviour leaves detectable environmental imprints that could offer insights into the underlying swarm dynamics. However, this task is complex due to the hidden and interconnected relationships between the rules governing agent interactions, the emergent swarm behaviour, and the environmental changes generated by this behaviour. In this work, we propose a method for extracting human-readable controllers from demonstrations showing only observable environmental imprints caused by the swarm. This approach explores whether these environmental imprints can reveal the swarm's actions, even when the individual agents are challenging to track. Our approach eliminates the need for prior knowledge about the controller or its structure, enabling the successful learning of controllers from a single demonstration. We provide a novel method for understanding and managing both natural and engineered swarms by utilising the environmental imprints left by swarm behaviours, even when direct observation of the swarm's actions is not feasible.

Keywords: Swarm Behaviour · Imitation Learning · Environmental Imprints

1 Introduction

Swarm behaviour emerges when individual agents within a swarm interact with their neighboring agents and the surrounding environment, guided by rules that each agent follows independently, without knowledge of the resulting collective behaviour at the swarm level. Examples of such behaviours from nature, including bird flocking, fish schooling, and bee foraging, have inspired the design of robot swarms valued for their robustness, adaptability, and scalability [5,26].

Z. S. Abdallah, S. Hauert—Both authors have contributed equally to the work.

Swarm robotics can be applied in various scenarios, including search and rescue operations, environmental surveillance, logistics management, and infrastructure inspection [6,29]. Previous research in this area has primarily focused on understanding the link between an agent's local rules and the emergent collective behaviour [3,5,8,9,12,23]. However, what if we cannot directly observe the swarm's behaviour and only have access to its impact on the environment? Could these environmental imprints serve as indicators of the swarm behaviour? This question is relevant in cases where it is challenging or impossible to track individual agents within the swarm. In biological systems, such as cells, ants, and bees, or applications involving micro- and nano-robots, monitoring each agent's behaviour is often unfeasible. However, swarm behaviour typically leaves observable environmental imprints or changes in the surroundings [4,11,15,31,33,34]. These environmental imprints could provide focused, informative metrics that reveal aspects of the swarm's underlying collective behaviour. Imitation learning enables the extraction of swarm controllers by observing demonstrations of the desired emergent swarm behaviour. If we have a demonstration where the swarm is like a ghost, visible only through the traces it leaves in the environment as shown in Fig. 1, we can attempt to extract its underlying rules from this demonstration. Extracting rules from environmental imprints offers a valuable approach to understanding both natural swarms and developing robotic swarms. When the environmental imprints of natural swarms are observable, it can provide insights into how these complex behaviours emerge from simple local interactions. This is vital when studying behaviours where individual observa-

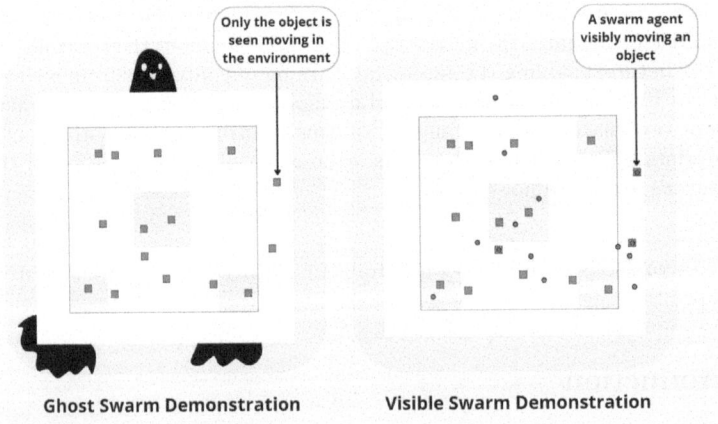

Fig. 1. An illustration of a demonstration showing a swarm behaving like a "ghost", visible only through the traces it leaves in the environment, alongside a normally visible swarm. In this environment, swarm agents are illustrated as purple circles, with five green squares representing areas and 15 small blue squares representing objects in the environment. This visualisation highlights how environmental imprints can be used to infer the underlying rules of the swarm. (Color figure online)

tion of each agent is impossible. This method is also beneficial for robotic swarms since it can be used to design autonomous swarms that rely on environmental feedback to adapt to novel situations without requiring specific programming for every possible scenario. Especially when designing robot swarms for situations where direct control or communication between agents is limited, such as in remote or hazardous environments. Since the extracted swarm controllers are human-readable, they can further help us better understand both natural and artificial swarms and enable swarm engineers to understand and control swarm robots in challenging environments. This work introduces a method for learning swarm controllers from demonstrations where only environmental imprints are observable. In a prior study, we developed a method to extract understandable controllers from a single swarm behaviour demonstration [2]. We now develop this method to learn from a demonstration of a "ghost swarm" in an environment containing both static elements (areas) and movable elements (objects). The method does not require prior knowledge of the controller or its structure; instead, it constructs controllers using a list of primitive nodes. While this list does not cover all possible swarm behaviours, it represents a significant portion. The environmental imprints from the original demonstration are compared to those generated by the method, and the difference between these imprints is minimized using a genetic programming algorithm. Moreover, the framework is scalable, allowing the extraction of additional behaviours by expanding its set of leaf nodes and metrics.

The paper is organised as follows. Section 2 introduces an overview of the related works. Section 3 presents the components of the proposed learning framework. Our results are analysed and discussed in Sect. 4.

2 Related Works

Tracking individual agents within a swarm can be challenging, and biological examples, such as ants, bees, and cells, demonstrate how emergent behaviour from local interactions can leave observable environmental imprints without direct monitoring. For example, in biological swarms like ant colonies or bee hives, individual agents follow simple rules based on local interactions, and the collective result can be seen in how they organise their environments. Ants, for instance, exhibit emergent behaviours such as foraging or nest-building, which can be observed in the paths they create through pheromone trails or the structures they build [11,33]. Similarly, bees communicate and make collective decisions about tasks like food foraging, which affects their environment, leaving traces such as the formation of hives [31]. This coordination often involves stigmergy, a form of indirect communication where agents leave environmental traces that trigger subsequent behaviours by themselves or others [24]. At the cellular level, bacteria are a collective group of cells that move and grow together, resulting in patterns such as floral structures. These patterns can be seen as environmental imprints created by the interactions within the bacterial colony [34]. In the context of micro/nano swarms, individual agents sometimes operate in ways that are difficult to track accurately. Tracking is essential feedback for controlling these systems, as inaccurate feedback can lead to incorrect control inputs, potentially destabilising the swarm or causing navigation failures. To enhance swarm control, incorporating information about the surrounding environment can help improve both the efficiency and accuracy of control [4,15]. These examples from biology and robotics illustrate that swarms can imprint observable changes in their environments, even when individual agents may be challenging to track directly. Thus, the environmental imprints they leave can provide essential insight about the swarm dynamics. This information can be leveraged in various ways, including imitation learning. For instance, if we treat a demonstration as a "ghost swarm", where only environmental changes are visible without the agents themselves, imitation learning could use these cues to infer and replicate swarm behaviour. In imitation learning, a demonstration of swarm behaviour is used as an input to train a controller to imitate the demonstrated behaviour [20]. This can be achieved through behaviour cloning, which employs supervised learning to connect sensory data with corresponding actions as shown in a demonstration [14,21,28,36]. Alternatively, inverse reinforcement learning can be used, where a reward function is constructed from the swarm demonstration to guide the learning process [10,32,35]. Another approach, Turing learning, aims to reproduce the original behaviour by co-evolving a population of controllers and discriminators, with the discriminators' goal being to distinguish between original and imitated behaviours [22]. In a biological context, video observations are utilised to study and interpret swarm behaviours by analysing individual trajectories across various species, providing an understanding of individual actions and their significance in shaping emergent behaviours [3,8,12,23]. For instance, trajectories of ants are examined to explore the obstacle navigation strategies they use during cooperative transportation [25]. These methods are employed to study or imitate observed swarm behaviours. Behaviour trees can serve as controllers for both purposes due to their readability and executable nature [13,17]. Extracting controllers in this format can offer valuable insights for human operators of robot swarms, boosting their ability to adapt and manage control in complex environments [19]. This work focuses on learning controllers only from tracking environmental changes in an environment with two interacting elements (objects and areas). It aims to extract high-level, modular, and easily adaptable rules that can be applied to external systems. Our method employs a set of primitives to construct 50 simulated behaviour trees representing a diverse range of swarm behaviors, encompassing tasks such as object assembly, foraging, and adaptive behavior shifts based on the presence or absence of specific areas. These trees are used to generate demonstrations of swarm behaviours. In these demonstrations, only the environmental imprints left by the swarm serve as input for a learning process, which then reconstructs the corresponding behaviour trees. Our contributions are as follows: first, we present a method to extract behaviour trees from environments with two interacting elements, second, we introduce metrics that capture the co-change of two environmental components driven by the swarm's activity. Finally, our approach is lightweight, requiring only a single demonstration of the ghost swarm, even if the demonstration does not fully capture the complete ghost swarm behaviour.

3 Methodology

This work presents a method for extracting swarm controllers from demonstrations in which only the environmental imprints caused by swarm activities are visible. At the core of the method is genetic programming designed to evolve behaviour trees, which serve as the swarm controllers. The process begins with the random generation of a population of behaviour trees, followed by simulations to execute each behaviour tree and compute its corresponding metrics. The fitness of each behaviour tree is determined by the distance between its metrics and those of the original demonstration. Through iterative selection, mutation, and crossover, the algorithm evolves behaviour trees to closely replicate the demonstrated environmental imprints. Metrics capture interactions among environmental elements (areas and objects). Behaviour trees that serve as controllers have their structure evolved from the leaf nodes list up to depth 3. The general framework of the proposed method is depicted in Fig. 2. The metrics list includes object-specific metrics and metrics that describe the co-change between two environmental components: objects and areas. The behaviour tree is constructed from a set of leaf nodes consisting of motion nodes, nodes that respond to neighboring robots, and nodes reacting to objects or areas in the environment. This section details the core components of the proposed framework for extracting the swarm controller. We then explore the dataset and the metrics employed to evaluate our method.

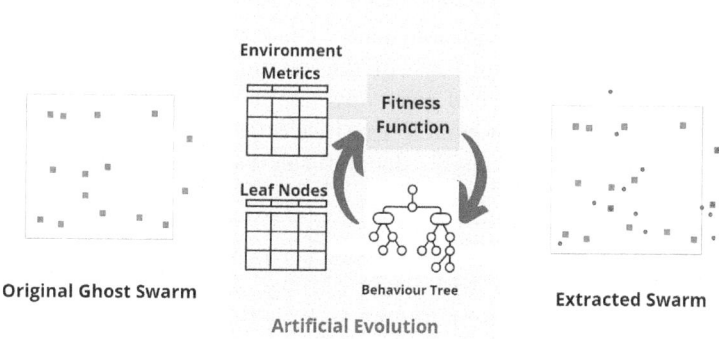

Fig. 2. The proposed method extracts swarm controllers from demonstrations where only environmental imprints are visible as if the swarm were a ghost. Behaviour trees are evolved using metrics that capture these environmental imprints, with the trees constructed from leaf nodes, including motion and response nodes.

3.1 Behaviour Tree Controller

In this work, swarm controllers are designed as behaviour trees. These behaviour trees are implemented to have two types of root nodes: sequence or selection. A sequence node prompts each leaf node from left to right to execute unless any failure happens in any of the leaf nodes. In contrast, a selection node ticks its leaf nodes until success is reported from any of them [17]. To initialise a population for the evolutionary method, we used Ramped Half and Half to randomly generate a list of behaviour trees with varying structures, each with a maximum depth of three [27]. A library of leaf nodes is constructed with two different types of sub-modules. Sub-modules that control the agent's action (action leaf nodes) and sub-modules where the agent senses the local surroundings (condition leaf nodes). Swarm agents use a behaviour tree as their controller to make decisions by selecting an action leaf node. The action leaf nodes provide choices like moving in specific directions, executing random motion, aligning positions with neighbouring robots, and picking or dropping objects. Alternatively, they may decide not to select any action. These decisions are reactive, influenced by sensing the local surroundings, as the behaviour tree integrates condition leaf nodes to assess neighbouring robots and nearby environmental elements such as objects, areas, and boundaries. Through these reactions, we anticipate the emergence of diverse swarm behaviours. Leaf nodes used to form behaviour trees in this work are listed as follows:

- **Motion nodes:** has five action nodes that direct motion in the random, southeast, southwest, northeast, and northwest directions.
- **Neighbour nodes:** has two action nodes (Aggregation and Repulsion) and one condition node (No Neighbour?) that check neighbors' existence in the local radius.
- **Boundary force node:** one action node that implements a force to remain within the defined boundaries.
- **Area nodes:** has two condition nodes: (Inside area?), which checks if the robot is located within any area of the environment, and (Sensing area?), which reveals if the robot can sense any area of the environment.
- **Object nodes:** has two action nodes (Pick object and Drop object) and one condition node (Picked object?) that returns true if the robot has picked up and moved an object.

Examples of behaviour trees that can be generated from leaf node options are shown in Fig. 3. As behaviour trees evolve through generations, they can become complex and difficult to read or interpret. To address this, we implement tree pruning both during and after evolution. At the end of each fitness computation, we remove tree nodes that were never ticked since they do not contribute to the produced swarm behaviour. Additionally, to avoid overly complex trees, we introduce a penalty to the fitness score, calculated at 0.003 per node. Post-evolution, trees undergo a pruning process to reduce them to their essential form.

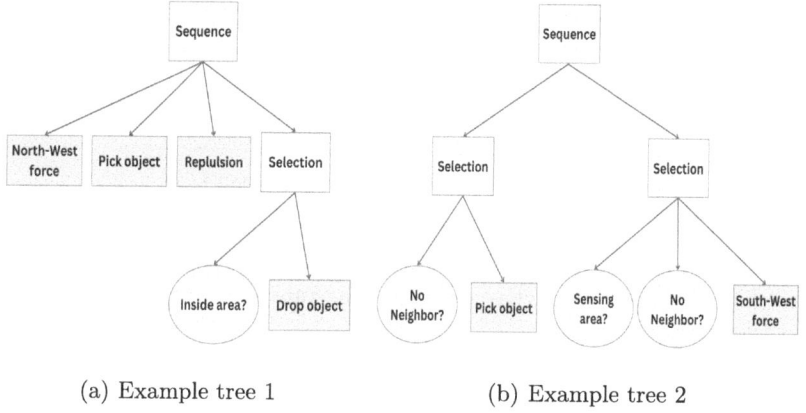

(a) Example tree 1 (b) Example tree 2

Fig. 3. Examples behaviour trees generated from leaf node options, which serve as controllers for swarm agents. These trees enable collective behaviours within an environment with objects and designated areas. The root node is illustrated as a green square, action nodes as blue squares, and condition nodes as yellow circles. (Color figure online)

Simulation Environment. A 2D simulation environment is built using C++ and Python Matplotlib to generate swarm behaviour and track its traces in the environment for each behaviour tree. Robots in this simulation are modeled off of the DOTS robots in our laboratory [16]. The simulation environment includes a square area (5 m × 5 m). Swarm robots are simulated as a circle with a radius of 12.5 cm and a sensory range of 22.5 cm. Robots can move in any direction based on the velocity vector resulting from the execution of the behaviour tree, with a speed equal to 25 cm/s. The update frequency of behaviour tree is once every 0.1 s. Each swarm controller is simulated for 600 time steps (60 s), where at each time step, the behaviour tree is ticked to update the swarm behaviour. The environmental components are set in advance to consist of five square areas: four located at the corners of the arena, each with a width of 100 cm, and one positioned centrally with a width of 150 cm. Furthermore, the environment also contains 15 square objects, each with a width of 14 cm, distributed within, near, and outside of the designated areas. The original ghost swarm demonstration was produced using 20 swarm robots. To evaluate each behaviour tree in the population, we vary the swarm size randomly between 15 and 25 and assign random initial robot positions. This variation reflects the assumption that we lack access to exact swarm details, such as the number of robots or their starting positions, and can only observe the resulting environmental changes.

3.2 Learning Controllers from Demonstrations via Genetic Programming

Genetic programming (GP) has been used to evolve behaviour trees using operations that consider their hierarchical structure [13,17]. In this work, our goal is to minimise the fitness.

Fitness Function. The fitness function evaluates how closely the produced environmental imprints match the original ones by calculating the Euclidean distance between their respective metrics. The maximum of these distances is used as the fitness score 1.

$$Fitness = \max\left\{\text{distance}(i) = \sqrt{\sum_t (\text{Original SM}_{it} - \text{Assessed SM}_{it})^2} : i = 1,\ldots,n\right\} \quad (1)$$

where:

- **Original SM:** A set of metrics describing the environmental imprints in the original demonstration caused by the invisible swarm.
- **Assessed SM:** A set of metrics describing the environmental imprints produced by simulating a behavior tree for evaluation.
- i: The index of a specific metric from the metrics set, represented as a time-series.
- t: The time-step. The demonstration consists of snapshots of the environment taken every 2 s rather than continuous data. Therefore, each metric is compared at intervals of $t = 20$ time-steps.

Environmental imprints refer to modifications in the surrounding environment resulting from the collective behaviour of the swarm. These imprints are captured using a set of metrics. A description of these metrics is provided in the following list. In this description, each object is defined as o where o is a composite of two vectors that store the object's location in the x and the y direction, and s is the total number of objects in the environment. The *distance* is computed using Euclidean distance. Each area is defined as a where a is a composite of two vectors that store the location of the center of the area a in the x and the y direction, and m is the total number of areas in the environment. In our assessment, we utilised two categories of metrics:

1. **Object metrics** assess the swarm behaviour based on the recorded trajectories of objects in the environment.

    1- The average object shift represents the mean positional change of all objects, measured at each time step t.

    $$\frac{1}{n}\sum_{i=1}^{n}\{(o_{i,t} - o_{i,t-1}) : i = 1...n\} \quad (2)$$

    2- Nearest object distance: calculate each object's distance to its nearest neighbour. The metric is the average of these nearest distances for all objects.

    $$\frac{1}{n}\sum_{i=1}^{n}\min\{\text{distance}(o_i, o_j) : j = 1,\ldots,n, j \neq i\} \quad (3)$$

3- The number of moving objects monitored at each time-step.
2. **Object-area metrics** describe the co-change of two environmental components (objects and areas) due to the swarm's activity within the environment. They include:
4- The connectivity index is determined by the number of paths between pairs of areas and objects. A pair of an object and an area is considered to have a connecting path if their distance is less than the average distance of all other pairs.

$$\frac{\sum_{i=1}^{m}\sum_{j=1}^{s}\mathbb{1}_{\{\text{distance}(a_{i,t},o_{j,t})<(\frac{1}{m*s}\sum_{i=1}^{m}\sum_{j=1}^{s}\text{distance}(a_{i,0},o_{j,0}))\}}}{m+s} \quad (4)$$

5- The shortest of the longest paths metric is measured by identifying the shortest distance within a set of path lengths, representing the greatest distance between each area and all objects.

$$min\left\{max\left\{distance(o_i, a_j) : i = 1....s\right\} : j = 1....m\right\} \quad (5)$$

6- The percentage of objects within all areas monitored at each time-step.

All metrics are structured as time-series data, with calculations performed every 20 time-steps. The metric distances are normalised using min-max normalisation to standardise the scales across all metrics. For each metric, the maximum and minimum distances are recorded across the entire population in the first generation. These values are then used to normalise the metric distances throughout the evolutionary process.

Genetic Programming: Core Operators. The elitism strategy is used to copy the best three behaviour trees to the next generation without any change. The remaining individuals are selected using tournament selection with a tournament size of 3. Afterwards, three operators are applied, including single-point crossover, sub-tree mutation, and node mutation. In the single-point crossover, the cross point is chosen randomly, and the two behaviour trees swap their sub-trees. In subtree mutation, a random node in the tree is selected, and the entire subtree rooted at that node is replaced with a randomly generated tree (maximum depth = 2). Node mutation involves selecting a random leaf node and replacing it with another leaf node of the same type. The evolution runs for 50 generations, and the behaviour tree with the best fitness score in the final generation is chosen as the extracted swarm controller. The three operator rates are as shown in Table 1. The evolutionary process here is designed to favor broad exploration while preserving high-quality solutions for future iterations. By refraining from re-evaluating elite solutions, the algorithm treats them as stable benchmarks even as the system explores new alignment with environmental imprints. This balance between exploration and elite retention is essential when the goal is to evolve a swarm controller that can mirror the environmental changes observed in the original demonstration.

Table 1. Genetic Parameters

Parameters	Value
Population size	85
Generation number	50
Elitism size	3
Tournament size	3
Crossover rate	0.4
Sub-tree mutation rate	0.3
Node mutation rate	0.3

3.3 Performance Evaluation

Dataset of Original Behaviour Trees. To evaluate the performance of our learning method, we start by generating 50 random behaviour trees. Each tree is executed to produce a swarm behaviour that causes environmental changes. These changes, represented by simulated object trajectories and area locations, serve as the sole input to the learning algorithm. The method aims to learn a behaviour tree capable of generating swarm behaviour that replicates these observed environmental changes. Notably, our method operates without prior knowledge of the original behaviour tree or its structure.

Performance Evaluation. Learning is considered as successful if it reproduces the original behaviour tree accurately. Since our goal is to learn an understandable controller from which the swarm behaviour emerges that leads to the observed changes in the environment, we evaluate the performance of our method by measuring the similarity between the original controller and the learned controller. Assessing the similarity of the extracted controller to the original behaviour tree in most cases is not a trivial task and requires human reasoning. The final extracted behaviour tree is post-processed to reduce it, making it easier to understand and compare to the original tree [18]. To count the number of identically extracted behaviour trees, we decompose a tree as a set of actions and conditioned actions. We then used the Jaccard index to compute the similarity between the two trees [7]. If tree similarity is one, the two trees are the same, whereas zero indicates they are entirely different. Behaviour trees with a tree similarity below 0.5 fall into the low similarity group, indicating that less than half of the tree is similar. Conversely, those with a similarity of 0.5 or higher are categorized into the high similarity group. This metric serves as ground truth for evaluating the similarity between behaviour trees. However, it cannot be used directly in the fitness function, which relies on indirect metrics derived from demonstrations.

4 Results

4.1 Performance of the Controller Extraction Method

We tested the method with 50 environmental observations generated from 50 behaviour trees. Our method achieved a perfect extraction rate for 26% of the controllers. Additionally, 34% of the controllers were extracted with high similarity, while the remaining 40% were extracted with lower similarity. These results stress the potential of utilising environmental metrics alone to learn behaviour tree controllers from single demonstrations. The results are noteworthy, given the inherent complexity of the problem. Table 2 summarizes the method's performance. We then executed more experimental runs (4 in total) to analyse the robustness of our learning method. Across the runs, the average percentage of behaviour trees extracted perfectly was 20%, with a standard deviation of 3.08 extractions, indicating moderate variability across runs. The average for high-similarity behaviour trees was 38.5%, with a more consistent performance indicated by a lower standard deviation of 2.8 extractions. Behaviour trees extracted with low similarity demonstrated an average of 41.5% and a standard deviation of 1.30 extractions, showing minimal variation across runs. This analysis shows that our experimental setup is reproducible, considering the noise introduced by randomness in the number and initial locations of the swarm agents. The best fitness and average fitness during the learning process for all controllers in the dataset are illustrated in Fig. 4. The figure reveals that the evolution began converging around the 30th generation mark for the best fitness. While the average fitness continued to explore a more extensive solution space, it also demonstrated a notable convergence trend over time. This behaviour highlights the interplay between exploitation and exploration in the optimization process.

Table 2. Performance measures of the controller learning method

Controller Similarity	Rate
Perfect Extraction	26%
High similarity	34%
Low similarity	40%

4.2 Discussion

Learning behaviour tree controllers from demonstrations with only environmental imprints is inherently challenging due to the variability in swarm behaviours and the unknown relationships between individual agents and their impact on the environment. Despite these challenges, our approach demonstrates promising results, highlighting the potential of leveraging environmental metrics as a foundation for learning behaviour tree controllers. Figure 5 illustrates the variance of metrics over time in the original demonstrations, averaged across all

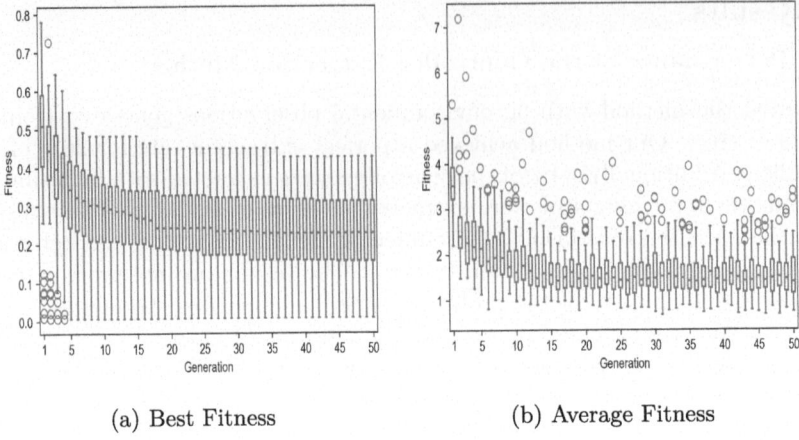

(a) Best Fitness   (b) Average Fitness

Fig. 4. Evolution of best fitness and average fitness during the learning process across all controllers in the dataset. The best fitness begins converging around the 30th generation, indicating early identification of high-quality solutions. The average fitness shows broader exploration initially, with gradual convergence.

controllers and grouped by the extraction categories: perfect and low similarity. The figure shows that low-similarity extractions have environmental metrics with lower variance in the original demonstrations. The figure also illustrates the decrease in distances between the six environmental metrics used in the fitness function and those of the original demonstrations from the first to the last generation, showing that all the metrics were minimised in perfect extractions.

We ran a multinomial logistic regression analysis using average temporal variance across all metrics as the sole predictor for each behaviour tree to predict if it belongs to one of the extraction categories [1,30]. The test shows that temporal variance significantly predicts classification into high similarity and perfect extraction relative to the baseline (low similarity category). Temporal variance has a positive effect, with coefficients of 13.1602 ($p = 0.001$) and 8.8338 ($p = 0.019$) for the high similarity extraction category and perfect extraction category, respectively, indicating that higher temporal variance increases the likelihood of classification into these categories. Conversely, when we ran the test with the decrease in fitness from the first to last generation as the sole predictor, it shows a negative relationship, with coefficients of -7.3111 ($p = 0.012$) for the high similarity extraction category and -8.2891 ($p=0.010$) for the perfect extraction category, suggesting that the higher minimisation of the distance between original and produced metrics, the higher the probability of the controller to be extracted with perfect and high similarity.

These results indicate that extracting controllers with low similarity can be due to demonstrations having minimal motion and remaining primarily in static states. Thus, they provide limited environmental imprints, making it challenging for the method to learn the controllers. On the other hand, when we have more dynamic environmental observations, they offer us richer informative data. Consequently, the method can better utilise environmental metrics, leading to more accurate extractions. Since temporal variance reveals richer information, utilising the variance as a weight for each metric in the fitness function could improve the learning method. However, this addition must be carefully balanced between low-variance metrics with low informative value and very high-variance metrics that lead to controllers that overfit noise.

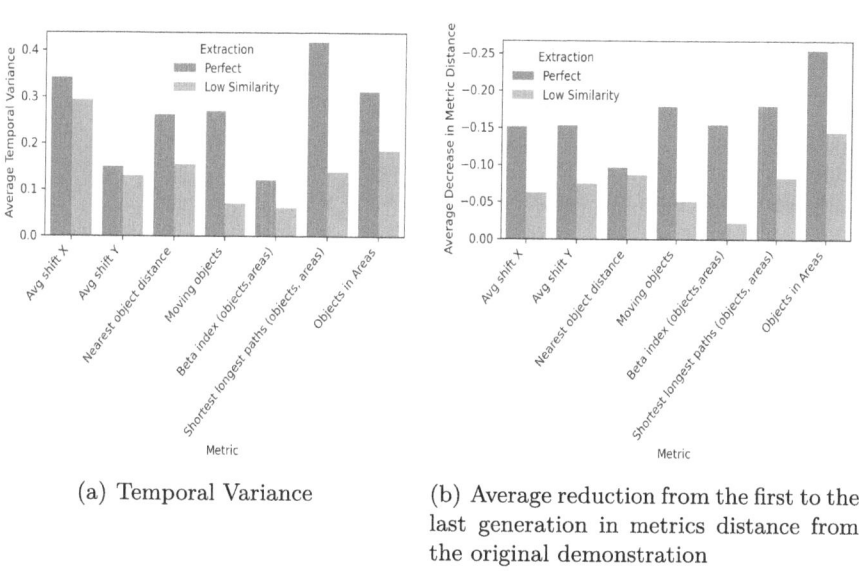

(a) Temporal Variance

(b) Average reduction from the first to the last generation in metrics distance from the original demonstration

Fig. 5. Temporal variance of metrics from original demonstrations grouped by perfect and low similarity extraction categories. Low-similarity extractions show low variance, indicating static states with limited information. Perfect extractions exhibit higher variance, and a more accurate controller can be extracted with better optimization of metrics' alignment.

Figure 6 presents an example of a behaviuor tree, guiding a swarm to move randomly and pick up objects. The object-picking action is associated with a double condition: it occurs only when agents have neighbors and are not sensing any surrounding areas. This behaviour was perfectly extracted, which is promising, considering it is based solely on environmental imprints. The method, in this case, can infer not only when objects are picked but also that the agent's actions are conditioned to have neighbors, an aspect not immediately apparent from the environment alone. Our extraction method relies solely on a single demonstration, yet it may not capture the complete range of behaviour. For instance, in the input demonstration of the extraction case in Fig. 6, object picking occurs at a rate of only 0.03 across all agents throughout the entire duration. More examples are featured in the supplementary video[1].

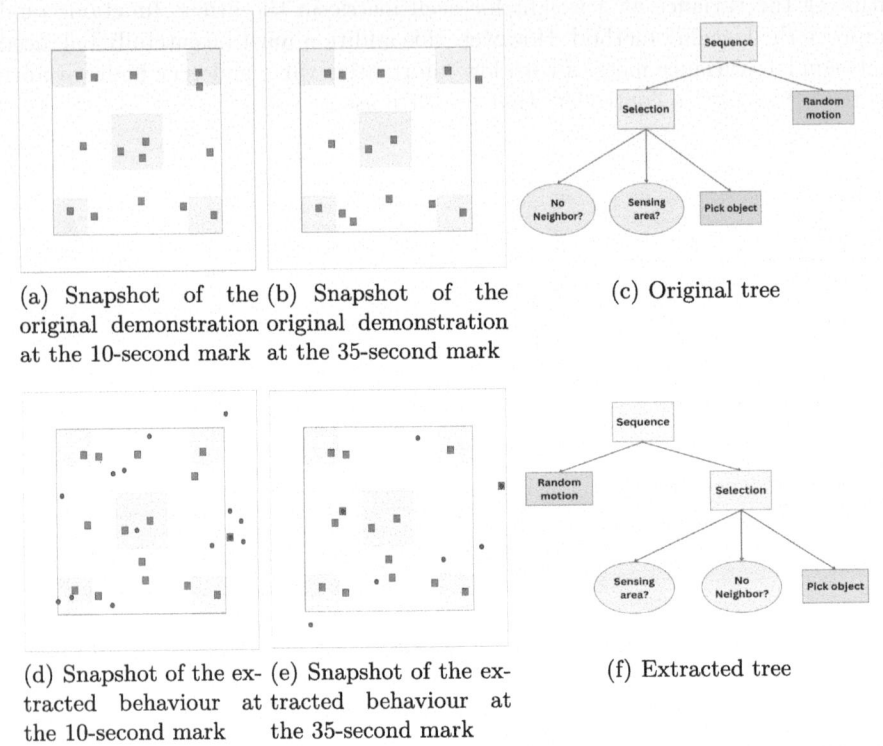

Fig. 6. Example of a behaviour tree producing a swarm that moves randomly and picks objects when the agent has neighbors and does not sense any area zones in their surroundings. The behaviour was perfectly extracted, showing the ability of the method to learn complex interactions from environmental imprints.

5 Conclusion

Extracting the rules from swarm demonstrations is crucial for both the engineering of robot swarms and the interpretability of the swarm dynamics, whether natural or artificial. This study is motivated by the difficulty in understanding swarm behaviour in scenarios where tracking individual agents is impractical, such as with biological cells, ants, and bees, or in high-risk environments. Traditional research on swarm behaviour has focused on linking local agent rules to the resulting collective behaviour. However, this research proposes a different approach: examining the environmental changes to uncover the hidden collective

---

[1] https://youtu.be/8YXzNJjYwv4.

behaviours driving those changes. We present a learning method for extracting understandable swarm controllers from a single demonstration that only includes environmental imprints. Our experiments, using a dataset of 50 simulated behaviour trees as ground truth, show that the method can extract 60% of the behaviours perfectly or with high similarity using a single demonstration of the environmental imprints caused by the swarm. These results demonstrate the method's potential in diverse applications, from robotics to biology.

References

1. Agresti, A.: Analysis of Ordinal Categorical Data, 2nd edn. Wiley, Hoboken (2010)
2. Alharthi, K., Abdallah, Z.S., Hauert, S.: Automatic extraction of understandable controllers from video observations of swarm behaviors. In: Swarm Intelligence, pp. 41–53. Springer, Cham (2022)
3. Amornbunchornvej, C., Berger-Wolf, T.: Framework for inferring following strategies from time series of movement data. ACM Trans. Knowl. Discov. Data **14**(3), 35:1–35:22 (2020)
4. Bijli, M.K., Verma, P., Singh, A.P.: A systematic review on the potency of swarm intelligent nanorobots in the medical field. Swarm Evol. Comput. 101524 (2024)
5. Brambilla, M., Ferrante, E., Birattari, M., Dorigo, M.: Swarm robotics: a review from the swarm engineering perspective. Swarm Intell. **7**(1), 1–41 (2013)
6. Carrillo-Zapata, D., et al.: Mutual shaping in swarm robotics: user studies in fire and rescue, storage organization, and bridge inspection. Front. Robot. AI **7**, 53 (2020)
7. Chung, N., Miasojedow, B., Michał, S., Gambin1, A.: Jaccard/tanimoto similarity test and estimation methods for biological presence-absence data. BMC Bioinform. **20** (2019)
8. Eriksson, A., Nilsson Jacobi, M., Nyström, J., Tunstrøm, K.: Determining interaction rules in animal swarms. Behav. Ecol. **21**(5), 1106–1111 (2010)
9. Francesca, G., Birattari, M.: Automatic design of robot swarms: achievements and challenges. Front. Robot. AI **3**, 29 (2016)
10. Gharbi, I., Kuckling, J., Ramos, D.G., Birattari, M.: Show me what you want: inverse reinforcement learning to automatically design robot swarms by demonstration. In: 2023 IEEE International Conference on Robotics and Automation (ICRA), pp. 5063–5070 (2023)
11. Gordon, D.M.: The ecology of collective behavior. PLoS Biol. **12**(3), e1001805 (2014)
12. Herbert-Read, J.E., Perna, A., Mann, R.P., Schaerf, T.M., Sumpter, D., Ward, A.: Inferring the rules of interaction of shoaling fish. Proc. Natl. Acad. Sci. **108**(46), 18726–18731 (2011)
13. Hogg, E., Hauert, S., Harvey, D., Richards, A.: Evolving behaviour trees for supervisory control of robot swarms. Artif. Life Robot. **25**(4), 569–577 (2020). https://doi.org/10.1007/s10015-020-00650-2
14. Hu, T.K., Gama, F., Chen, T., Wang, Z., Ribeiro, A., Sadler, B.M.: VGAI: end-to-end learning of vision-based decentralized controllers for robot swarms. In: ICASSP 2021 - 2021 IEEE International Conference on Acoustics, Speech and Signal Processing (ICASSP), pp. 4900–4904 (2021)

15. Jiang, J., Yang, L., Zhang, L.: An overview of micro/nanorobot swarm control: from fundamental understanding to autonomy. IEEE/ASME Trans. Mechatron. (2024)
16. Jones, S., Milner, E., Sooriyabandara, M., Hauert, S. DOTS: an open testbed for industrial swarm robotic solutions (2022). https://arxiv.org/abs/2203.13809
17. Jones, S., Studley, M., Hauert, S., Winfield, A.: Evolving behaviour trees for swarm robotics. In: Distributed Autonomous Robotic Systems, pp. 487–501. Springer (2018)
18. Jones, S.W.: Onboard evolution of human-understandable behaviour trees for robot swarms. Ph.D. thesis, University of Bristol (2020)
19. Kolling, A., Sycara, K., Nunnally, S., Lewis, M.: Human swarm interaction: an experimental study of two types of interaction with foraging swarms. J. Hum.-Rob. Interact. **2**(2) (2013)
20. Kuckling, J.: Recent trends in robot learning and evolution for swarm robotics. Front. Robot. AI **10** (2023)
21. Li, J., Tan, Y.: A two-stage imitation learning framework for the multi-target search problem in swarm robotics. Neurocomputing **334**, 249–264 (2019)
22. Li, W., Gauci, M., Groß, R.: Turing learning: a metric-free approach to inferring behavior and its application to swarms. Swarm Intell. **10**(3), 211–243 (2016). https://doi.org/10.1007/s11721-016-0126-1
23. Mann, R.P.: Bayesian inference for identifying interaction rules in moving animal groups. PLoS ONE **6**(8), e22827 (2011)
24. Marsh, L., Onof, C.: Stigmergic epistemology, stigmergic cognition. Cogn. Syst. Res. **9**(1–2), 136–149 (2008)
25. McCreery, H.F., Dix, Z.A., Breed, M.D., Nagpal, R.: Collective strategy for obstacle navigation during cooperative transport by ants. J. Exp. Biol. **219**(21), 3366–3375 (2016)
26. Nedjah, N., Junior, L.S.: Review of methodologies and tasks in swarm robotics towards standardization. Swarm Evol. Comput. **50**, 100565 (2019)
27. Poli, R., Langdon, W.B., McPhee, N.F.: A field guide to genetic programming (2008). http://lulu.com, http://www.gp-field-guide.org.uk. With contributions by JR Koza
28. Schilling, F., Lecoeur, J., Schiano, F., Floreano, D.: Learning vision-based flight in drone swarms by imitation. IEEE Robot. Autom. Lett. **4**(4), 4523–4530 (2019)
29. Schranz, M., Umlauft, M., Sende, M., Elmenreich, W.: Swarm robotic behaviors and current applications. Front. Robot. AI **7**, 36 (2020)
30. Seabold, S., Perktold, J.: Statsmodels: econometric and statistical modeling with python. In: Proceedings of the 9th Python in Science Conference, vol. 57 (2010). https://www.statsmodels.org/
31. Siefert, P., Buling, N., Grünewald, B.: Honey bee behaviours within the hive: insights from long-term video analysis. PLoS ONE **16**(3), e0247323 (2021)
32. Šošić, A., KhudaBukhsh, W.R., Zoubir, A.M., Koeppl, H.: Inverse reinforcement learning in swarm systems. In: Proceedings of the 16th Conference on Autonomous Agents and MultiAgent Systems, AAMAS 2017, pp. 1413–1421. International Foundation for Autonomous Agents and Multiagent Systems, Richland (2017)
33. Werfel, J., Petersen, K., Nagpal, R.: Designing collective behavior in a termite-inspired robot construction team. Science **343**(6172), 754–758 (2014)
34. Xiong, L., Cao, Y., Cooper, R., Rappel, W.J., Hasty, J., Tsimring, L.: Flower-like patterns in multi-species bacterial colonies. eLife **9** (2020)

35. Yu, X., Wu, W., Feng, P., Tian, Y.: Swarm inverse reinforcement learning for biological systems. In: 2021 IEEE International Conference on Bioinformatics and Biomedicine (BIBM), pp. 274–279 (2021)
36. Zhou, S., Phielipp, M.J., Sefair, J.A., Walker, S.I., Amor, H.B.: Clone swarms: learning to predict and control multi-robot systems by imitation. In: 2019 IEEE/RSJ International Conference on Intelligent Robots and Systems (IROS), pp. 4092–4099 (2019)

A Systematic Evaluation of Evolving Highly Nonlinear Boolean Functions in Odd Sizes

Claude Carlet[1,2], Marko Đurasević[3], Domagoj Jakobović[3(✉)], Stjepan Picek[4], and Luca Mariot[5]

[1] Department of Mathematics, Université Paris 8, 2 Rue de la Liberté, 93526 Saint-Denis Cedex, France
[2] University of Bergen, Bergen, Norway
[3] Faculty of Electrical Engineering and Computing, University of Zagreb, Unska 3, Zagreb, Croatia
{marko.durasevic,domagoj.jakobovic}@fer.hr
[4] Digital Security Group, Radboud University, PO Box 9010, Nijmegen, The Netherlands
stjepan.picek@ru.nl
[5] Semantics, Cybersecurity and Services Group, University of Twente, Drienerlolaan 5,, 7522 NB Enschede, The Netherlands
l.mariot@utwente.nl

Abstract. Boolean functions are mathematical objects used in diverse applications. Different applications also have different requirements, making the research on Boolean functions very active. In the last 30 years, evolutionary algorithms have been shown to be a strong option for evolving Boolean functions in different sizes and with different properties. Still, most of those works consider similar settings and provide results that are mostly interesting from the evolutionary algorithm's perspective. This work considers the problem of evolving highly nonlinear Boolean functions in odd sizes. While the formulation sounds simple, the problem is remarkably difficult, and the related work is extremely scarce. We consider three solutions encodings and four Boolean function sizes and run a detailed experimental analysis. Our results show that GP outperforms other EA in evolving highly nonlinear functions. Nevertheless, the problem is challenging, and finding optimal solutions is impossible except for the smallest tested size. However, once we added local search to the evolutionary algorithm, we managed to find a Boolean function in nine inputs with nonlinearity 241, which, to our knowledge, had never been accomplished before with evolutionary algorithms.

Keywords: Boolean functions · nonlinearity · genetic programming · evolutionary algorithms · odd dimension · encodings

1 Introduction

Boolean functions are used in diverse applications. For instance, some of the domains with long history and still active research developments include combinatorics [13], coding theory [20], computational complexity theory [1], and cryptography [11]. Since Boolean functions are used in various contexts, the conditions they need to fulfill are different. For instance, depending on the application, it may be relevant to know the Boolean function size, whether it is even or odd, or the value for some specific property. Thus, since there are multiple scenarios to consider, the research on Boolean functions is an active research domain. Considering how to construct Boolean functions with specific properties, two main directions are to either use algebraic constructions or heuristics [4].[1] Algebraic constructions have the advantage of clear mathematical formulation, commonly working for multiple sizes, both even and odd. On the other hand, heuristics need to be tested for every dimension of interest, but one could construct functions with properties that are not attainable with algebraic constructions (or, at least, with the currently known algebraic constructions). Unfortunately, heuristics commonly struggle when considering larger Boolean functions due to the super-exponential growth of the search space with respect to the number of variables of the function. It has been observed that Genetic Programming (GP) is usually able to evolve better Boolean functions than other Evolutionary Algorithms (EA), such as Genetic Algorithms (GA) [22, 28, 31].

This paper considers EA to construct Boolean functions with high nonlinearity and odd sizes. While this problem sounds simple, it is far from it. For instance, it was not known for more than 30 years if a Boolean function of nine variables could have a nonlinearity larger than 240. This question was solved positively in 2007 by Kavut et al. using simulated annealing [18]. However, the authors needed to use custom heuristics and limit the search space to the class of rotation symmetric (RS) Boolean functions, which is considerably smaller than the search space size for general Boolean functions (see Table 2).

Considering the works done with evolutionary algorithms, we can informally divide them into those relevant from both mathematical and EA perspectives and those relevant from the EA side only. In the first category, we can single out the result from Kavut et al. since it solved a long-standing open problem [18]. Another example is the work from Carlet et al., where the authors used evolutionary algorithms to improve the algebraic construction of Boolean functions [7]. Finally, some works consider evolving Boolean functions with properties that do not achieve good values when Boolean functions are constructed with algebraic constructions [27, 37]. On the other hand, multiple papers have relevance from the evolutionary perspective only (e.g., in the sense of benchmarking) since the results are either negative (such as those reported in [23]) or there is no analysis that would show if results are new (for example, finding Boolean functions with desired properties without evaluating if they are achievable by known algebraic constructions). Naturally, there is nothing wrong with considering only the evolutionary perspective. Still, one needs to be careful not to consider problems that also hold little relevance from the EA perspective anymore. For instance, while evolving bent Boolean functions was challenging up to a few years ago,

---

[1] It is also possible to use a hybrid approach where both algebraic constructions and heuristics are used, see, e.g., [33].

more recent EA results show the problem to be rather readily solvable [15,29]. Since we also know multiple algebraic constructions that produce bent Boolean functions, the problem relevance becomes less clear from both the mathematical and EA sides. Thus, while we can conclude that evolving Boolean functions is still very interesting, one must take care to find relevant problem instances. We concentrate on one such problem relevant to mathematical and EA research.

In this work, we systematically evaluate the problem of evolving highly nonlinear Boolean functions in odd dimensions (and small size). We emphasize that we do not constrain functions to be balanced, and we do not consider even sizes, making it impossible to obtain bent Boolean functions. The constraint on high nonlinearity and odd dimension makes the problem mathematically relevant, as we do not know algebraic constructions capable of reaching upper bounds for a number of Boolean function sizes. Since the problem is difficult (due to the large search space), we limit our attention to small function sizes, allowing us to reach more relevant conclusions. More precisely, we evaluate three solution encodings (bitstring, symbolic, and floating-point) and four Boolean function sizes, namely from $n = 7$ to $n = 13$ input variables. The results show that GP outperforms other EA in consistently evolving highly nonlinear functions. However, the main finding is that this problem is difficult: we can find optimal results for certain sizes, but such solutions are rare. Already for nine inputs, none of the algorithms, including GP, can reach the optimal value. Then, we add several local search variants to our algorithms, making the results somewhat better, especially for the solution representation performing suboptimally before. Interestingly, with one evolutionary algorithm and local search combination, we even find a Boolean function with nonlinearity 241 and size 9. As far as we know, this is the first time EAs have found such a function.

The rest of this paper is organized as follows. Section 2 provides relevant information on Boolean functions. Section 3 discusses related works. In Sect. 4, we provide information about the algorithms, parameters, fitness functions, and encodings. Section 5 gives experimental results and discussion. Finally, in Sect. 6, we provide conclusions and possible future research directions.

2 Preliminaries on Boolean Functions

Let $\mathbb{F}_2 = \{0, 1\}$ denote the finite field with two elements. Given $n \in \mathbb{N}$, by $\mathbb{F}_2^n$ we denote the n-dimensional vector space over $\mathbb{F}_2$. The inner product of $a, b \in \mathbb{F}_2^n$ equals $a \cdot b = \bigoplus_{i=1}^{n} a_i b_i$ in $\mathbb{F}_2^n$. A mapping $f : \mathbb{F}_2^n \to \mathbb{F}_2$ is called a Boolean function of n variables.

In the next sections we concisely present the basic representations and properties of Boolean functions used throughout the paper, referring the reader to [4] for a more thorough discussion of the topic.

2.1 Representations

One common way to uniquely represent a Boolean function f on $\mathbb{F}_2^n$ is by using its truth table. The truth table of a Boolean function f is the list of pairs of function inputs (in $\mathbb{F}_2^n$) and function outputs, with the size of the value vector being 2^n. The output vector is the binary vector composed of all $f(x), x \in \mathbb{F}_2^n$, with a certain order selected on $\mathbb{F}_2^n$. Usually, one uses a vector $(f(0), \ldots, f(1))$ that contains the function values of f, ordered lexicographically [4].

The Walsh-Hadamard transform W_f is another common representation of a Boolean function f. The Walsh-Hadamard transform measures the correlation between $f(x)$ and the linear functions $a \cdot x$, defined for all $a \in \mathbb{F}_2^n$ as:

$$W_f(a) = \sum_{x \in \mathbb{F}_2^n} (-1)^{f(x) \oplus a \cdot x}, \qquad (1)$$

with the sum calculated in $\mathbb{Z}$.

2.2 Properties and Bounds

Balancedness. A Boolean function f is balanced if its truth table vector is composed of an equal number of zeros and ones.

Nonlinearity. The minimum Hamming distance between a Boolean function f and all affine functions is the nonlinearity of f. The nonlinearity nl_f of a Boolean function f can be calculated from the Walsh-Hadamard values [4]:

$$nl_f = 2^{n-1} - \frac{1}{2} \max_{a \in \mathbb{F}_2^n} |W_f(a)|. \qquad (2)$$

By the Parseval relation, it holds that $\sum_{a \in \mathbb{F}_2^n} W_f(a)^2 = 2^{2n}$ for any Boolean function $f : \mathbb{F}_2^n \to \mathbb{F}_2$, which implies that the nonlinearity of any n-variable Boolean function is bounded above by the so-called covering radius bound:

$$nl_f \leq 2^{n-1} - 2^{\frac{n}{2}-1}. \qquad (3)$$

Equation (3) cannot be tight when n is odd. The functions whose nonlinearity equals the maximal value from Eq. (3) are called bent, and they exist only for n even.

For n odd, a slightly better bound is $2\lfloor 2^{n-2} - 2^{\frac{n}{2}-2} \rfloor$ [14]. The nonlinearity $2^{n-1} - 2^{\frac{n-1}{2}}$ is called the quadratic bound[2] since for n odd, it is a tight upper bound on the nonlinearity of Boolean functions with algebraic degree at most two. Note also that it is called a bent concatenation bound since it is a tight upper bound on the nonlinearity of the concatenation of two bent functions f, g in $n-1$ variables. The quadratic bound is the best value of the nonlinearity that can be reached for $n \leq 7$ while for $n \geq 9$ better nonlinearity values exist, see [4]. We report the best-known values for nonlinearity in Table 1.

---

[2] Note that when we speak of a quadratic bound concerning general Boolean functions, it is not strictly speaking a bound but rather a value that we can try to exceed with the nonlinearity of certain functions.

Table 1. Nonlinearities of Boolean functions in odd dimensions [4].

condition	n			
	7	9	11	13
quadratic bound	56	240	992	4032
best-known	56	242	996	4040
upper bound	58	244	1000	4050

2.3 Rotation Symmetric Functions

A Boolean function over $\mathbb{F}_2^n$ is called rotation symmetric (RS) if it is invariant under any cyclic shift of input coordinates. The number of rotation symmetric Boolean functions is less than the number of Boolean functions, as the output value remains the same for certain input vectors. Stănică and Maitra used the Burnside lemma to deduce that the number of rotation symmetric functions is 2^{g_n}, where g_n equals [36]:

$$g_n = \frac{1}{n} \sum_{t|n} \phi(t) 2^{\frac{n}{t}}, \qquad (4)$$

and ϕ is Euler totient function that counts the number of positive integers less than n that are relatively prime to it. Thus, g_n represents the number of orbits where an orbit is a rotation symmetric partition composed of vectors equivalent under rotational shifts. Considering rotation symmetric functions allows an exhaustive search for larger Boolean function sizes, or at least a "simpler" problem for heuristics. We list the search space sizes in Table 2.

Table 2. The number of (rotation symmetric) Boolean functions.

criterion	n			
	7	9	11	13
# general	2^{128}	2^{512}	2^{2048}	2^{8192}
# RS	2^{20}	2^{60}	2^{188}	2^{632}

3 Related Work

As already discussed, most of the research works on evolutionary algorithms and Boolean functions considers two cases: 1) evolving bent Boolean functions (thus, considering only even dimensions and imbalanced Boolean functions) or 2) evolving balanced, highly nonlinear Boolean functions (plus maybe some other cryptographic properties) [10]. On the other hand, evolving maximally nonlinear

Boolean functions in odd dimensions is a relatively unexplored topic. We provide a brief overview of works considering highly nonlinear Boolean functions in odd sizes, novel EA techniques, or constructing rotation symmetric functions. The survey [10] gives a more comprehensive outlook of the literature related to metaheuristics for the design of Boolean functions with good cryptographic properties.

To our knowledge, Millan et al. were the first to apply a genetic algorithm (GA) to evolve Boolean functions with good cryptographic properties [25]. The authors used a genetic algorithm (and hill climbing) to evolve Boolean functions with high nonlinearity. The authors considered sizes from 8 to 16 inputs, and the best result for nine inputs is achieved with a combination of GA and hill climbing and it equals 236. Clark and Jacob used a combination of simulated annealing and hill-climbing with a cost function motivated by the Parseval theorem to find functions with high nonlinearity and low autocorrelation [8]. Most of the work considers functions with eight inputs, but the authors also report results for sizes 4 to 12. The best result for the nine inputs and genetic algorithms equals 236, and simulated annealing is 238. Burnett et al. used custom heuristics to generate Boolean functions with good cryptographic properties [3]. They reported a nonlinearity of 240 for the Boolean function in nine inputs.

Picek et al. were the first to use genetic programming (GP) to find Boolean functions with high nonlinearity (alongside more properties) [30]. The authors considered only Boolean functions with eight inputs, but already from there, it was clear that GP could easily outperform GA. Mariot and Leporati proposed using Particle Swarm Optimization to find Boolean functions with good trade-offs of cryptographic properties [24]. The authors consider sizes between 7 and 12 variables and report the best nonlinearity of 236 for functions of $n = 9$ variables. Picek et al. proposed to use Cartesian Genetic Programming to evolve Boolean functions with good cryptographic properties [32]. While the results were good, the authors analyzed Boolean functions with eight inputs only.

Stănică et al. used simulated annealing to evolve rotation symmetric Boolean functions [35]. By reducing the search space this way, the authors constructed Boolean functions in nine variables with nonlinearity 240. Kavut et al. used a steepest descent-like iterative algorithm to construct highly nonlinear Boolean functions [18]. The authors found imbalanced Boolean functions in nine variables with a nonlinearity of 241. This represented a breakthrough as, before this result, it was not known if one could obtain a function in nine variables with nonlinearity larger than 240. Kavut et al. conducted an efficient exhaustive search of rotation symmetric Boolean functions in nine variables having nonlinearity greater than 240 [17]. They showed there are 1512 functions with a nonlinearity of 241 and no rotation symmetric Boolean function with a nonlinearity greater than 241. Kavut and Yucel used a steepest-descent-like iterative algorithm to construct imbalanced Boolean functions in 9 variables with nonlinearity 242 [19]. For this result, the authors considered the generalized rotation symmetric Boolean functions class to allow nonlinearity to potentially reach above 241 while, at the same time, making the search space size significantly smaller than for general Boolean functions. Liu and Youssef used simulated annealing in combination with some algebraic techniques to construct balanced rotation symmetric Boolean functions in 10 inputs with nonlinearity equal to 488 [21]. Wang et al. employed genetic algorithms (GAs) to construct rotation symmetric Boolean functions [38]. The authors reported constructing balanced, highly nonlinear rotation symmetric functions in 8, 10, and 12 inputs. Recently, Carlet et al. investigated evolutionary algorithms for the evolution of rotation symmetric Boolean functions [6]. The authors report finding balanced Boolean functions in nine variables with nonlinearity 240. Moreover, they achieve it in two ways: either evolving general Boolean functions with the tree encoding or evolving rotation symmetric Boolean functions with the bitstring encoding. Unfortunately, they do not consider the evolution of imbalanced Boolean functions in odd sizes.

A different research line investigated the evolution of algebraic constructions of Boolean functions. The idea is to employ GP (or a variant thereof) to evolve a tree that combines seed functions with good cryptographic properties with new variables, to obtain a larger function with analogous good properties. While the usual target in this optimization problem is to evolve constructions for bent functions (see e.g. [29]), Carlet et al. considered the use of GP to evolve constructions for balanced functions with high nonlinearity [5]. The authors remarked that GP can easily generate constructions yielding highly nonlinear functions (i.e., 240 for nine variables), but most of them are actually equivalent to the well-known indirect sum construction.

To conclude, a large part of the related works either completely ignore the Boolean functions in odd sizes or impose constraints that the functions need to be balanced. From the remaining works, most achieve suboptimal results (i.e., not reaching the upper bound). The best results are achieved with custom heuristics, and evolutionary algorithms do not seem to be able to compete.

4 Experimental Settings

4.1 EC Representations

Bitstring Encoding. The most common option for encoding a Boolean function is the bitstring representation [10]. The bit string represents the truth table of the function. For a Boolean function with n inputs, the truth table is coded as a bit string with a length of 2^n. For rotationally symmetric Boolean functions, the number of truth table entries to be coded is significantly lower, and is specified by the number of orbits g_n in Eq. (4). For each evaluation, the bitstring genotype is first decoded into the full Boolean truth table (which is trivial since we know the orbits), and the desired property is computed (since we consider the nonlinearity property, we must first translate the truth table representation into the Walsh-Hadamard spectrum).

Symbolic Encoding. The second approach in our experiments uses tree-based GP to represent a Boolean function in its symbolic form. In this case, we represent a candidate solution by a tree whose leaf nodes correspond to the input variables $x_1, \cdots, x_n \in \mathbb{F}_2$. The internal nodes are Boolean operators that combine the inputs received from their children and forward their output to the respective parent nodes.

The output of the root node is the output value of the Boolean function. The corresponding truth table of $f : \mathbb{F}_2^n \to \mathbb{F}_2$ is determined by evaluating the tree over all possible 2^n assignments of the input variables at the leaf nodes. Each GP individual is evaluated according to the truth table it generates.

Floating Point Encoding. The last approach to representing a Boolean function is the floating-point genotype, which is defined as a vector with continuous variables. This requires defining the translation of a vector of floating point numbers into the corresponding genotype, which is then translated into a complete truth table (binary values). The idea behind this translation is that each continuous variable (a real number) of the floating point genotype represents a subsequence of bits in the genotype. All real values in the floating point vector are restricted to the interval $[0, 1]$. If the genotype size is g_n, the number of bits represented by a single continuous variable of the floating point vector can vary:

$$decode = \frac{g_n}{dimension}, \quad (5)$$

where the parameter $dimension$ denotes the floating point vector size (number of real values). This parameter can be modified if the genotype size is divisible by its value. The first step of the translation is to convert each floating point number to an integer value. As each real value must represent $decode$ bits, the size of the interval decoding to the same integer value is given as:

$$interval = \frac{1}{decode}. \quad (6)$$

To obtain a distinct integer value for a given real number, every element d_i of the floating point vector is divided by the calculated interval size, generating a sequence of integer values:

$$int\_value_i = \left\lfloor \frac{d_i}{interval} \right\rfloor. \quad (7)$$

The final translation step involves decoding the integer values into a binary string that can be used for evaluation. For further details on using floating point representation for evolving rotation symmetric Boolean functions, see [6].

4.2 Fitness Function

Several objective functions can be defined to optimize Boolean function nonlinearity regardless of the representation and search algorithm. The fitness function used here was selected based on the literature study of common choices in related works [10] and the authors' previous experience. Apart from maximizing the nonlinearity value, the applied fitness function considers the whole Walsh-Hadamard spectrum and not only its extreme value (see Eq. (2)). Here, we count the number of occurrences of the maximal absolute value in the spectrum, denoted as $\#max\_values$. As higher nonlinearity corresponds to a lower maximal absolute value, we aim for as few occurrences of the maximal value as possible, hoping it would be easier for the algorithm to reach the next nonlinearity value. In this way, we provide the algorithm with additional information, making the objective space more gradual. The fitness function is thus defined as:

$$fitness = nl_f + \frac{2^n - \#max\_values}{2^n}. \qquad (8)$$

The second term never reaches the value of 1 since, in that case, we effectively reach the next nonlinearity level.

4.3 Algorithm Parameters

Bitstring Encoding. The corresponding variation operators we use are the simple bit mutation and the shuffle mutation. The simple bit mutation inverts a randomly selected bit. The shuffle mutation shuffles the bits within a randomly selected substring. For the crossover operators, we use the one-point crossover and uniform crossover. The one-point crossover combines a new solution from the first part of one parent and the second part of the other parent with a randomly selected breakpoint. The uniform crossover randomly selects one bit from both parents at each position in the child bitstring that is copied. Each time the evolutionary algorithm invokes a crossover or mutation operation, one of the previously described operators is randomly selected.

Symbolic Encoding. In our experiments, we use the following function set: OR, XOR, AND, AND2, XNOR, IF, and function NOT that takes a single argument. The function AND2 behaves the same as the function AND but with the second input inverted. The function IF takes three arguments and returns the second one if the first one evaluates to true and the third one otherwise. This function set is common when dealing with the evolution of Boolean functions with cryptographic properties [6,10].

The genetic operators used in our experiments with tree-based GP are simple tree crossover, uniform crossover, size fair, one-point, and context preserving crossover [34] (selected at random), and subtree mutation. The option to use multiple genetic operators was based on the initial experiments.

We employ the same evolutionary algorithm for both bitstring and symbolic encoding: a steady-state selection with a 3-tournament elimination operator (denoted SST). In each iteration of the algorithm, three individuals are chosen at random from the population for the tournament, and the worst one in terms of fitness value is eliminated. The two remaining individuals in the tournament are used with the crossover operator to generate a new child individual, which then undergoes mutation with individual mutation probability $p_{mut} = 0.5$. Finally, the mutated child replaces the eliminated individual in the population.

Floating Point Encoding. When FP encoding is used, one can vary the number of bits a single FP value will represent (*decode*, Eq., (5)). Based on related work [6], all FP-based algorithms use the same setting with *decode* = 3. The floating point representation can be used with any continuous optimization algorithm, which increases its versatility. We investigated the following algorithms: Artificial Bee Colony (ABC) [16], Clonal Selection Algorithm (CLONALG) [2], CMA-ES [12], Differential Evolution (DE) [26], Optimization Immune Algorithm (OPTIA) [9], and a GA-based algorithm with steady-state selection (GA-SST), which is also used with TT and GP and whose behavior is described above. Due to lack of space, we do not provide algorithms' parameters, but note we used the ECF software framework[3] with default parameter values. The termination condition is set at 10^6 evaluations for all configurations.

5 Experimental Evaluation

5.1 Evolutionary Algorithms

The results for the three representations and the different optimization methods are outlined in Table 3 for the four selected problem sizes. Each experiment was executed 30 times, with descriptive statistics like the maximum, average, and standard deviation being outlined in the table. The best results are given in bold for each problem size. The results of all methods are also outlined in Fig. 1. For $n = 13$, the figure does not include the results for FP/DE due to the poor solutions the methods obtain, making them unreadable.

The table and figure evidently show that the best results are obtained by GP across all the problem sizes. The differences between it and the other methods become more pronounced as the size of the problem increases. Regarding the TT and FP representations, it is not possible to say whether each one is consistently better than the other since it depends on the problem size. Regarding the optimization algorithms that were used with the FP representation, we see that the choice of the algorithm had a significant influence on the obtained results. Again, no single optimization method consistently achieved the best results for all problem sizes. However, the SST algorithm seems to be the most stable because it achieved good performance for most problem sizes. An additional benefit of GP against the other methods is that it achieved a very small standard deviation value, in several cases the smallest among all the methods. Thus, the GP results are not dispersed, and the algorithm is rather stable. Regarding the comparison with the best-known solutions, for size 7, each representation obtained the best-known result, a solution with a nonlinearity of 56. However, this was not the case for larger sizes, and the obtained solutions are worse than the best-known solutions for those sizes (see Table 1).

To determine whether a statistically significant difference between the results obtained by different methods exists, we used the Kruskal-Wallis test with a significance value of 0.05. Since a p-value of 0 was obtained for all problem

---

[3] http://solve.fer.hr/ECF/.

Table 3. Summary of the results of the various representations and optimization algorithms (obtained fitness values).

Enc.	Algorithm	7			9			11			13		
		max.	avg.	std.	max.	avg.	std.	max.	avg.	std.	max.	avg.	std.
FP	ABC	55.85	55.05	0.31	234.95	233.84	0.42	971.00	970.09	0.61	3827.00	3810.97	6.71
	CLONALG	56.63	56.62	0.01	235.98	235.01	0.18	969.00	967.76	0.57	3888.00	3853.40	24.37
	CMAES	54.93	54.65	0.50	231.98	231.02	0.18	964.00	963.00	0.52	3938.00	3934.23	1.41
	DE	54.93	54.90	0.03	231.98	230.79	0.48	960.00	958.50	1.01	2836.00	2701.12	58.23
	OPTIA	56.64	56.57	0.19	232.99	232.85	0.34	967.00	965.43	0.57	3918.00	3894.17	18.22
	SST	56.63	56.46	0.30	236.95	236.80	0.38	978.97	976.78	1.42	3923.00	3911.70	7.20
GP	SST	56.69	**56.64**	0.03	240.72	**240.64**	0.03	992.69	**992.63**	0.02	4032.69	**4030.52**	11.62
TT	SST	56.63	56.60	0.02	236.91	236.55	0.74	978.96	974.44	1.88	3980.99	3977.22	2.51

sizes, we conclude that there is a significant difference in the performance of the tested algorithms and representations. The post-hoc Dunn test with the Bonferroni correction method was used to determine these differences.

For problem size 7, the results demonstrate that although GP achieves the best result, it does not perform significantly better than the FP representation with the CLONALG or OPTIA methods. Instead, the TT representation was significantly worse than GP but equally good as FP for some algorithms (again, CLONALG and OPTIA). For problem sizes 9 and 11, GP achieves significantly better results than all other methods, except for FP with SST, which came second. Furthermore, in those cases, there is no difference between the TT and FP representation (when considering the result obtained by the best algorithm). For size 13, GP achieves significantly better results than all other methods, except for TT, against which the differences are not significant. Furthermore, TT achieves equally good results as the FP representation with CMAES.

Based on the previously outlined observations and analyses, we can conclude that GP is the most appropriate method for tackling the considered problem. It consistently achieved the best results, which were significantly better than most of the results obtained by any other tested method. Unfortunately, even these best results fall short of the result with, e.g., custom heuristics.

5.2 Evolutionary Algorithms + Local Search

Since the case with nine variables was the smallest size where evolutionary methods did not obtain the best-known value, we investigated the possibilities of improving the efficiency in this particular case. The first modification adds local search, which was applied in two forms. The first form is a mutation-based local search operator: the operator acts on a single solution and performs a number of mutations. If a better solution is found, the new solution immediately replaces the original one, and the operator is applied again. If no better solution is found after a predefined number of mutations, the operator terminates. The operator

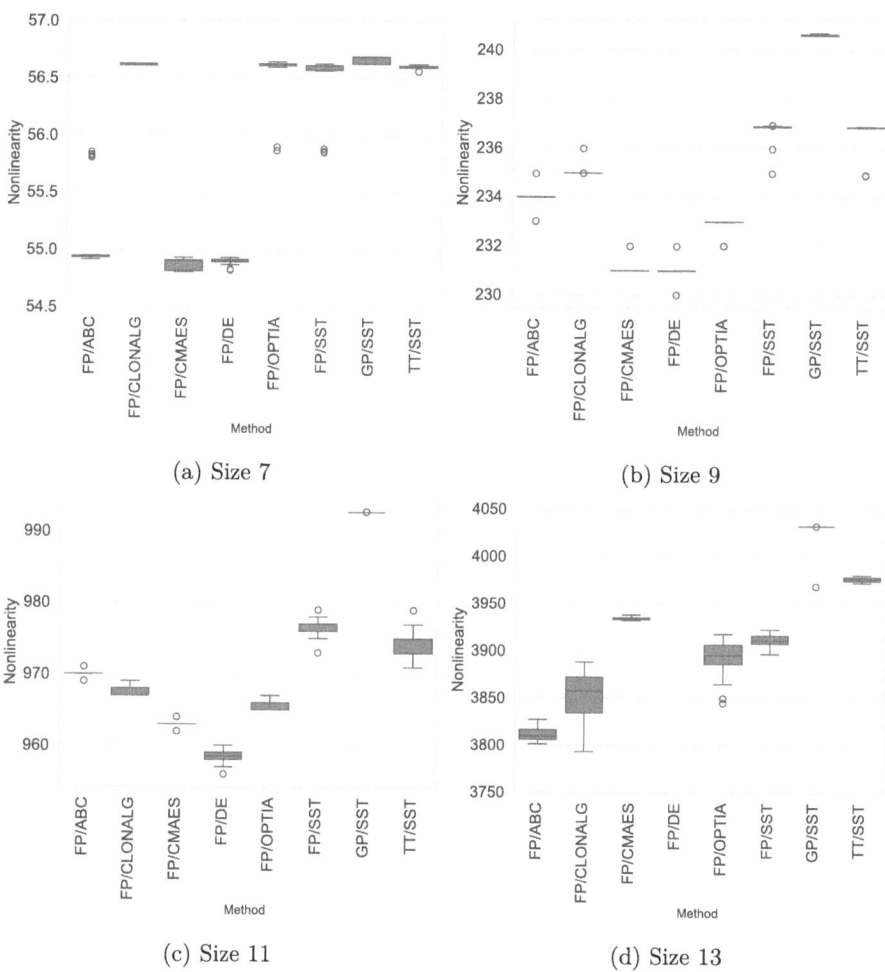

Fig. 1. Boxplot representation of the results obtained for various sizes.

is applied after each generation and acts upon the current best solution and a number of random solutions. In our experiments, the number of solutions undergoing local search was set to 5% of the population size, and the number of trials (random mutations per individual) was set to 25. This operator is general in that it can be applied to any encoding.

However, for the bitstring representation, we included the second form of the local operator that performs individual bit flips instead of random mutations. The operator is exhaustive, meaning it will perform all possible bit flips and terminate only if there is no improvement. We applied the local search operators only to GP and bitstring encodings as the most efficient variants; in the bitstring case, three combinations were tested with either the mutation (denoted as "-LS1") or bit flip operator (denoted as "-LS2"), or both (denoted as "-LS3"). The results with these modifications were not encouraging, despite the experimental design which included a thousand runs for every combination with a time limit of 2000 s, which amounts to approximately 300 million evaluations per run. The GP efficiency was not altered since GP always found the same nonlinearity value of 240 in every run, with or without local search. The results for the bitstring encoding were slightly improved and are shown in Table 4.

Following these experiments, we introduced another modification: the use of rotation invariant encoding. In this encoding, we only consider rotation-symmetric Boolean functions, limiting the number of possible solutions with high nonlinearity while drastically reducing the search space. For instance, in the nine variable case, the representation for rotation symmetric functions consists of only 60 bits (as opposed to 512 in the general case). These results are also included in Table 4 and denoted with "-RI".

To better outline the effect of different LS operators on the results, Fig. 2 provides the boxplot representation of the results. Clearly, applying LS operators affects the results, especially the application of the one based on the bit flip operator. To determine whether the improvement in the results is statistically significant, the Kruskal-Wallis test was again applied. Since a p-value of 0 was obtained, we deduce that the results are significantly different. The post hoc analysis demonstrates that by using LS, it is possible to improve the results significantly compared to the basic algorithm. Furthermore, the rotation invariant algorithm variant leads to significant improvements in the results. Remarkably, it also archives the nonlinearity of 241 for the TT-RI-LS1 combination. To our knowledge, this is the first time EA reached nonlinearity 241 for Boolean functions with nine inputs. Regarding the different LS operators, the analysis demonstrates that there is no significant difference between them.

Let us consider how our results compare with [18]. The authors reported that among 200 million RSBFs in nine inputs evaluated with the steepest descent-like algorithm, five have nonlinearity 241. Clearly, this is more successful than what we achieved, as we found only one such function. The question is from where the better performance arises. First, there is a difference in the objective function, although Kavut et al. also considered the sum of square errors of the Walsh-Hadamard values. Next, the difference is in the steepest descent-like nature of their algorithm since they introduced a step where, once the cost cannot be minimized further, there is a deterministic step in the reverse direction corresponding to the smallest possible cost increase. Finally, we did not use exclusively local search but a combination with EA. Based on the results, we conclude that local search is a crucial step, which indicates that EA operators are either 1) too disruptive or 2) reach local optima and cannot produce a small change required to improve the fitness value. Increasing nonlinearity from 240 to 241 requires—in the best case scenario—only a single change in the truth table representation.

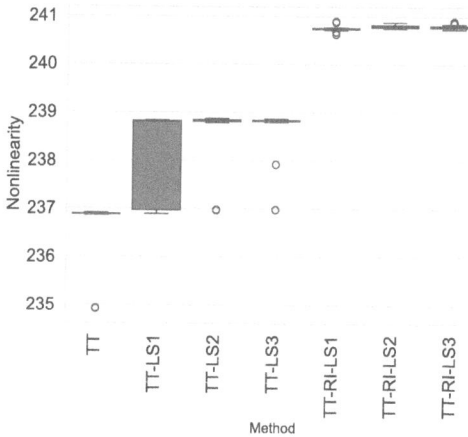

Fig. 2. Boxplot representation of the results for the application of LS operators on the TT representation

Table 4. Results for additional runs with TT representation and LS (size 9)

	max.	avg.	std.
TT	236.91	236.55	0.74
TT-LS1	238.83	237.98	0.95
TT-LS2	238.87	238.58	0.64
TT-LS3	238.87	238.69	0.48
TT-RI-LS1	**241.75**	240.75	0.05
TT-RI-LS2	240.88	240.80	0.03
TT-RI-LS3	240.90	240.79	0.04

6 Conclusions and Future Work

This paper systematically evaluates the evolution of Boolean functions with high nonlinearity and in odd sizes. The experiments with EAs and three solution encodings indicate GP to be the best option, regardless of the fact that GP works with general Boolean functions, while FP and TT consider only a subspace of rotation symmetric Boolean functions. Unfortunately, even such best results fall short of the best-known results reached with custom heuristics (except for the smallest Boolean function size). Next, we added several local search variants to our best EAs. Those modifications did not help GP but improved the TT results. Moreover, one combination of the TT encoding and local search operators even resulted in a nine variable Boolean function with nonlinearity 241, which is the best possible value within the rotation symmetric Boolean function class, and something previously not achieved with EAs.

Our results indicate several future research directions. GP again shows to be the best general encoding, but with it, we cannot consider solutions belonging to a specific subclass of Boolean functions. It would be interesting to explore how to circumvent this problem. On the other hand, TT and FP are not very successful in the general case, but they benefit from constraining to subclasses and offer a larger choice of local search operators. Finally, different types of selection, such as lexicase variants, may improve the convergence.

References

1. Arora, S., Barak, B.: Computational Complexity: A Modern Approach, 1st edn. Cambridge University Press, USA (2009)
2. Brownlee, J., et al.: Clonal selection algorithms. Complex Intelligent Systems Laboratory, Swinburne University of Technology, Australia (2007)
3. Burnett, L., Millan, W., Dawson, E., Clark, A.: Simpler methods for generating better Boolean functions with good cryptographic properties. Australas. J. Combin. **29**, 231–248 (2004)
4. Carlet, C.: Boolean Functions for Cryptography and Coding Theory. Cambridge University Press, Cambridge (2021). https://doi.org/10.1017/9781108606806
5. Carlet, C., Djurasevic, M., Jakobovic, D., Mariot, L., Picek, S.: Evolving constructions for balanced, highly nonlinear Boolean functions. In: Proceedings of the Genetic and Evolutionary Computation Conference, GECCO 2022, pp. 1147–1155. Association for Computing Machinery, New York (2022). https://doi.org/10.1145/3512290.3528871
6. Carlet, C., Durasevic, M., Gasperov, B., Jakobovic, D., Mariot, L., Picek, S.: A new angle: on evolving rotation symmetric boolean functions. In: Smith, S.L., Correia, J., Cintrano, C. (eds.) Applications of Evolutionary Computation - 27th European Conference, EvoApplications 2024, Held as Part of EvoStar 2024, Aberystwyth, UK, 3–5 April 2024, Part I. Lecture Notes in Computer Science, vol. 14634, pp. 287–302. Springer (2024)
7. Carlet, C., Jakobovic, D., Picek, S.: Evolutionary algorithms-assisted construction of cryptographic Boolean functions. In: Proceedings of the Genetic and Evolutionary Computation Conference, GECCO 2021, pp. 565–573. Association for Computing Machinery, New York (2021). https://doi.org/10.1145/3449639.3459362
8. Clark, J.A., Jacob, J.L.: Two-stage optimisation in the design of Boolean functions. In: Dawson, E.P., Clark, A., Boyd, C. (eds.) Information Security and Privacy, pp. 242–254. Springer, Heidelberg (2000)
9. Cutello, V., Nicosia, G., Pavone, M.: Real coded clonal selection algorithm for unconstrained global optimization using a hybrid inversely proportional hypermutation operator. In: Proceedings of the 2006 ACM symposium on Applied computing, pp. 950–954 (2006)
10. Djurasevic, M., Jakobovic, D., Mariot, L., Picek, S.: A survey of metaheuristic algorithms for the design of cryptographic Boolean functions. Cryptogr. Commun. **15**(6), 1171–1197 (2023)
11. Dobbertin, H.: Construction of bent functions and balanced Boolean functions with high nonlinearity. In: Preneel, B. (ed.) Fast Software Encryption, pp. 61–74. Springer, Heidelberg (1995)
12. Hansen, N., Müller, S.D., Koumoutsakos, P.: Reducing the time complexity of the derandomized evolution strategy with covariance matrix adaptation (CMA-ES). Evol. Comput. **11**(1), 1–18 (2003)

13. Helleseth, T., Kholosha, A.: Bent Functions and Their Connections to Combinatorics. London Mathematical Society Lecture Note Series, pp. 91–126. Cambridge University Press (2013)
14. Hou, X.D.: On the norm and covering radius of the first-order Reed-Muller codes. IEEE Trans. Inf. Theory **43**(3), 1025–1027 (1997). https://doi.org/10.1109/18.568715
15. Husa, J., Dobai, R.: Designing bent Boolean functions with parallelized linear genetic programming. In: Proceedings of the Genetic and Evolutionary Computation Conference Companion, GECCO 2017, pp. 1825–1832. Association for Computing Machinery, New York (2017). https://doi.org/10.1145/3067695.3084220
16. Karaboga, D., Gorkemli, B., Ozturk, C., Karaboga, N.: A comprehensive survey: artificial bee colony (ABC) algorithm and applications. Artif. Intell. Rev. **42**, 21–57 (2014)
17. Kavut, S., Maitra, S., Sarkar, S., Yücel, M.D.: Enumeration of 9-variable rotation symmetric Boolean functions having nonlinearity $\geq$ 240. In: Barua, R., Lange, T. (eds.) INDOCRYPT 2006. LNCS, vol. 4329, pp. 266–279. Springer, Heidelberg (2006). https://doi.org/10.1007/11941378_19
18. Kavut, S., Maitra, S., Yucel, M.D.: Search for Boolean functions with excellent profiles in the rotation symmetric class. IEEE Trans. Inf. Theory **53**(5), 1743–1751 (2007). https://doi.org/10.1109/TIT.2007.894696
19. Kavut, S., Yücel, M.D.: 9-variable Boolean functions with nonlinearity 242 in the generalized rotation symmetric class. Inf. Comput. **208**(4), 341–350 (2010). https://doi.org/10.1016/j.ic.2009.12.002, https://www.sciencedirect.com/science/article/pii/S0890540109002454
20. Kerdock, A.: A class of low-rate nonlinear binary codes. Inf. Control **20**(2), 182–187 (1972)
21. Liu, W.M., Youssef, A.: On the existence of (10, 2, 7, 488) resilient functions. IEEE Trans. Inf. Theory **55**(1), 411–412 (2009). https://doi.org/10.1109/TIT.2008.2008140
22. Manzoni, L., Mariot, L., Tuba, E.: Balanced crossover operators in genetic algorithms. Swarm Evol. Comput. **54**, 100646 (2020)
23. Mariot, L., Jakobovic, D., Leporati, A., Picek, S.: Hyper-bent Boolean functions and evolutionary algorithms. In: Sekanina, L., Hu, T., Lourenço, N., Richter, H., García-Sánchez, P. (eds.) Genetic Programming, pp. 262–277. Springer, Cham (2019)
24. Mariot, L., Leporati, A.: Heuristic search by particle swarm optimization of Boolean functions for cryptographic applications. In: Proceedings of the Companion Publication of the 2015 Annual Conference on Genetic and Evolutionary Computation, GECCO Companion 2015, pp. 1425–1426. Association for Computing Machinery, New York (2015). https://doi.org/10.1145/2739482.2764674
25. Millan, W., Clark, A., Dawson, E.: An effective genetic algorithm for finding highly nonlinear Boolean functions. In: Han, Y., Okamoto, T., Qing, S. (eds.) Information and Communications Security, vol. 1334, pp. 149–158. Springer, Heidelberg (1997)
26. Pant, M., Zaheer, H., Garcia-Hernandez, L., Abraham, A., et al.: Differential evolution: a review of more than two decades of research. Eng. Appl. Artif. Intell. **90**, 103479 (2020)
27. Picek, S., Batina, L., Jakobovic, D.: Evolving DPA-resistant Boolean functions. In: Bartz-Beielstein, T., Branke, J., Filipič, B., Smith, J. (eds.) PPSN 2014. LNCS, vol. 8672, pp. 812–821. Springer, Cham (2014). https://doi.org/10.1007/978-3-319-10762-2_80

28. Picek, S., Carlet, C., Guilley, S., Miller, J.F., Jakobovic, D.: Evolutionary algorithms for Boolean functions in diverse domains of cryptography. Evol. Comput. **24**(4), 667–694 (2016)
29. Picek, S., Jakobovic, D.: Evolving algebraic constructions for designing bent Boolean functions. In: Proceedings of the Genetic and Evolutionary Computation Conference 2016, GECCO 2016, pp. 781–788. Association for Computing Machinery, New York (2016). https://doi.org/10.1145/2908812.2908915
30. Picek, S., Jakobovic, D., Golub, M.: Evolving cryptographically sound Boolean functions. In: Proceedings of the 15th Annual Conference Companion on Genetic and Evolutionary Computation, GECCO 2013 Companion, pp. 191–192. Association for Computing Machinery, New York (2013). https://doi.org/10.1145/2464576.2464671
31. Picek, S., Jakobovic, D., Miller, J.F., Batina, L., Cupic, M.: Cryptographic Boolean functions: One output, many design criteria. Appl. Soft Comput. **40**, 635–653 (2016). https://doi.org/10.1016/j.asoc.2015.10.066, http://www.sciencedirect.com/science/article/pii/S1568494615007103
32. Picek, S., Jakobovic, D., Miller, J.F., Marchiori, E., Batina, L.: Evolutionary methods for the construction of cryptographic Boolean functions. In: Machado, P., et al. (eds.) EuroGP 2015. LNCS, vol. 9025, pp. 192–204. Springer, Cham (2015). https://doi.org/10.1007/978-3-319-16501-1_16
33. Picek, S., Marchiori, E., Batina, L., Jakobovic, D.: Combining evolutionary computation and algebraic constructions to find cryptography-relevant Boolean functions. In: Bartz-Beielstein, T., Branke, J., Filipič, B., Smith, J. (eds.) PPSN 2014. LNCS, vol. 8672, pp. 822–831. Springer, Cham (2014). https://doi.org/10.1007/978-3-319-10762-2_81
34. Poli, R., Langdon, W.B., McPhee, N.F.: A Field Guide to Genetic Programming (2008). lulu.com
35. Stanica, P., Maitra, S., Clark, J.A.: Results on rotation symmetric bent and correlation immune Boolean functions. In: Roy, B.K., Meier, W. (eds.) Fast Software Encryption, 11th International Workshop, FSE 2004, Delhi, India, 5–7 February 2004, Revised Papers. Lecture Notes in Computer Science, vol. 3017, pp. 161–177. Springer, Heidelberg (2004)
36. Stănică, P., Maitra, S.: Rotation symmetric Boolean functions—count and cryptographic properties. Discrete Appl. Math. **156**(10), 1567–1580 (2008). https://doi.org/10.1016/j.dam.2007.04.029, https://www.sciencedirect.com/science/article/pii/S0166218X07001734
37. Wang, Q., Stănică, P.: Transparency order for Boolean functions: analysis and construction. Designs Codes Cryptogr. **87**(9), 2043–2059 (2019). https://doi.org/10.1007/s10623-019-00604-1
38. Wang, Y., Gao, G., Yuan, Q.: Searching for cryptographically significant rotation symmetric Boolean functions by designing heuristic algorithms. Secur. Commun. Netw. **2022**, 1–6 (2022). https://doi.org/10.1155/2022/8188533

Exploring the Impact of Data Scale on Mutation Step Size in SLIM-GSGP

Davide Farinati[1]($\boxtimes$), Gloria Pietropolli[2], and Leonardo Vanneschi[1]

[1] NOVA Information Management School (NOVA IMS), Universidade Nova de Lisboa, Campus de Campolide, 1070-312 Lisbon, Portugal
{dfarinati,lvanneschi}@novaims.unl.pt
[2] Department of Mathematics, Informatics, and Geosciences, University of Trieste, Trieste, Italy
gloria.pietropolli@units.it

Abstract. The Semantic Learning algorithm based on Inflate and deflate Mutation (SLIM) is a promising recent variant of Geometric Semantic Genetic Programming (GSGP) that introduces a new Deflate Geometric Semantic Mutation (DGSM). This operator maintains the key feature of the standard Geometric Semantic Mutation (GSM), inducing a unimodal error surface for any supervised learning problem, while generating smaller offspring than their parents, and thus allowing SLIM to generate compact, and potentially interpretable, final solutions. A key parameter controlling the evolution process in both GSGP and SLIM is the Mutation Step (MS), which regulates the extent of perturbation to the parent semantics. While it is intuitive that the optimal value of MS has a relationship with the scale of the dataset features, to the best of our knowledge no prior research has extensively explored this relationship. In this work, we provide the first comprehensive investigation into this topic. First, we hypothesize a general rule by analyzing results from artificial datasets, and then we confirm these findings with more complex, real-world datasets. This approach offers a solid alternative to the typical hyperparameter tuning approach.

Keywords: Genetic Programming · Geometric Semantic Genetic Programming · Geometric Mutation · Mutation Step · Symbolic Regression

1 Introduction

Geometric Semantic Genetic Programming (GSGP) is a variant of Genetic Programming (GP) [16,26] that uses Geometric Semantic Operators (GSOs) to replace the traditional (syntax-based) crossover and mutation and induce precise and known geometric properties in the semantic space [20,34]. The key property of GSOs is that they induce a unimodal error surface for any supervised learning problem. However, GSGP is also known for producing extremely large final solutions, that are usually not interpretable by humans. A new variant of GSGP was

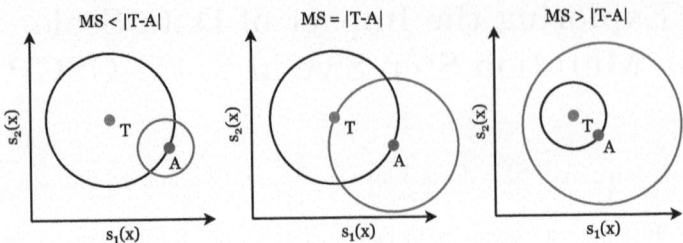

Fig. 1. Visual representation of the impact of the MS in a bidimensional feature space($s_1(x)$ and $s_2(x)$). The target point T is shown, with the black circle representing the space of solutions fitter than an individual, whose semantic is point A. The blue circle illustrates the space of solutions that can be generated by mutating that individual. Three conditions are shown: when the MS is smaller than, equal to, or greater than the distance between the target (T) and the individual.

recently introduced, which preserves the property of inducing a unimodal error surface, while also producing models that are compact enough to be interpreted by humans [30,33]. This variant, called Semantic Learning algorithm based on Inflate and deflate Mutation (SLIM), relies on two types of mutation: besides the traditional Geometric Semantic Mutation (GSM) of GSGP, called Inflate Geometric Semantic Mutation (IGSM) as it creates offspring are larger than their parents, SLIM introduces a Deflate Geometric Semantic Mutation (DGSM), so called because of its property of producing offspring smaller than their parents. A crossover for SLIM has not been defined yet and a significant amount of literature indicates that GSGP using only mutation is often competitive with, or even better than, GSGP using both crossover and mutation [4,21,35]. For these reasons, we will not consider crossover in this paper as well.

Like the traditional mutation of GSGP, both IGSM and DGSM in SLIM adjust an individual's semantic coordinates by perturbing them of an amount contained within a specific range $[-\text{MS}, \text{MS}]$, where MS (mutation step) is a user-defined parameter. Geometric mutation shifts individuals within the solution space, with MS setting the maximum shift amplitude and significantly influencing the optimization process. Figure 1 visually illustrates the effect of MS. As widely shown in the literature, the ideal MS value depends on the problem at hand and is typically determined through hyperparameter tuning, often requiring time-consuming trial and error [6,7]. Despite its importance in guiding the evolution process, few studies have specifically explored optimal configuration of the MS parameter.

For instance, [13] introduced an adaptive MS method, demonstrating that the optimal MS for each GSM event can be computed deterministically using the Moore-Penrose pseudoinverse [9]. However, while effective on training data, this approach often leads to overfitting, limiting its generalizability to new data. Similarly [17] employs an optimal-step one-tree GSGP mutation operator. This operator determines the optimal step for each mutation event by calculating the value that minimizes the distance of the resulting tree from the optimum.

Although this approach did not systematically outperform the standard application of GSGP.

Currently, hyperparameter tuning remains the most widely used approach, even if it is resource-intensive and often fails to identify the true optimal value, leading to less-than-ideal results and inefficient use of resources without guaranteeing generalization.

Selecting an appropriate MS value is closely tied to the dataset's specific characteristics, particularly the distribution and range of its input and output values. This is crucial since GSM involves exploring the solution space, and its effectiveness depends on how well the choice aligns with the dataset's properties. This study investigates how the scale of a dataset-whether in its input features, output features, or both-affects the selection of an appropriate MS for SLIM. Finding a general rule to automatically determine the optimal MS, or at least a close approximation, would eliminate the laborious and time-consuming process of manual hyperparameter tuning.

The paper is structured as follow: Sect. 2 introduces GSGP and SLIM. Section 3 describes the adopted methodology. Section 4 describes the datasets (artificial and real-world) used for the experimental study, as well as the used experimental settings. Section 5 provides a detailed analysis of the obtained results. Finally, Sect. 6 summarizes the main contributions of the paper and provides directions for further research.

2 Background

This section introduces the algorithms used in this study. First, in Sect. 2.1, we present GSGP, with a particular focus on GSM. Then, in Sect. 2.2, we describe the recently introduced variant SLIM.

2.1 Geometric Semantic Genetic Programming

When we tackle supervised learning problems, GP individuals are typically computer programs that map inputs into outputs. Given a GP individual P and a set of inputs $X = \{x_1, x_2, \ldots, x_n\}$, the output vector $s_P = \{P(x_1), P(x_2), \ldots, P(x_n)\}$ is referred to as the *semantics* of the individual P [20]. Traditionally, GP uses operators like mutation and crossover, which manipulate the syntactic structure of individuals. For example, crossover swaps subprograms between individuals, while mutation replaces subprograms with random programs. Though these operations are easy to describe, their effects on the semantics can be unpredictable. For instance, small syntactic changes can lead to significant semantic differences, compromising locality and thus complicating the search for optimal solutions [23]. To address this, semantic awareness was integrated into GP [32]. Among other approaches, GSGP was introduced by [20] as a variant of GP that replaces traditional syntax-based operators with GSOs. These operators, although modifying the syntax of the individuals, induce known geometric properties in the semantic space and, for any supervised learning problem, they generate a unimodal error surface.

Given a parent function $T : \mathbb{R}^n \to \mathbb{R}$, *Geometric Semantic Mutation* (GSM) generates the function $T_M = T + \text{MS} \cdot (T_{R1} - T_{R2})$, where MS is the mutation step and T_{R1} and T_{R2} are random real functions whose output ranges in the interval $[0, 1]$. This is usually obtained wrapping two random expressions with a sigmoid function. GSM generates an individual contained in the hyper-sphere of radius MS centred in the semantics of the parent in the semantic space. Building on the success of GSGP, researchers have explored many variations in recent years [3,5,22,25]. Its major drawback is that it generates offspring larger than their parents, leading to final solutions that are hard to interpret [20].

2.2 Semantic Learning algorithm based on Inflate and deflate Mutation

As previously mentioned, one major drawback of GSGP is that the standard GSM causes a rapid increase in the size of individuals throughout the evolutionary process, making the final solutions difficult to interpret. Several solutions have been proposed to address this issue, such as dynamic population approaches [12] and the integration of gradient descent techniques [24,25]. However, most of the existing methods still depend on the GSO introduced by [20] or utilize new operators that continue to produce offspring larger than their parents [2]. Recently, [33] introduced a new GSM that maintains the same properties as the traditional semantic mutation (renamed Inflate Geometric Semantic Mutation, or IGSM), while also producing smaller individuals than their parents. This new operator, called Deflate Geometric Semantic Mutation (DGSM), is based on two ideas. The first idea is that the IGSM definition from the previous section can be rewritten as follows:

$$IGSM(T) = T + \text{MS} \cdot (T_{R1} - T_{R2}) = T - \text{MS} \cdot (T_{R2} - T_{R1}),$$

given that, trivially, $(T_{R1} - T_{R2}) = -(T_{R2} - T_{R1})$. The second idea is that the two random programs, T_{R1} and T_{R2}, are interchangeable because they are independent random variables with the same probability distribution. Therefore, IGSM can also be rewritten as:

$$IGSM(T) = T - \text{MS} \cdot (T_{R1} - T_{R2}).$$

Now, consider an individual T to which the IGSM has been applied, say, three times, resulting in the new individual:

$$T + \text{MS} \cdot (T_{R1} - T_{R2}) + \text{MS} \cdot (T_{R3} - T_{R4}) + \text{MS} \cdot (T_{R5} - T_{R6}).$$

Given that the two definitions are equivalent, we can now apply IGSM again using subtraction, resulting in the individual:

$$T + \text{MS} \cdot (T_{R1} - T_{R2}) + \text{MS} \cdot (T_{R3} - T_{R4}) + \text{MS} \cdot (T_{R5} - T_{R6}) - \text{MS} \cdot (T_{R7} - T_{R8}).$$

Clearly, the size of the individual continues to grow with each application of IGSM. However, if we apply IGSM with subtraction and reuse a pair of random programs from a previous mutation event such as, for instance, T_{R3} and T_{R4}[1], we can simplify the expression to:

$$T + \text{MS} \cdot (T_{R1} - T_{R2}) + \cancel{\text{MS} \cdot (T_{R3} - T_{R4})} + \text{MS} \cdot (T_{R5} - T_{R6}) - \cancel{\text{MS} \cdot (T_{R3} - T_{R4})}.$$

The new individual has now a smaller genotype than its parent. This idea inspires the DGSM, which consists in the simple removal of a previously added term, and, as shown, is simply an alternative way of applying GSM by using subtraction and a previously used pair of random programs. DGSM retains the same semantic properties as GSM, perturbing each semantic coordinate by $[-\text{MS}, \text{MS}]$, and inducing a unimodal error surface for all supervised learning problems.

As the name SLIM suggests, this algorithm uses both IGSM and DGSM, as two independent operators, each with its own probability of being applied, with the only restriction that whenever a program contains only one term, only IGSM can be applied. More specifically, the number of times each GSM is applied is controlled by a parameter, which regulates the probability of applying DGSM instead of IGSM. When this parameter is set to zero, only IGSM is applied, making SLIM completely identical to traditional GSGP.

2.3 Different SLIM variants

Several variants of SLIM have been introduced, corresponding to as many ways to define the GSM operators. In this work, we consider three variants of SLIM, each using a different definition of GSM.

In the standard GSM defined in Sect. 2.1, a function with codomain $[-\text{MS}, \text{MS}]$ is *added* to the parent tree. Following [33], we refer to this mutation as (+2SIG), since it involves two random trees, both wrapped in a sigmoid function. Recent studies [2,30,33] propose new ways to introduce ball mutations with range $[-\text{MS}, \text{MS}]$. Among the different variants, we focus on those that performed best in [33]. These are called *1SIG (representing one random expression wrapped in a sigmoid) and *ABS (representing absolute value).

Given a parent function $T : \mathbb{R}^n \to \mathbb{R}$, the *1SIG mutation with radius MS returns the function: $GSM(T) = T \cdot (1 + (\text{MS} \cdot (2 \cdot T_R - 1)))$, where T_R is a random function with codomain $[0, 1]$.

For the second variant, given a parent function $T : \mathbb{R}^n \to \mathbb{R}$, the *ABS mutation with radius MS returns the function: $GSM(T) = T \cdot \left(1 + \left(\text{MS} \cdot \left(1 - \frac{2}{1+|T_R|}\right)\right)\right)$, where T_R is a random function.

While the traditional GSM achieves a ball mutation by adding a positive or negative quantity centered at 0, the *1SIG and *ABS mutations achieve the same effect by multiplying by a quantity centered at 1. Unlike the +2SIG mutation, which uses two random functions, both (*1SIG) and (*ABS) use only one random function. As shown in [33], this modification reduces the growth rate of

---

[1] Reusing random material is not new; it was done, for example, in [31].

individuals during evolution while maintaining comparable performance to the standard GSM.

These two new mutation definitions can be applied in both GSGP and SLIM, creating two new variants of these algorithms. In this work, we use the varinats that were proven to be have the best overall performance in [33]. Specifically we use the SLIM variants that employ the +2SIG, *1SIG, and *ABS GSM, denoted as SLIM+2SIG, SLIM*1SIG, and SLIM*ABS, respectively.

2.4 Linked-List Implementation of SLIM

As presented by [33], SLIM can be implemented by representing the genotype of each individual as a linked list of subexpressions. The IGSM appends one element at the end of the list, while the DGSM removes one element from the list. According to the variant used, the semantics of the individual can be obtained by accumulating the semantics of all the blocks using the sum (for (SLIM+2SIG)) or the product (for (SLIM*1SIG) and (SLIM*ABS)). A visual representation of the linked list implementation and the application of the IGSM and DGSM is presented in Fig. 2.

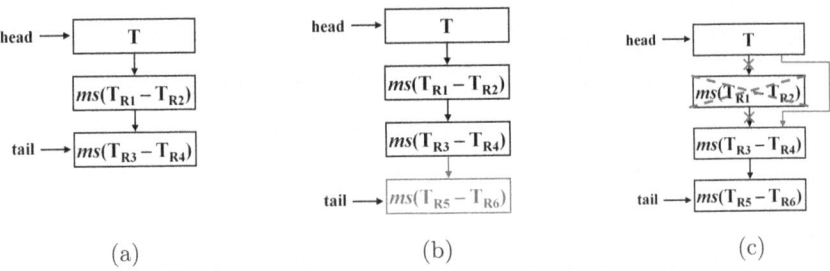

Fig. 2. An example genotype of a SLIM individual, with a visual representation of the effect of the IGSM and DGSM on its genotype. When the IGSM is applied, a block is appended to the genotype of the individual, while when the DGSM is used, a block is removed.

3 Experimental Methodology

This study investigates how the intrinsic characteristics of a dataset affect the choice of an appropriate MS for GSMs, including both IGSM and DGSM. While different datasets require different MS values for effective evolution, this need likely depends on variation in input and/or output ranges or data distribution. For example, even if data may cover a large domain, if the majority of values are concentrated in a small subrange, this could influence the effectiveness of the MS more than the range of the domain itself. To determine which factors (input/output range and/or data distribution) are more influential, and based on that to develop a general rule to set the optimal MS, we present an experimental study consisting of two phases.

The first phase focuses on synthetic datasets, where we can control the input and output ranges. We generate different versions with varied ranges and test multiple MSs on them. These synthetic functions are designed to cover a wide range of complexities (non-linearity, multimodality, etc.) that commonly occur in real-world datasets. Comparing these results allows us to derive a general rule, which will later be tested on real-world datasets where ranges cannot be controlled. In this phase, we do not impose any specific structure on the input distribution; instead, we generate random input values from a uniform distribution and compute target values using corresponding functions. After generating the datasets, we scale the input and output features using the following formula to map feature x to the desired $[a, b]$ range:

$$x_{scaled} = a + \left((b - a) \cdot \frac{(x - x_{min})}{x_{max} - x_{min}} \right)$$

where x_{min} and x_{max} are the minimum and maximum values of x. We vary both the input range $[a_{in}, b_{in}]$ and the output range $[a_{out}, b_{out}]$, and create datasets with small-scale inputs and large-scale outputs, large-scale inputs and small-scale outputs, and cases where both input and output are either on a small or large scale. We then evolve SLIM using different MSs values, then rank performance to identify which dataset characteristics influence optimal MS selection and derive a general rule. More specifically, we will use synthetic datasets consisting of 1000 rows and 10 input columns, where the input and output values are constrained to a range of $[0, max]$, with max selected from 1, 10, 100, 1000. For each combination, we evaluated five MSs, corresponding to the ranges $[0, 0.1]$, $[0, 1]$, $[0, 10]$, $[0, 100]$, and $[0, 1000]$. At each mutation event, the mutation step is randomly drawn, with uniform probability, from this range.

The second phase of our study validates this hypothesis using real-world benchmark datasets. We analyze the distribution of input and output features and then evolve SLIM with different MSs to see if the best-performing ones match the rule inferred in the first phase. Unlike synthetic datasets, real-world data distributions are typically not uniform, so we need to determine whether the shape of the distribution affects the performance of different MSs. To address this, we test the MSs from the first phase, while also introducing new ones based on the statistical properties of the data distributions, such as the median, minimum, and maximum values. The objective of this second phase is to confirm or refine the rule established in the first phase and ultimately provide a definitive optimal value for the MS, that can be determined a priori by analyzing the data distribution, enabling an informed choice before training.

In summary, this work seeks to answer the following Research Questions (RQs):

- **RQ1:** Is the optimal MS influenced by any dataset feature? If so, is it affected by the output range, input range, or both?

- **RQ2:** Does the data distribution also influence the MS, or is it solely dependent on the scale of the input and output values?
- **RQ3:** Can we define a general rule that ensures a good MS without the need for hyperparameter tuning?

4 Material and Experimental Setup

This section describes the employed datasets (both synthetic and real-world) (Sect. 4.1) and provides all the experimental settings (Sect. 4.2) to make the experiments completely reproducible. The code used for this experiment is available in the slim library.

4.1 Test Problems

As synthetic benchmarks, we select a set of functions widely used in the literature. These functions represent a variety of mathematical challenges and are commonly used to test the robustness and effectiveness of optimization algorithms. Table 1 outlines the functions used in this study, including their corresponding references. The chosen functions cover a broad range of characteristics, from multimodality and non-linearity (e.g., Ackley and Rastrigin) to simpler, unimodal problems (e.g., Sphere), allowing us to assess the impact of different MSs values across different landscapes.

Table 1. Benchmark functions used in the first phase of our experimental study.

Function Name	Formula	Reference
Ackley	$f(x) = -20\exp\left(-0.2\sqrt{\frac{1}{d}\sum_{i=1}^{d} x_i^2}\right) - \exp\left(\frac{1}{d}\sum_{i=1}^{d}\cos(2\pi x_i)\right) + 20 + e$	[1]
Alpine1	$f(x) = \sum_{i=1}^{d} |x_i \sin(x_i) + 0.1 x_i|$	[29]
Alpine2	$f(x) = \prod_{i=1}^{d} \sqrt{x_i}\sin(x_i)$	[11]
Michalewicz	$f(x) = -\sum_{i=1}^{d} \sin(x_i)\left(\sin\left(\frac{ix_i^2}{\pi}\right)\right)^{2m}$	[19]
Rastrigin	$f(x) = 10d + \sum_{i=1}^{d}\left(x_i^2 - 10\cos(2\pi x_i)\right)$	[27]
Rosenbrock	$f(x) = \sum_{i=1}^{d-1}\left[100(x_{i+1} - x_i^2)^2 + (x_i - 1)^2\right]$	[28]
Sphere	$f(x) = \sum_{i=1}^{d} x_i^2$	[15]

For the second phase, we conduct experiments on a range of real-world datasets from various domains. These datasets have often been used as benchmarks for GP, and their characteristics have been extensively analyzed in the literature [4,12,14,33]. Table 2 provides an overview of their key features, including the number of instances and variables. For a more detailed description of these datasets, the reader is referred to [18].

Table 2. Dataset specifications including the number of observations, features, input range, and target feature statistics.

Dataset	Observations	Features	Input Range	Target Range	Target Median	Target Std
yatch	307	6	[−5, 5]	[0, 62]	3	15.1
airfoil	1502	5	[0, 2000]	[103, 141]	125.7	6.9
slump	102	9	[0, 1050]	[0, 30]	21.5	8.7
strength	1029	8	[0, 1145]	[2.3, 83]	33.8	16.3
ppb	131	626	[−999, 2016472]	[0.5, 100]	85	31.1
bioav	359	241	[−71, 62651]	[0.4, 100]	75	30.5
ld50	234	626	[−999, 2289926]	[0.25, 89000]	775	2023.8

Table 3. Parameter values used in the evolutionary algorithm.

Parameter	Value
Generations	1000
Population Size	100
Elitism	True
Maximum Initial Depth	6
Function Set	$\{+, -, \times, \div\}$
Selection Algorithm	Tournament
Tournament Size	2
Initialization Method	Ramped Half-and-Half

4.2 Experimental Settings

Table 3 provides a detailed overview of the parameters utilized for the SLIM algorithm. Following [30], the probability of utilizing IGSM was set to 0.3 for all test datasets, with two exceptions: the LD50 dataset, where the probability was set to 0.1, and the concrete strength dataset, where it was increased to 0.5. The probability of utilizing DGSM was always set to 1 minus the probability of utilizing IGSM. At each application of the GSM, the mutation step is always generated from a uniform distribution in range [0, MS], where MS is the parameter studied in this work. The investigation of the most appropriate value for MS is performed for all the three distinct SLIM versions (defined in Sect. 2), to determine whether different mutation definitions lead to varying optimal MS values. Specifically, the variants considered were SLIM+2SIG, SLIM*1SIG, and SLIM*ABS. The first, SLIM+2SIG, was selected as it is based on the standard mutation definition, while the others were included based on their excellent performance demonstrated in previous studies [30,33].

We conducted 10 runs for each SLIM algorithm and each artificial dataset in the first phase, and 30 runs in the second phase. The Monte Carlo cross-validation [10,34] was applied with each train-test partition randomly drawn at each different run, and with 70% of the observations forming the training set and the remaining 30% the test set. The partitions were consistent across all runs. We evaluated the results by computing the median of the Root Mean Squared Error (RMSE) across all runs.

5 Experimental Results

This section presents the results of the experiments described in Sect. 3. The goal is to analyze these results and determine how the range of the values in the dataset impacts the choice of an appropriate MS value, starting by inferring a general rule using artificial datasets (in Sect. 5.1) and then confirming the validity of our hypothesis with more complex real-world datasets (in Sect. 5.2). All the presented results are relative to the best individual in the population on the training set at each generation.

5.1 Experiments on the Artificial Datasets

Table 4 and Table 5 show the results from the first phase of the experimental study. RQ1 investigates whether input, output range, or both influence the optimal mutation step (MS). Table 4 explores this through three experiments: equal input-output ranges, a fixed input range with varying output, and a fixed output range with varying input. Variants of SLIM with different MSs were tested, and median rankings from best (1) to worst (5) are shown. For example, with SLIM*1SIG and a range of [0,1], the rankings for MS are: (1) [0, 0.1], (2) [0, 1], (3) [0, 10], (4) [0, 10^2], (5) [0, 10^3]. The left side of the table shows that, with equal input-output ranges, small ranges favor lower MSs, and larger ranges favor higher MSs, suggesting that the optimal MS typically matches or is slightly below the range's order of magnitude. This pattern holds across all algorithms tested. However, when only the output range varies (middle section of the table), rankings remain similar, implying that output range primarily drives MS performance. Fixing the output range at [0, 1] (right side of the table) shows the best MS remains in the [0, 1] range regardless of input range, reinforcing that input range does not affect the best MS selection.

Table 4. The results, as median rankings across all the datasets, of the first three experiments for all the studied algorithms.

In & Out Range Equal						Fix In Range						Fix Out Range					
*1SIG		\multicolumn{5}{c	}{Mutation Step}	*1SIG		Mutation Step				*1SIG		Mutation Step			
In	Out	0.1	1	10	10^2	10^3	In	Out	0.1	1	10	10^2	10^3	In	Out	0.1	1	10	10^2	10^3
1	1	1	2	3	4	5	1	1	1	2	3	4	5	1	1	1	2	3	4	5
10	10	1	2	2	3	5	1	10	2	3	2	4	5	10	1	3.5	1	2	4	4.5
10^2	10^2	4	5	2	3	2	1	10^2	5	4	2	1	3	10^2	1	3.5	1	2	4	4
10^3	10^3	4	5	3	1	3	1	10^3	4	5	3	2	1	10^3	1	3	1	3	4	5
*ABS		Mutation Step					*ABS		Mutation Step					*ABS		Mutation Step				
In	Out	0.1	1	10	10^2	10^3	In	Out	0.1	1	10	10^2	10^3	In	Out	0.1	1	10	10^2	10^3
1	1	1	2	3	4	5	1	1	1	2	3	4	5	1	1	1	2	3	4	5
10	10	1.5	2	2	3.5	5	1	10	1	2	3	4	5	10	1	3.25	1	2.5	4	4
10^2	10^2	4	5	2	2	2.5	1	10^2	5	3	2	1	4	10^2	1	3.5	1	2	4	4
10^3	10^3	4	4	3	1	2	1	10^3	5	3.5	3.5	2	1	10^3	1	3	1	2	4	5
+2SIG		Mutation Step					+2SIG		Mutation Step					+2SIG		Mutation Step				
In	Out	0.1	1	10	10^2	10^3	In	Out	0.1	1	10	10^2	10^3	In	Out	0.1	1	10	10^2	10^3
1	1	1	2	3	4	5	1	1	1	2	3	4	5	1	1	1	2	3	4	5
10	10	2	2	2	4	5	1	10	2	2	2	4	5	10	1	3	1	3	4	4
10^2	10^2	4	4	2	2	3	1	10^2	5	4	2	1	3	10^2	1	3	1	2	4	5
10^3	10^3	4	4	3	1	2	1	10^3	5	4	3	2	1	10^3	1	2	1	3	4	5

Table 5. The results, as median rankings across all the datasets, of the last three experiments for all the studied algorithms.

		Exp 1							Exp 2							Exp 3				
*1SIG		Mutation Step					*1SIG		Mutation Step					*1SIG		Mutation Step				
In	Out	0.1	1	10	10^2	10^3	In	Out	0.1	1	10	10^2	10^3	In	Out	0.1	1	10	10^2	10^3
10^2	1	3.5	1	2	4	4	10^3	1	3	1	3	4	5	1	1001	5	4	3	2	1
10^2	10	1	2	3	3	5	10^3	10	1	2	3	4	5	1	1010	5	4	3	2	1
10^2	10^2	4	5	2	3	2	10^3	10^2	5	2	2	1	3	1	1100	5	4	3	1	2
10^2	10^3	2	4	5	3	2	10^3	10^3	4	5	3	1	3	1	2000	5	4	3	1	2
*ABS		Mutation Step					*ABS		Mutation Step					*ABS		Mutation Step				
In	Out	0.1	1	10	10^2	10^3	In	Out	0.1	1	10	10^2	10^3	In	Out	0.1	1	10	10^2	10^3
10^2	1	3.5	1	2	4	4	10^3	1	3	1	2	4	5	1	1001	4	3.5	4	2	1
10^2	10	1	2	3	3	5	10^3	10	1	2	3	3	5	1	1010	5	3.5	3.5	2	1
10^2	10^2	4	5	2	2	2.5	10^3	10^2	5	3	2	2	3	1	1100	5	3.5	3.5	1	2
10^2	10^3	2	4	5	3	1.5	10^3	10^3	4	5	3	1	2	1	2000	5	3	4	1	2
+2SIG		Mutation Step					+2SIG		Mutation Step					+2SIG		Mutation Step				
In	Out	0.1	1	10	10^2	10^3	In	Out	0.1	1	10	10^2	10^3	In	Out	0.1	1	10	10^2	10^3
10^2	1	3	1	2	4	5	10^3	1	2	1	3	4	5	1	1001	5	4	3	2	1
10^2	10	2	2	3	3	5	10^3	10	1	2	3	4	5	1	1010	5	4	3	1	2
10^2	10^2	4	4	2	2	3	10^3	10^2	5	4	2	2	3	1	1100	5	4	3	1	2
10^2	10^3	3	4	5	3	2	10^3	10^3	4	4	3	1	2	1	2000	5	4	3	1	2

This supports that MS operates within output space, where its magnitude must align with the output range. To further confirm, additional tests are presented in Table 5, which verify that optimal MSs match the output range's order of magnitude across varying input ranges. For example, with increased input ranges ($[0, 100]$ and $[0, 1000]$), optimal MSs are still determined by the output range alone. Finally, a test with the output range scaled by 1000 shows larger optimal MSs ($[0, 100], [0, 1000]$), confirming that the MS depends on the output range scale rather than absolute values, emphasizing the importance of matching the order of magnitude.

5.2 Experiments on Real-World Datasets

The first phase of our study on synthetic datasets suggested that output scale largely determines the optimal mutation step (MS). Here, we test this on real-world datasets, which present additional challenges like noise and non-uniform distributions. Along with confirming our hypothesis for RQ1, we address RQ2 by examining whether the data distribution itself significantly influences MS selection. The results, shown in Fig. 3, illustrate test fitness progression across generations for different SLIM versions (columns) and datasets (rows). We test MSs from Sect. 5.1 alongside two additional values based on output distribution: $[0, \text{median}]$ and $[\text{min}, \text{max}]$. As concluded from prior results, these values are based on the output range alone. The conclusions drawn from these plots are statistically validated by a Mann-Whitney U test (not shown here due to space limitations). Figure 4 presents the distributions of target variable for all the studied datasets.

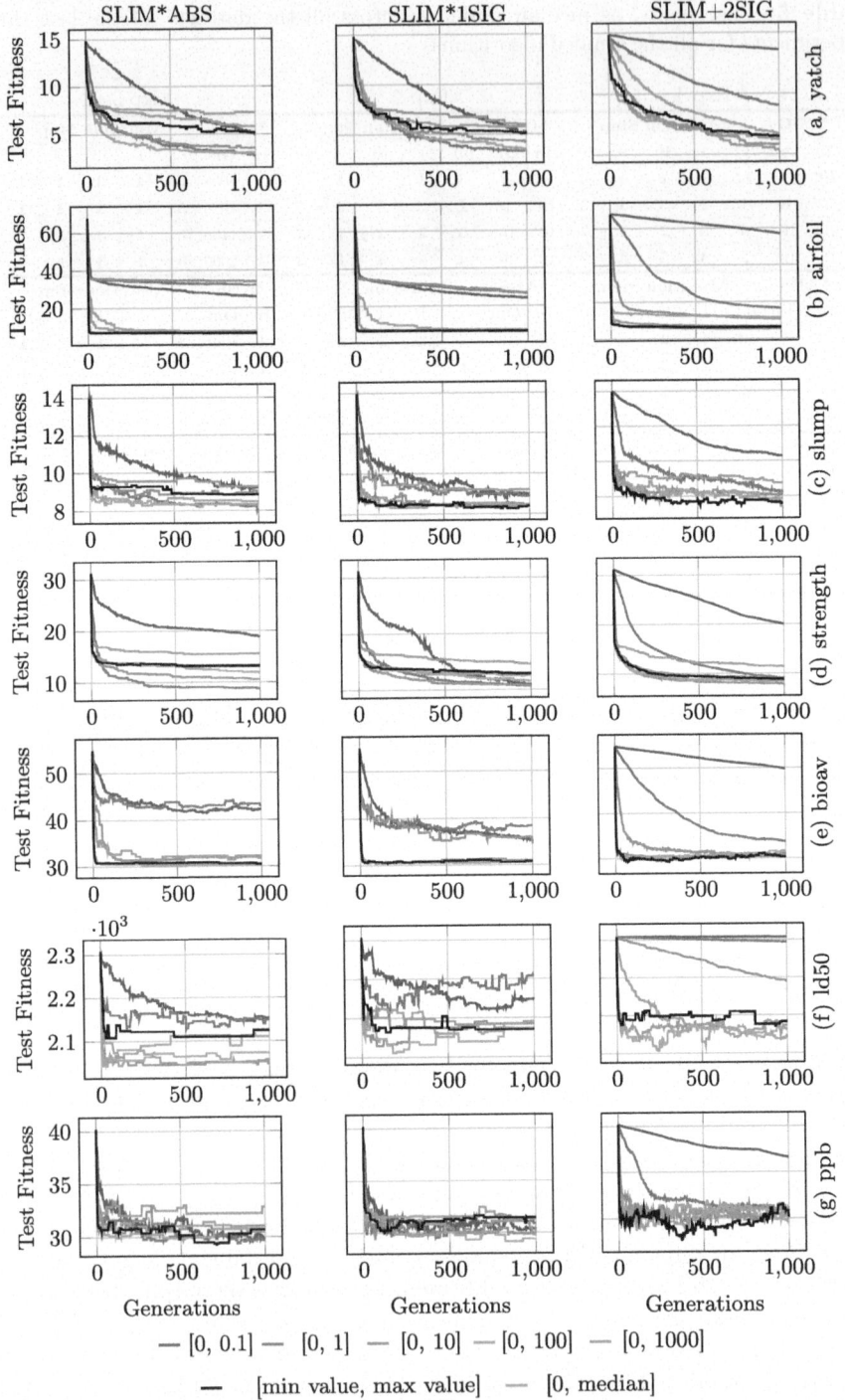

Fig. 3. Test Fitness of various datasets across three algorithms: SLIM*ABS, SLIM*1SIG, and SLIM+2SIG

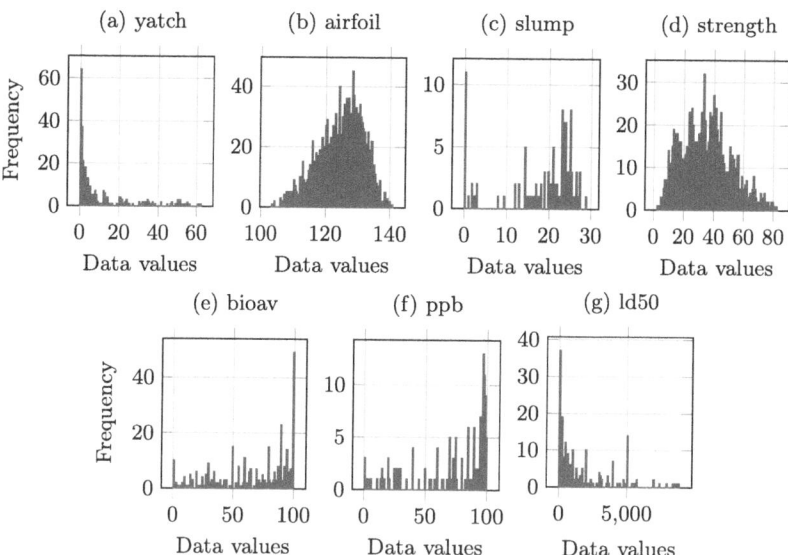

Fig. 4. Distribution of Target Variables for Various Datasets

Analyzing the results, the hypothesis that an optimal MS should match the order of magnitude of the output range holds on real-world datasets. When MS values are significantly lower than the target scale, models converge slowly to suboptimal solutions, as seen in the airfoil dataset (target range [103, 141]). Here, models with smaller MS values ([0, 0.1], [0, 1]) yield higher test fitness, while larger values closer to the target scale ([0, 100], [0, 1000]) show better performance. Conversely, a too-large MS relative to the output scale also hinders convergence, as seen in the strength dataset (target range [2.3, 80]). A MS of [0, 1000] leads to quick but suboptimal convergence. These patterns confirm that setting MS to align with the target scale supports optimal performance. However, real-world data reveal nuances beyond output range alone, particularly in datasets with skewed distributions. For instance, datasets like strength, bioav, and ppb have similar ranges, yet MS [0, 1] performs well only in strength due to differences in data skewness. In the yacht dataset (target range [0, 62]), where values concentrate near the lower end, MS [0, 1] performs well despite being smaller than the target range. These observations indicate that skewed distributions benefit from MSs that better reflect data density. The consistency of [0, median] as a top-performing MS across datasets, as seen in Fig. 3, underscores the importance of incorporating distribution properties. In contrast, the performance of [min, max], which matches target scale but ignores data skew, varies with distribution. For instance, in the left-skewed ld50 dataset (range [0, 89000], median 775), [min, max] underperforms due to its disproportionate maximum. Thus, while [min, max] performs well for symmetric or right-skewed distributions, left-skewed distributions benefit from MSs like [0, median] that align with data density. In summary, both range and distribution shape inform MS selection. A MS smaller than the target range slows convergence to suboptimal solutions, while a too-large MS accelerates convergence but misses optimal solutions. For symmetric or right-skewed distributions, [min, max] performs well, whereas [0, median] emerges as a robust choice across diverse data shapes, yielding optimal results across all datasets tested.

6 Conclusion and Future Work

This study presents an effective approach to optimizing the mutation step (MS) parameter in the Semantic Learning algorithm based on Inflate and deflate Mutation (SLIM) by basing its selection on dataset-specific characteristics. SLIM, a variant of Geometric Semantic Genetic Programming (GSGP), uses Inflate Geometric Semantic Mutation (IGSM) and Deflate Geometric Semantic Mutation (DGSM) to build compact, interpretable models, inducing a unimodal error surface. The MS parameter, crucial for SLIM's performance, traditionally requires extensive hyperparameter tuning to control offspring perturbation. Our findings streamline this process by establishing a practical selection rule for MS that eliminates manual tuning. Experiments on synthetic and real-world datasets showed that MS values aligning with the target output's magnitude, rather than input scale, improved convergence rate as well as accuracy. Real-world datasets confirmed that the target output scale is the main factor for effective MS selection. Skewed distributions performed best with MS centered on the target median, which improved model stability. Our selection rule based on [0, median(target)] offers a practical alternative to hyperparameter tuning, especially for complex datasets.

This work simplifies SLIM configuration, advancing its accessibility for machine learning and symbolic regression. Future research could investigate dynamic MS adjustments during the evolution or even the introduction of an individual-based MS, where each individual adapts MS differently from the others, based on its fitness, data distribution, diversity or other characteristics. Another future work scenario should focus on how feature standardization, as proposed in [8], affects the MS parameter choice. Finally, a Geometric Semantic Crossover (GSC) crossover, leveraging SLIM's linked-list structure, and generating individuals of a compact size, remains a valuable avenue for future investigation.

Acknowledgments. This work was supported by national funds through FCT (Fundação para a Ciência e a Tecnologia), under the project - UIDB/04152/2020 - Centro de Investigação em Gestão de Informação (MagIC)/NOVA IMS (https://doi.org/10.54499/UIDB/04152/2020).

References

1. Ackley, D.H.: A connectionist machine for genetic hillclimbing (1987)
2. Bakurov, I., et al.: Geometric semantic genetic programming with normalized and standardized random programs. Genetic Program. Evolvable Mach. **25**(1) (2024). https://doi.org/10.1007/s10710-024-09479-1
3. Bonin, L., Rovito, L., De Lorenzo, A., Manzoni, L.: Cellular geometric semantic genetic programming. Genet. Program Evolvable Mach. **25**(1), 8 (2024)
4. Castelli, M., Manzoni, L., Gonçalves, I., Vanneschi, L., Trujillo, L., Silva, S.: An analysis of geometric semantic crossover: a computational geometry approach, pp. 201–208 (2016). https://doi.org/10.5220/0006056402010208
5. Castelli, M., Trujillo, L., Vanneschi, L., Silva, S., Z-Flores, E., Legrand, P.: Geometric semantic genetic programming with local search. In: Proceedings of the 2015 Annual Conference on Genetic and Evolutionary Computation, pp. 999–1006 (2015)
6. Castelli, M., Vanneschi, L., Popovič, A.: Parameter evaluation of geometric semantic genetic programming in pharmacokinetics. Int. J. Bio-Inspir. Comput. **8**(1), 42–50 (2016). https://doi.org/10.1504/IJBIC.2016.074634, https://www.inderscienceonline.com/doi/abs/10.1504/IJBIC.2016.074634, pMID: 74634
7. Dick, G.: An ensemble learning interpretation of geometric semantic genetic programming. Genetic Programm. Evolvable Mach. **25**(1) (2024). https://doi.org/10.1007/s10710-024-09482-6
8. Dick, G., Owen, C.A., Whigham, P.A.: Feature standardisation and coefficient optimisation for effective symbolic regression. In: ACM Conferences, pp. 306–314. Association for Computing Machinery, New York (2020). https://doi.org/10.1145/3377930.3390237
9. Dresden, A.: The fourteenth western meeting of the American Mathematical Society. Bull. Am. Math. Soc. **26**(9), 385–396 (1920)
10. Dubitzky, W., Granzow, M., Berrar, D.P.: Fundamentals of Data Mining in Genomics and Proteomics. Springer (2006)
11. Epitropakis, M.G., et al.: Benchmark problems for CEC 2013 special session on real-parameter optimization. In: Proceedings of the 2013 IEEE Congress on Evolutionary Computation, pp. 1096–1103 (2013)
12. Farinati, D., Bakurov, I., Vanneschi, L.: A study of dynamic populations in geometric semantic genetic programming. Inf. Sci. **648**, 119513 (2023). https://doi.org/10.1016/j.ins.2023.119513, https://www.sciencedirect.com/science/article/pii/S0020025523010988
13. Gonçalves, I., Silva, S., Fonseca, C.M.: On the generalization ability of geometric semantic genetic programming. In: Machado, P., et al. (eds.) Genetic Programming, pp. 41–52. Springer, Cham (2015)
14. Gonçalves, I., Silva, S., Fonseca, C.M., Castelli, M.: Unsure when to stop? In: Proceedings of the Genetic and Evolutionary Computation Conference. ACM (2017). https://doi.org/10.1145/3071178.3071328
15. Jong, K.D.: An analysis of the behavior of a class of genetic adaptive systems. Doctoral dissertation, University of Michigan (1975)
16. Koza, J.R.: Genetic Programming: On the Programming of Computers by Means of Natural Selection. MIT Press, Cambridge (1992)
17. McDermott, J., Agapitos, A., Brabazon, A., O'Neill, M.: Geometric semantic genetic programming for financial data. In: Esparcia-Alcázar, A.I., Mora, A.M. (eds.) EvoApplications 2014. LNCS, vol. 8602, pp. 215–226. Springer, Heidelberg (2014). https://doi.org/10.1007/978-3-662-45523-4_18

18. McDermott, J., et al.: Genetic programming needs better benchmarks. In: Proceedings of the 14th Annual Conference on Genetic and Evolutionary Computation, GECCO 2012, pp. 791–798. ACM, New York (2012). https://doi.org/10.1145/2330163.2330273
19. Michalewicz, Z.: Genetic Algorithms + Data Structures = Evolution Programs. Springer (1994)
20. Moraglio, A., Krawiec, K., Johnson, C.G.: Geometric Semantic Genetic Programming. In: Coello, C., Cutello, V., Deb, K., Forrest, S., Nicosia, G., Pavone, M. (eds.) Parallel Problem Solving from Nature. LNCS, vol. 7491, pp. 21–31. Springer (2012)
21. Moraglio, A., Mambrini, A.: Runtime analysis of mutation-based geometric semantic genetic programming for basis functions regression. In: Proceedings of the annual international conference on Genetic and Evolutionary Computation, GECCO 2013, pp. 989–996. ACM, New York (2013)
22. Nadizar, G., Sakallioglu, B., Garrow, F., Silva, S., Vanneschi, L.: Geometric semantic GP with linear scaling: darwinian versus Lamarckian evolution. Genet. Program Evolvable Mach. **25**(2), 1–24 (2024)
23. Nguyen, Q.U.: Examining semantic diversity and semantic locality of operators in genetic programming. Ph.D. thesis, University College Dublin, Ireland (2011). http://ncra.ucd.ie/papers/Thesis_Uy_Corrected.pdf
24. Pietropolli, G., Manzoni, L., Paoletti, A., Castelli, M.: Combining geometric semantic GP with gradient-descent optimization. In: Medvet, E., Pappa, G., Xue, B. (eds.) Genetic Programming, pp. 19–33. Springer, Cham (2022)
25. Pietropolli, G., Manzoni, L., Paoletti, A., Castelli, M.: On the hybridization of geometric semantic GP with gradient-based optimizers. Genet. Program Evolvable Mach. **24**(2), 16 (2023)
26. Poli, R., Langdon, W.B., Mcphee, N.F.: A field guide to genetic programming (2008)
27. Rastrigin, L.A.: Systems of extremal control. Moscow University (1974)
28. Rosenbrock, H.H.: An automatic method for finding the greatest or least value of a function. Comput. J. **3**, 175–184 (1960)
29. Tang, K.S., et al.: Overlapping decomposition for scalability of GP-based function optimization. In: Proceedings of the 2007 IEEE Congress on Evolutionary Computation, pp. 2044–2051 (2007)
30. Vanneschi, L.: SLIM_GSGP: the non-bloating geometric semantic genetic programming. In: Giacobini, M., Xue, B., Manzoni, L. (eds.) Genetic Programming, pp. 125–141. Springer, Cham (2024)
31. Vanneschi, L., Castelli, M., Manzoni, L., Silva, S.: A new implementation of geometric semantic GP and its application to problems in pharmacokinetics. In: Krawiec, K., Moraglio, A., Hu, T., Uyar, A.S., Hu, B. (eds.) Proceedings of EuroGP. LNCS, pp. 205–216. Springer (2013)
32. Vanneschi, L., Castelli, M., Silva, S.: A survey of semantic methods in genetic programming. Genetic Program. Evolvable Mach. **15**(2), 195–214 (2014). https://doi.org/10.1007/s10710-013-9210-0
33. Vanneschi, L., Farinati, D., Rasteiro, D., Rosenfeld, L., Pietropolli, G., Silva, S.: Exploring non-bloating geometric semantic genetic programming. Genetic Program. Theory Pract. (to appear)
34. Vanneschi, L., Silva, S.: Lectures on Intelligent Systems. Natural Computing Series. Springer (2023). https://doi.org/10.1007/978-3-031-17922-8

35. Vanneschi, L., Silva, S., Castelli, M., Manzoni, L.: Geometric semantic genetic programming for real life applications. In: Riolo, R., Vladislavleva, K., Moore, J. (eds.) Genetic Programming Theory and Practice XI. Genetic and Evolutionary Computation. Springer (2013)

Multi-objective Evolutionary Design of Explainable EEG Classifier

Martin Hurta[✉][iD], Anna Ovesna[iD], Vojtech Mrazek[iD], and Lukas Sekanina[iD]

Faculty of Information Technology, Brno University of Technology,
Brno, Czech Republic
{ihurta,mrazek,sekanina}@fit.vut.cz, xovesn03@stud.fit.vut.cz

Abstract. Deep neural networks (DNNs) have achieved impressive results in many fields. However, the use of black-box solutions based on DNNs in medical applications poses challenges, as understanding the rationale behind decisions is crucial for application in healthcare. For those reasons, we propose a new method for the evolutionary multi-objective design (MOD) of small and potentially explainable EEG (Electroencephalography) signal classifiers. We evaluate a combination of genetic algorithm (GA) for feature selection with multiple algorithms for the automated design of the classifier, including Support Vector Machine, k-Nearest Neighbors, and Naive Bayes. To further improve the classification quality and obtain less complex solutions, we compare three different MOD scenarios targeting the accuracy, specificity, sensitivity, and the number of used features. In addition, we evaluate the use of Cartesian Genetic Programming (CGP) as a way to achieve smaller and more interpretable solutions and combine it with the compositional co-evolution of selected features to improve computational requirements and find solutions in a reasonable time. The proposed methods are experimentally evaluated on tasks of alcohol use disorder and major depressive disorder classification. Experimental results show that newly proposed MOD scenarios lead to significantly better trade-offs between the accuracy and the number of features compared to the state-of-the-art method employing the NSGA-II algorithm. The proposed co-evolution of features (evolved by GA) and classifier (evolved by CGP) allowed the design of small and potentially explainable solutions and led to 20–100 times faster convergence than the baseline CGP-based approach.

Keywords: Multi-objective design · Classification · Explainability · Co-evolution · Genetic algorithm · Cartesian genetic programming · EEG

1 Introduction

Electroencephalography (EEG) is a medical method for measuring brain response to different stimuli and state of mind based on electrical signals produced within the brain [3]. EEG is non-invasive, widely available, mobile, and cost-effective and is used in a growing number of fields. The use of EEG signals has already been successfully evaluated in tasks such as decoding brain dynamics during different cognitive tasks, emotion recognition, neuromarketing research, brain stimulation, measuring the effect of brain treatments and interventions, or diagnosis and tracking of neurological disorders, including epilepsy and Alzheimer's disease, Parkinson's disease, dementia, and sleep disorders [3,14].

Some essential diagnostic methods, such as the classification of *alcohol use disorder* (AUD) or *major depressive disorder* (MDD), benefit especially from the use of EEG data. Classification of AUD, one of the most common mental illnesses in the world, usually depends either on trustworthiness in questionary or complete physical and psychological evaluation. However, this approach is time-consuming, laborious, and subjective [10]. MDD is another of the most prevalent mental health issues, with approximately more than 264 million affected people worldwide [22]. Diagnosis of MDD is usually made through clinical questionnaire-based assessment, which is determined by patients' responses and behavioral activities [11]. Again, these are highly subjective metrics that make diagnosis challenging [11]. This work aims to create a method that would help physicians with the classification of different health conditions that manifest in changes in brain activity and possibly provide new information about how these conditions affect the human brain.

Traditional solutions for automated classification of AUD and MDD based on *machine learning* (ML) algorithms require manual selection of features by medical experts [6,10]. The selection can often be difficult and makes scaling and use in new tasks problematic [10]. The use of *deep learning* algorithms can overcome the need for manual feature selection, but it is usually at a price of highly complex and uninterpretable solutions, which makes their use in the medical field problematic [14].

Mrazek et al. [14] proposed a method for the automated selection of features for interpretable classifiers in the task of MDD classification. The method extracts features by *Welch's method*, a *spectral density* estimation approach and uses ML algorithms *Support Vector Machine* (SVM) and *K-Nearest Neighbors* (k-NN) to evaluate the classification accuracy achieved for different candidate feature sets. The evolution of candidate feature sets is guided by the NSGA-II algorithm [4] as a way to achieve a good trade-off between *sensitivity*, *specificity* and the number of features. However, their study did not evaluate any other multi-objective design (MOD) algorithms published in the literature. The NSGA-II algorithm was used without any analysis of its effectivity in the task of the EEG signal classification. Evaluation of only the SVM and k-NN classifiers also leaves space for additional experiments with other algorithms, potentially leading to even better interpretability of the results.

The main contributions of this work are as follows. We evaluate not only the NSGA-II algorithm but also convert the MOD to a single-objective design by (i) using a weighted sum of objectives and (ii) introducing a constraint on the number of features. Despite of potential simplification, the notion of explainability adopted in our work is straightforward - smaller classifiers composed of simple operators and utilizing a few features show better explainability than complex classifiers [2]. For that reason, we aim to evaluate not only common ML algorithms (i.e., SVM, k-NN, and *Naive Bayes*) but also an *evolutionary algorithm* called *Cartesian Genetic Programming* (CGP) [13], which was successfully applied to the classification of various health conditions including Parkinson's and Alzheimer's disease [20] and can provide simple and potentially explainable classifiers [17]. As the evolution of CGP is computationally expensive, we employ *compositional co-evolution* of CGP classifiers together with features. To ensure applicability to the classification of new health conditions, we apply the proposed method to EEG data from two separate data sets of AUD and MDD classification and present the most significant features in their classification, as determined by our method.

2 Proposed Method

The proposed method aims to design small and potentially explainable solutions by utilizing a small number of features and less complex classifiers. To evaluate several MOD methods and classifier design algorithms, and choose the best setup for our method, we propose the following methodology depicted in Fig. 1.

The proposed method first preprocesses EEG data using *Wavelet transformation* [12] for AUD data and *Welch's method* [21] for MDD data to reduce noise and decompose signal at different channels. The proposed method works generally and with an arbitrary feature extraction algorithm. A wide number of features is then extracted and prepared for selection by GA. GA is initialized with a population of random subsets of extracted features. Evaluation of candidate feature sets is performed by means of the selected classifier. The classifier is trained utilizing the candidate features on the training data. The accuracy, sensitivity, and specificity achieved on the validation data are together with the number of used features utilized as the optimization objectives in the MOD method. This method aims to, over multiple generations, find a suitable set of features for the currently solved task and provide a classifier working with these features. Individual parts of the proposed method are further described in the following subsections.

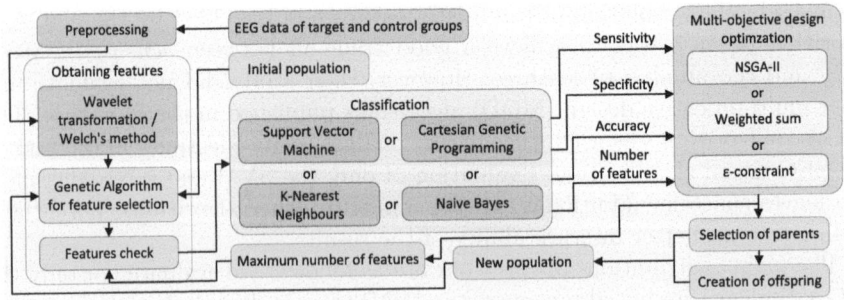

Fig. 1. Overview of the methodology for the MOD of EEG classifiers using a small number of features. Green represents the part used in traditional ML algorithms. Yellow represents newly proposed parts introduced for the optimization of the chosen feature set. (Color figure online)

2.1 Data Preprocessing

The original EEG data represent the values of the EEG signal for the captured channels in the time domain. The data are cleaned and filtered. Since AUD and MDD are not event-based (such as epilepsy), we transformed the data to the frequency domain. AUD data are first constricted to selected 16 channels based on the results of Ong et al. [16]. Wavelet transformation with Daubechies 4 wavelet is performed on the selected channels to receive six separate coefficients at different frequencies (i.e., Delta 0–4 Hz, Theta 4–8 Hz, Alpha 8–16 Hz, Beta 16–32 Hz, Gamma 32–64 Hz). The first coefficient with a frequency of 64–128 Hz is in accordance with the common methodology discarded as it contains unwanted artifacts. The second coefficient with the frequency of 32–64 Hz was kept despite its usual removal, as Rodrigues et al. [19] has shown that it is beneficial in the task of AUD classification.

The vectors representing the coefficients of different frequencies are further preprocessed into statistical values of minimum, maximum, median, mean, standard deviation, variance, coefficients of skewness and kurtosis, spectral power, entropy, and peak frequency. In total, this means 16 EEG channels, 5 selected coefficients per channel, and 11 features of each coefficient, thus making it a total of 880 different extracted features describing the AUD data.

MDD data are preprocessed in accordance with findings in Mrazek et al. [14], where *Power Spectrum Density* (PSD) is used as a feature for the classification of MDD. PSD is calculated from the data using the *Fast Fourier Transform* (FFT) with Welch's method and Hamming window to estimate the power spectrum of time series with 2-second segments, 50% overlapping, and the non-equidistant FFT. With 19 used EEG channels, the preprocessing results in a total of 950 extracted features.

2.2 Feature Selection and Classifier Design

Feature selection is performed by GA, with the *chromosome* represented as a 1D array of length equal to the total number of features in the data set (i.e., 880 for AUD and 950 for MDD). The chromosome has binary encoding: a single bit determines whether a given feature is used (1) or not (0). The first generation of candidate solutions is generated randomly. The quality of candidate feature sets is expressed as the sensitivity, specificity, accuracy (achieved by the classifier utilizing selected features on training data) and the number of features used. The MOD method then selects parent solutions with good trade-offs among the criteria. Offspring are created using one-point crossover and a bit flip at random positions.

Four algorithms are evaluated in the task of classifier design. These include common ML algorithms (SVM, k-NN, Naive Bayes), and evolutionary algorithm CGP. The common ML algorithms are considered because they represent, in today's standard, less complex approaches and potentially allow for better explainability than deep neural networks. Nevertheless, SVM still provides rather black-box solutions and does not allow for full explainability. Naive Bayes can provide information about the probability of features being connected with the classified disorder but does not provide information about how they function. The k-NN algorithm lacks a traditional model that could be interpreted and the mapping of data into higher dimensions makes the explainability of similarity between data points non-trivial. CGP was selected as it was already successfully applied in the tasks of classifying Parkinson's disease, Alzheimer's disease, breast cancer [20] or Levodopa Induced Dyskinesia [9]. As CGP generates small explainable classifiers in the form of arithmetic expressions (as shown in Fig. 2), it can provide new information about features important in the recognition of different health conditions to neuroscientists.

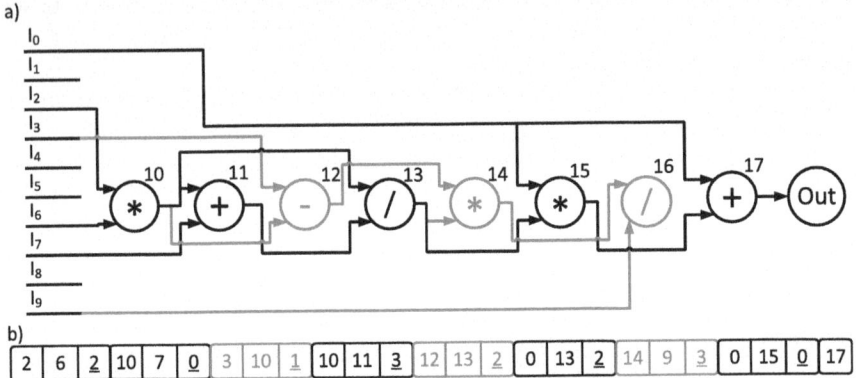

Fig. 2. An example of (a) CGP computational graph and (b) corresponding encoding. Displayed CGP has parameters of 10 inputs, 1 output, 8 columns, 1 row, and a set of functions $\Gamma = \{+^{(0)}, -^{(1)}, *^{(2)}, /^{(3)}\}$. Computational graph corresponds to arithmetic expression $y = I_0 + I_0 * ((I_2 * I_6)/(I_2 * I_6 + I_7))$. Grayed-out parts represent parts not utilized in the final arithmetic expression.

2.3 Multi-objective Design Scenarios

In this study, we consider three possible scenarios for the evolutionary multi-objective design:

Scenario S1: NSGA-II. The first scenario uses the NSGA-II algorithm and corresponds to the method from the recent work [14]. NSGA-II aims to sustain a diverse set of candidate solutions throughout the evolution process. This is achieved by maintaining Pareto fronts of candidate solutions up to the last one that can still fully fit into the number of required parent solutions of the next generation. If the remaining number of required parents is smaller than the size of the next Pareto front, the crowding distance between its solutions is calculated, and the most diverse candidate solutions are selected until the required number of parents is met.

Scenario S2: SO-W (Single Objective Weighted Sum). The second design scenario implements MOD through its transformation into a single objective design by use of a weighted sum. This approach requires a setting of suitable weight for each objective. The fitness of the candidate solution s is calculated as:

$$f_s = w_1 * s_{accuracy} + w_2 * s_{sensitivity} + w_3 * s_{specificity} + w_4 * s_{num\_of\_features}, \quad (1)$$

where w_i are parts of the weight vector. This MOD method allows setting different importance levels for different objectives by setting their respective weights higher in comparison to the rest of the weights.

Scenario S3: SO-C (Single Objective Constraint). The third scenario solves MOD by transforming it into a single-objective design by introducing constraints. In this case, the number of extracted features is constrained with ε. The evolution is focused on achieving the highest accuracy possible for a given number of features ε. Combining solutions obtained from independent runs with different ε settings can, in an ideal case, lead to wide Pareto front of solutions. The fitness function implementing the ε constraint is as follows:

$$f_s = \begin{cases} s_{accuracy}, & \text{if } s_{num\_of\_features} \leq \varepsilon \\ 0, & \text{otherwise} \end{cases}. \quad (2)$$

2.4 Compositional Co-evolution of Features and Classifiers

The proposed method for the design of explainable classifiers requires training the classifier for every candidate feature set. This becomes problematic for algorithms with longer training times (e.g., CGP) and disables running the evolution process for a sufficient number of evaluations.

In the task of evolutionary design of preprocessing algorithm and classifier of Levodopa Induced Dyskinesia, Hurta et al. [8] proposed a three-population algorithm, which simultaneously develops one population of CGP to design a preprocessing function and the second population of CGP classifiers. The third population implemented Adaptive Size Fitness Predictors (ASFP) [5].

We propose to adapt the three-population algorithm from [8] to the form of a two-population co-evolutionary algorithm for the simultaneous feature selection by GA and classifier design by CGP. Both populations cooperate during the evaluation of fitness, where the fitness is calculated as the fitness of the composition of selected features and classifier. This method will eliminate the need for a complete run of the CGP algorithm for each of the candidate feature sets and thus accelerate the evolution process. An overview of the proposed co-evolutionary design method is shown in Fig. 3.

During the initialization phase, a random initial population of both GA for features and CGP for classifiers is created. All combinations of features and classifiers are then evaluated for their fitness (i.e., accuracy) to obtain the initial best functioning pair of feature set FE_{init} and classifier CL_{init}. To eliminate computationally expensive each-with-each evaluation in later generations, an archive of the top-ranked solutions FE_{best} and CL_{best} is created and initialized with candidate solutions FE_{init} and CL_{init}. During the evolution process, both

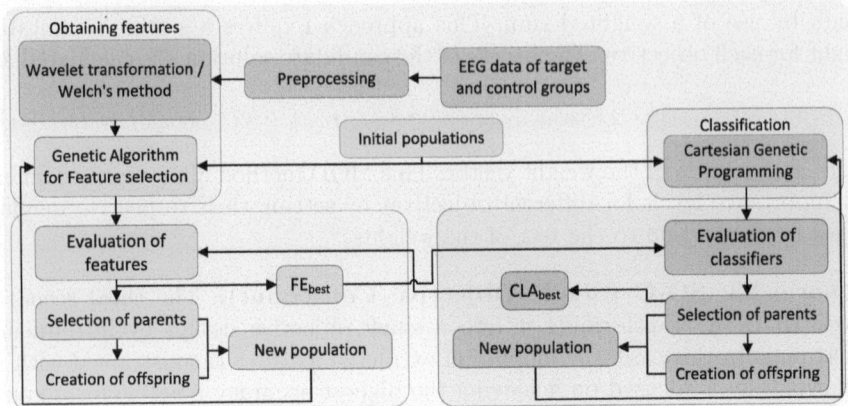

Fig. 3. Overview of a compositional co-evolutionary method utilizing GA for feature selection and CGP for design of classifier. Both populations take turns in executing the evolution. Candidate solutions of one population are evaluated in connection with the currently top-tanked solution from the second population. The fitness of candidate solutions is based on classification accuracy achieved by combining a feature set and classifier.

populations take turns. In each turn, candidate solutions of one population are evaluated in connection with the current top-ranked candidate solution from the other population, with fitness equal to the accuracy achieved by the created pair of features and classifier. Archives FE_{best} and CL_{best} are updated each time the corresponding population reaches the new best fitness. Archives FE_{best} and CL_{best} also contain final solutions at the end of the evolution process. The method is further explained in Algorithm 1.

The challenge of combining the variable size of feature sets with one continuous run of the CGP algorithm, which, by definition, has a fixed number of inputs, was solved by setting the number of features to a fixed size. The number of features is equal to the number of inputs of CGP. CGP can then utilize only a subset of the inputs, thus effectively resulting in MOD scenario S2: SO-C with ε equal to the number of CGP inputs.

3 Experimental Results

This section presents the clinical EEG data for tasks of classifying AUD and MDD, the experimental setup, and the experimental evaluation results of the proposed methods.

3.1 Major Depressive Disorder and Alcoholism Data Sets

EEG data used in this paper come from two clinical studies, both following the international standard for EEG sensor location placement called the 10–20

Algorithm 1: Pseudocode of co-evolution of Features and Classifiers

$FS \leftarrow$ Random initial population of GA (feature subsets)
$CL \leftarrow$ Random initial population of CGP (classifiers)
Evaluation of fitness of all compositions of FS_i and CL_j
$Fit_{best} \leftarrow$ Highest achieved fitness
$FS_{best} \leftarrow$ Feature set from composition with the highest fitness
$CL_{best} \leftarrow$ Classifier from composition with the highest fitness

while *terminating conditions are not met* **do**
    // Feature selection part of evolution
    for FS_i in FS **do**
        | Evaluate fitness of composition FS_i and CL_{best}
    end
    $best\_index \leftarrow$ Index of candidate FS with the best fitness
    if $FS_{best\_index}.fitness \geq Fit_{best}$ **then**
        $Fit_{best} \leftarrow FS_{best\_index}.fitness$
        $FS_{best} \leftarrow FS_{best\_index}$
    end
    $FS \leftarrow$ New generation of candidate feature sets
    // Classification part of evolution
    for CL_i in CL **do**
        | Evaluate fitness of composition CL_i and FS_{best}
    end
    $best\_index \leftarrow$ index of candidate CL with the best fitness
    if $CL_{best\_index}.fitness \geq Fit_{best}$ **then**
        $Fit_{best} \leftarrow CL_{best\_index}.fitness$
        $CL_{best} \leftarrow CL_{best\_index}$
    end
    $CL \leftarrow$ New generation of candidate classifiers
end

system. EEG data used for classifying MDD come from the study [15] and are publicly available. MDD data contain records from 34 persons with MDD (17 males and 17 females; the mean age is 40.3 ± 12.9) and 30 healthy controls (21 males and 9 females; the mean age is 38.3 ± 15.6). All persons with MDD met the diagnostic criteria for depression given by the Diagnostic and Statistical Manual-IV (DSM-IV). EEG recordings were performed during three different stimuli (i.e., eyes open, eyes closed, and performing a task), with only data recording during state *eyes open* being used in this paper. This results in the use of recordings from 30 persons with MDD and 32 healthy controls.

MDD data were randomly split into three groups: training data (60 %), validation data (20 %), and test data (20 %), while ensuring that all data from each person were always contained only in one group.

Data used in the task of classifying AUD come from the public data set [1], specifically from the version *The Large Data Set*. This data set contains records from 10 persons with AUD and 10 healthy controls. From each person, 20 recordings were created for each of the three different stimuli (i.e., shown single stimulus, shown two same stimuli, shown two different stimuli). Recordings from all subjects and stimuli are equally divided into two groups (i.e., training and test data).

Data in the AUD data set were originally split into two equally big groups of training data and test data. To allow the use of validation data during the evolution process and ensure comparability with other studies, we divided training data into two groups: training data (40%) and validation data (10%), while leaving the test data (50%) in the original form. The total number of input vectors is shown in Table 1.

Table 1. The number of input vectors in data groups of AUD and MDD data sets.

	AUD		MDD	
	Positive	Negative	Positive	Negative
Training	239	230	17	20
Validation	61	57	8	5
Test	300	278	5	7

3.2 Experimental Setup

The experimental setup of the proposed method involves: parameters of GA and classifier design algorithms, setting weights of MOD method SO-W, setting a set of constraints for MOD method SO-C, and parameters of a specific co-evolutionary variant of the proposed method. All parameter settings are based on either information from the literature or preliminary results.

The population size of GA is based on the selected MOD scenario: 10 for NSGA-II, 100 for SO-W, and 10 for SO-C. The number of generations is always selected so that a total of 10,000 fitness evaluations were made during the evolution process to allow for a fair comparison of evaluated methods. Parent solutions were selected using the tournament selection (tournament size = 2), with offspring created using single-point crossover with a probability of 0.7 and random mutation with a probability of 0.025. Naive Bayes, k-NN, and SVM methods were implemented using the Scikit-learn library [18]. Changes in parameter settings from default values include the selection of Gaussian Naive Bayes, the SVC variant of SVM with RBF kernel, and 18 neighbors of the k-NN method.

Parameters of CGP include the use of search strategy 1+4, 1 row, 8 columns, l-back of 8, use of Goldmann's Single mutation [7], and set of functions $\Gamma = \{+, -, *, \div, \frac{1}{X}, Sqrt, X^2, Abs, -X\}$. Additionally, due to a large number of inputs, the mutation of CGP was modified so that each node's input has a 50 % chance of being connected to one of the primary inputs and a 50 % chance of being connected to the output of one of the previous nodes.

Weights for the MOD scenario SO-W were selected based on preliminary results as 0.5 for accuracy, 0.5 for sensitivity, 0.5 for specificity, and $-\frac{1}{880}$ and $-\frac{1}{950}$ for the number of features of AUD and MDD, respectively. Results obtained using MOD scenario SO-C were achieved using ε constraints 10, 20, 30, 50, 100, and 200 features. Each constraint was used in $\frac{1}{6}$ of the total independent runs, and the resulting solutions that were obtained were merged together.

As GA is a stochastic method, all presented results are based on 50 independent runs of each setting. Classifiers, used for assessing the fitness of candidate feature sets, are trained using training data, and their accuracy, sensitivity, and specificity are calculated on validation data. The exception is the co-evolutionary

CGP algorithm, where switching between populations does not allow the use of training data for training and validation data for fitness evaluation. Here, both training and validation data are used in each step to evaluate the candidate solutions. The final results presented in this section are obtained from separate test data that were not used anywhere during evolution.

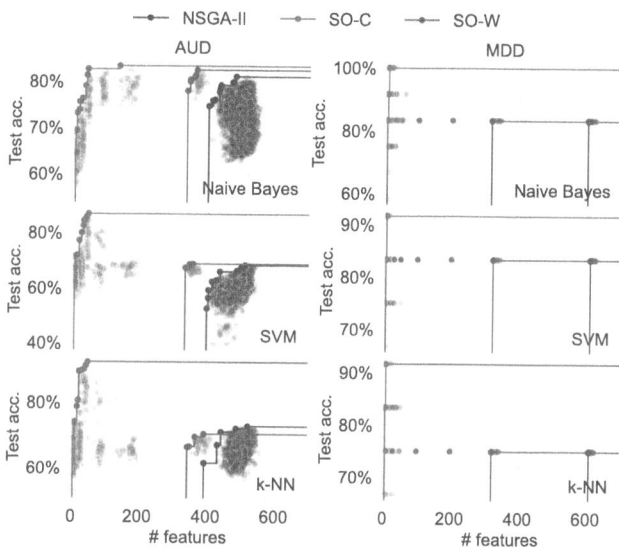

Fig. 4. Pareto fronts showing the trade-off between accuracy and the number of features obtained using different MOD scenarios (each row represents one classifier design algorithm). The left column shows results on the AUD task; the right column shows results on the MDD task. Colors represent different MOD scenarios.

3.3 Experiment 1: Evaluation of Multi-objective Design Algorithms

Evaluated MOD scenarios are compared based on their ability to design precise classifiers while maintaining a small number of used features. Fig. 4 shows Pareto fronts, with a trade-off between accuracy on test data and the number of used features achieved by individual scenarios on both data sets, using basic ML classifiers (i.e., SVM, k-NN, and Naive Bayes). Results show that both SO-W and SO-C scenarios achieved better accuracy than the NSGA-II algorithm on both datasets. The best accuracy was obtained using the SO-C scenario, with the k-NN classifier in the AUD task and the Naive Bayes classifier in the MDD task. SO-W scenario led to the highest accuracy in connection with the Naive Bayes classifier in both tasks.

SO-C scenario also leads to the use of the least number of extracted features, followed by SO-W and NSGA-II. Overall, the best trade-off between accuracy and the number of features used was achieved using the SO-C scenario for both tasks. Figure 5 shows a detail of the trade-off between accuracy on test data and the number of features achieved by the k-NN classifier using the SO-C scenario.

Results show that the most accurate solutions were obtained when $\varepsilon = 50$. Higher constraints of 100 and 200 lead to worse accuracy, which might be a result of overfitting that lower constraints might have prevented. Solutions obtained by $\varepsilon = 50$ surprisingly dominate the solutions obtained with $\varepsilon = 25$. Here, we speculate, that too low constraint might have limited explorable search space and led to the search being stuck in a local optimum. Table 2 shows the parameters of selected Pareto optimal solutions.

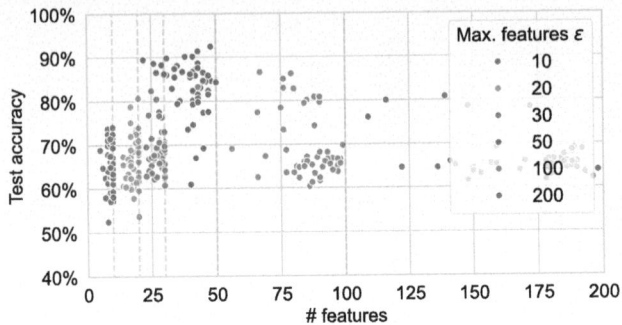

Fig. 5. Trade-off between accuracy on the AUD test data and the number of features obtained by the k-NN classifier using the SO-C scenario for ε constraints. Vertical dashed red lines represent individual constraints. (Color figure online)

Table 2. Parameters (i.e., the number of features, accuracy, sensitivity, and specificity) of selected Pareto optimal solutions.

Data Set	Classifier	MOD	Num. of features	Accuracy	Sensitivity	Specificity
AUD	Naive Bayes	NSGA-II	435	0.79	0.83	0.76
AUD	Naive Bayes	SO-C	10	0.73	0.69	0.77
AUD	Naive Bayes	SO-W	338	0.78	0.85	0.71
AUD	SVM	NSGA-II	402	0.57	0.77	0.37
AUD	SVM	SO-C	47	0.87	0.90	0.84
AUD	SVM	SO-W	335	0.67	0.60	0.74
AUD	k-NN	NSGA-II	522	0.73	0.83	0.63
AUD	k-NN	SO-C	48	0.92	0.96	0.89
AUD	k-NN	SO-W	393	0.70	0.77	0.64
MDD	Naive Bayes	NSGA-II	599	0.83	0.86	0.80
MDD	Naive Bayes	SO-C	10	1.00	1.00	1.00
MDD	Naive Bayes	SO-W	315	0.83	0.86	0.80
MDD	SVM	NSGA-II	609	0.83	0.86	0.80
MDD	SVM	SO-C	7	0.92	1.00	0.80
MDD	SVM	SO-W	322	0.83	0.86	0.80
MDD	k-NN	NSGA-II	608	0.75	0.71	0.80
MDD	k-NN	SO-C	8	0.92	1.00	0.80
MDD	k-NN	SO-W	321	0.75	0.71	0.80

3.4 Experiment 2: Co-evolution of Features with Cartesian Genetic Programming

In the second experiment, the classifier is co-evolved using CGP algorithm. The main motivation for the use of the proposed co-evolutionary method, which combines the design of feature selection and classification, is to accelerate the evolutionary process. However, it is crucial to ensure that the speedup of evolution will not be made at the expense of significantly worse results compared to the baseline method. To make the comparison as fair as possible, we compare the co-evolutionary method with the closest variant of the basic proposed method (the CGP classifier and SO-C scenario).

The left part of Fig. 6 shows boxplots of the accuracy of final solutions obtained on test data. Both variants achieved comparable results with identical medians. Performed Mann-Whitney U test did not reject the null hypothesis that the difference between variants' accuracy is significant (p-value = 0.705 for MDD and 0.043 for AUD).

The right part of Fig. 6 shows the time at which the convergence stopped and the final solution was found. The co-evolutionary method leads to comparable accuracy on test data while accelerating the design process 20x and 100x on average for AUD and MDD, respectively.

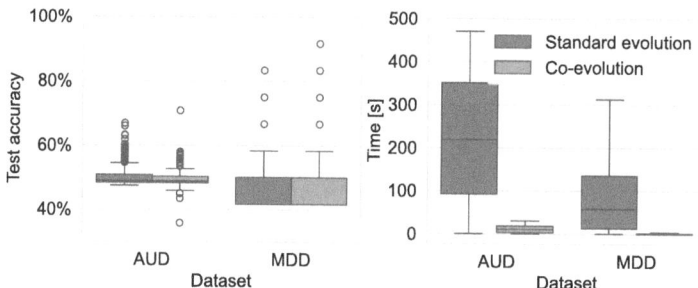

Fig. 6. A comparison of the basic variant of the proposed method utilizing evolved features in connection with the CGP classifier under SO-C scenario (blue) and the proposed co-evolutionary method (orange). The left part compares the accuracy of the final solutions on test data. The right part compares the time when the final solution was composed. (Color figure online)

Figure 7 shows the trade-off between the number of features used by the classifiers and achieved accuracy. Here, we can see that CGP achieves solutions with the smallest number of required features. In the task of AUD, CGP dominates space with an accuracy of up to 0.71. In the case of the MDD task, CGP dominates space with an accuracy of up to 0.92. CGP is thus able to design small, more explainable solutions (see Sect. 3.5), though they might achieve a lower accuracy than standard ML algorithms in some tasks.

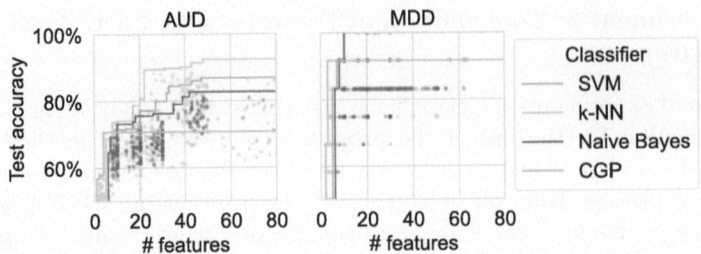

Fig. 7. Trade-off between accuracy on the test data and the number of features used by different classifiers in the SO-C scenario for ε constraints and co-evolutionary CGP method.

It is interesting to observe that CGP solutions use only 1 to 7 features in all cases, even though eight nodes of CGP allow connecting up to 9 inputs. Preliminary results show that larger CGP chromosomes do not lead to better results. Further analysis of evolved solutions has revealed that the most used functions in both tasks are multiplication, subtraction, addition, and division.

3.5 Selected Features and Resulting Classifiers

As we aim to create a method for the evolution of explainable classifiers, we present some of the evolved and relatively simple classifiers (i.e., potentially explainable) and key EEG channels and frequency bands for the classification of AUD and MDD. Selected small classifiers evolved by co-evolutionary CGP include the following classifier:

$$AUD = \frac{\sqrt{TP7\_\gamma(variance) + T8\_\alpha(max.) + AF8\_\gamma(max.)}}{FC2\_\beta(std.dev.)}, \qquad (3)$$

with reached the accuracy of 0.71 in the task of AUD classification. This classifier shows that AUD manifests by higher levels of variance and maximal values in alpha and gamma bands at temporal channels on both the left and right hemispheres and the anterior frontal channel of the right hemisphere. Higher variance in the beta band at the frontal central channel of the left hemisphere, on the other hand, strongly decreases the probability of AUD in the patients.

Classifiers evolved for MDD classification include the following solution:

$$MDD = (F4\_49\,Hz - F4\_46\,Hz) * (F4\_49\,Hz - F4\_46\,Hz + F7\_27\,Hz), \qquad (4)$$

whose accuracy is 0.92. This classifier shows that the level of MDD strongly corresponds to higher power levels at a frequency of 49 Hz at channel F4 in the frontal part of the right hemisphere. Interestingly, the probability of MDD decreases with higher power levels at 46 Hz at the same channel and 27 Hz at channel F7 in the frontal part of the left hemisphere.

Figure 8 shows a histogram of channels used in Pareto Optimal CGP classifiers visualized on the electrode placement map. The most used EEG channels are AF8, O1 and FC2 in the task of AUD and channels P3, P4 and F4 in the task of MDD classification. Additionally, the most prominent EEG band is the Gama band (32–50 Hz). Interestingly, the Gama band is the most prominent not only in the task of AUD (supporting the results of [19]) but also in the task of MDD.

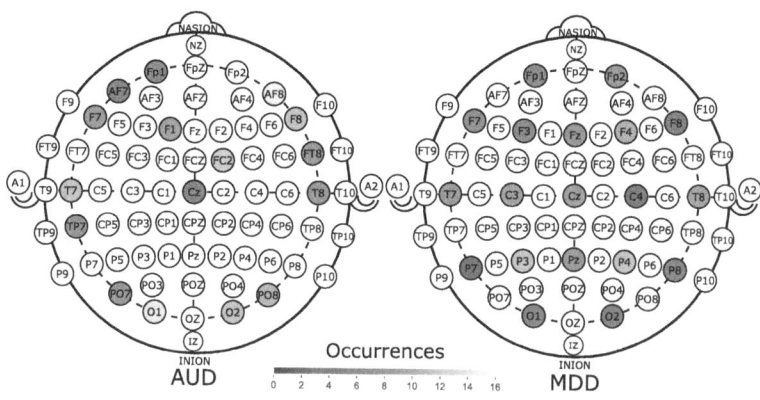

Fig. 8. Histograms of channels used in Pareto Optimal CGP classifiers visualized on electrode placement map.

4 Conclusion

In this paper, we proposed a method for multi-objective evolutionary design of explainable EEG classifiers by targeting small classifiers composed of simple operators and utilizing a few features. The proposed method aims to design accurate yet simple classifiers by combining traditional ML algorithms (e.g., SVM, k-NN, Naive Bayes) and GA for the automatic selection of features. To further improve the classification quality and obtain small and potentially explainable feature sets, we compared three multi-objective design scenarios with trade-offs among accuracy, specificity, sensitivity, and the number of used features. In addition, we evaluated the use of CGP as a classifier design algorithm, aiming to obtain a small and explainable classifier that would provide further information about the use of individual features. As CGP can be computationally demanding, we proposed a co-evolutionary variant of the method for simultaneously designing the feature set and classifier by CGP in two populations cooperating during fitness evaluation. All proposed methods were evaluated on the task of classifying alcohol use disorder and major depressive disorder.

Experimental results revealed that both newly proposed MOD scenarios lead to significantly better accuracy and the number of features than a state-of-the-art method based on the NSGA-II algorithm. The best results were achieved using the SO-C scenario, with the best results in AUD achieved in connection with the k-NN classifier and the best results in the MDD task with the Naive

Bayes classifier. The evolution process of CGP was significantly accelerated by the proposed co-evolutionary method, which led to 20–100× faster convergence. The co-evolutionary method using CGP achieved lower accuracy than other evaluated ML approaches but allowed the design of small and potentially explainable solutions using only a small subset of extracted features. Finally, we have presented the features with the highest impact on the classification of AUD and MDD and an example of classifiers designed using CGP.

In future work, we would like to focus on a deeper analysis of designed solutions from a medical point of view and the application of the proposed method to other disorders that cause permanent changes in the working of the human brain (i.e., non-event-based disorders). Additionally, we would like to evaluate additional algorithms for the MOD of classifiers and incorporate cross-validation into fitness calculation in the proposed method to ensure good generalization of designed solutions.

Acknowledgements. The work on this paper was supported by the Czech Science Foundation project No. 24-10990S. The computational resources were supported by the Ministry of Education, Youth and Sports of the Czech Republic through the e-INFRA CZ (ID:90254).

References

1. Begleiter, H.: EEG database. UCI Machine Learning Repository (1995). https://doi.org/10.24432/C5TS3D
2. Burkart, N., Huber, M.F.: A survey on the explainability of supervised machine learning. J. Artif. Int. Res. **70**, 245–317 (2021). https://doi.org/10.1613/jair.1.12228
3. Das, R.K., Martin, A., Zurales, T., Dowling, D., Khan, A.: A survey on EEG Data analysis software. Sci **5**(2), 23 (2023). https://doi.org/10.3390/sci5020023
4. Deb, K., Pratap, A., Agarwal, S., Meyarivan, T.: A fast and elitist multiobjective genetic algorithm: NSGA-II. IEEE Trans. Evol. Comput. **6**(2), 182–197 (2002). https://doi.org/10.1109/4235.996017
5. Drahosova, M., Sekanina, L., Wiglasz, M.: Adaptive fitness predictors in coevolutionary cartesian genetic programming. Evol. Comput. **27**(3), 497–523 (2019). https://doi.org/10.1162/evco_a_00229
6. Duan, L., et al.: Machine learning approaches for MDD detection and emotion decoding using EEG signals. Front. Hum. Neurosci. **14**, 284 (2020). https://doi.org/10.3389/fnhum.2020.00284
7. Goldman, B.W., Punch, W.F.: Reducing wasted evaluations in cartesian genetic programming. In: Krawiec, K., Moraglio, A., Hu, T., Etaner-Uyar, A.Ş, Hu, B. (eds.) EuroGP 2013. LNCS, vol. 7831, pp. 61–72. Springer, Heidelberg (2013). https://doi.org/10.1007/978-3-642-37207-0_6
8. Hurta, M., Drahosova, M., Mrazek, V.: Evolutionary design of reduced precision preprocessor for levodopa-induced dyskinesia classifier. In: Rudolph, G., Kononova, A.V., Aguirre, H., Kerschke, P., Ochoa, G., Tušar, T. (eds) PPSN 2022. LNCS, vol 13398, pp. 491–504. Springer, Cham (2022). https://doi.org/10.1007/978-3-031-14714-2_34

9. Hurta, M., Mrazek, V., Drahosova, M., Sekanina, L.: MODEE-LID: multiobjective design of energy-efficient hardware accelerators for levodopa-induced dyskinesia classifiers. In: 2023 26th International Symposium on Design and Diagnostics of Electronic Circuits and Systems (DDECS), pp. 155–160 (2023). https://doi.org/10.1109/DDECS57882.2023.10139399
10. Li, H., Wu, L.: EEG classification of normal and alcoholic by deep learning. Brain Sci. **12**(6), 778 (2022). https://doi.org/10.3390/brainsci12060778
11. Malik, A.S., Mumta, W.: EEG-Based Experiment Design for Major Depressive Disorder. Academic Press, London (2019). https://doi.org/10.1016/C2018-0-01657-6
12. Meyer, Y.: Wavelets and Operators: Volume 1. Cambridge Studies in Advanced Mathematics. Cambridge University Press (1993)
13. Miller, J.F.: Cartesian genetic programming: its status and future. Genetic Program. Evolvable Mach. **21**(1–2), 129–168 (2020). https://doi.org/10.1007/s10710-019-09360-6
14. Mrazek, V., Jawed, S., Arif, M., Malik, A.S.: Effective EEG feature selection for interpretable MDD (major depressive disorder) classification. In: Proceedings of the Genetic and Evolutionary Computation Conference, GECCO 2023, pp. 1427–1435. ACM, New York (2023). https://doi.org/10.1145/3583131.3590398
15. Mumtaz, W., Xia, L., Yasin, M., Ali, S., Malik, A.S.: A wavelet-based technique to predict treatment outcome for major depressive disorder. PLoS ONE **12**(2), e0171409–e0171409 (2017). https://doi.org/10.1371/journal.pone.0171409
16. Ong, K.M., Thung, K.H., Wee, C.Y., Paramesran, R.: Selection of a subset of EEG channels using PCA to classify alcoholics and non-alcoholics. In: Annual International Conference of the IEEE Engineering in Medicine and Biology - Proceedings, vol. 7, pp. 4195–4198 (2005). https://doi.org/10.1109/IEMBS.2005.1615389
17. Parziale, A., Senatore, R., Della Cioppa, A., Marcelli, A.: Cartesian genetic programming for diagnosis of Parkinson disease through handwriting analysis: performance vs. interpretability issues. Artif. Intell. Med. **111**, 101984 (2021). https://doi.org/10.1016/j.artmed.2020.101984
18. Pedregosa, F., et al.: Scikit-learn: machine learning in Python. J. Mach. Learn. Res. **12**, 2825–2830 (2011)
19. Rodrigues, J., Filho, P.R., Peixoto, E., Kumar, N.A., de Albuquerque, V.: Classification of EEG signals to detect alcoholism using machine learning techniques. Pattern Recognit. Lett. **125**, 140–149 (2019). https://doi.org/10.1016/j.patrec.2019.04.019
20. Smith, S.L., Walker, J.A., Miller, J.F.: Medical applications of cartesian genetic programming. In: Miller, J. (eds) Cartesian Genetic Programming. Natural Computing Series, pp. 309–336. Springer, Heidelberg (2011). https://doi.org/10.1007/978-3-642-17310-3_11
21. Welch, P.: The use of fast fourier transform for the estimation of power spectra: a method based on time averaging over short, modified periodograms. IEEE Trans. Audio Electroacoust. **15**(2), 70–73 (1967). https://doi.org/10.1109/TAU.1967.1161901
22. World Health Organization: Depression and other common mental disorders: global health estimates. Technical documents, World Health Organization (2017)

On the Effectiveness of Crossover Operators in Cartesian Genetic Programming

Mark Kocherovsky[✉][iD], Marzieh Kianinejad[iD], Illya Bakurov[iD], and Wolfgang Banzhaf[iD]

Department of Computer Science and BEACON Center for the Study of Evolution in Action, Michigan State University, East Lansing, MI 48824, USA
kocherov@msu.edu
https://banzhaf-lab.github.io/

Abstract. This study investigates the effectiveness of Cartesian Genetic Programming by analyzing numerous indicators of evolutionary dynamics when using different crossover operators and the canonical mutation-only (1 + 4) strategy. Specifically, we examine a traditional crossover operator which is based on the random selection of parental genes; Subgraph Crossover, where points in the range of active nodes are considered; and the recently-proposed Deep Neural Crossover (DNC) approach which utilizes a transformer network to learn correlations between genes and predict potentially beneficial crossover points. The performance of these different crossovers is evaluated on 11 standard and one real-world regression problem.

Keywords: Cartesian Genetic Programming · Crossover · Neural Crossover · Crossover Effects · Reinforcement Learning

1 Introduction

Introduced by Miller in the late 1990s [28,30], Cartesian Genetic Programming (CGP) is one of several Genetic Programming (GP) paradigms currently heavily used and under investigation in the field. This paradigm relies on a directed acyclic graph (DAG) representation, which allows the reuse of intermediate results and multiple outputs. Since its introduction, it was observed that traditional crossover operators were detrimental to search in CGP and the scientific community mostly avoids their usage [29]. Nevertheless, there were several attempts to design an effective crossover for CGP that could consistently outperform the canonical $(1 + \lambda)$ strategy [8–10,19–22,36,38], where there is no crossover, and only one parent is selected to be mutated several times. Despite numerous efforts, the scientific community shares the *assumption* that standard crossover is simply too destructive for CGP, and as we will see in Sect. 5, certain other assumptions. In summary, the literature skirts around the main question: *why is crossover so destructive in CGP in the first place?*

© The Author(s), under exclusive license to Springer Nature Switzerland AG 2025
B. Xue et al. (Eds.): EuroGP 2025, LNCS 15609, pp. 68–84, 2025.
https://doi.org/10.1007/978-3-031-89991-1_5

In an attempt to understand why traditional crossover is so destructive in CGP, the authors of [24] compared the performance of different crossover operators in CGP and Linear Genetic Programming (LGP). The latter is a GP paradigm that represents and executes solutions as a sequence of instructions, akin to a low-level programming language, where each instruction specifies an operation and the memory registers/locations [7]. Although it was shown that CGP and LGP can be mutually convertible [7], allowing direct comparison between their DAGs [40], crossover in LGP is frequently found to benefit search. In [24], this was linked to a more drastic node connectivity disruption in CGP when the genetic operators are applied, which results in children being far more genetically different from their parents than in LGP.

Previous research about the crossover in CGP relied upon methods that perform random selection of parental genes. For example, n-point crossover randomly selects n crossover points and swaps segments in between. Uniform Crossover swaps parent genes at each index with uniform probability. The real-valued crossover [9] performs a randomly weighted average of the two parents. Subgraph Crossover [19,21] essentially performs a one-point crossover between active function nodes of the two parents, aiming at recombining the subgraphs with semantic value.

In this work, we seek to deepen the community's understanding of crossover destructiveness in CGP following [24]. In particular, we adapt and utilize a recently proposed Deep Neural Crossover (DNC) [35] to CGP for the first time and compare it to the aforementioned crossover methods. DNC contains an encoder-decoder neural network trained in real-time to learn structural correlations between parent genes in order to select crossover points that maximize offspring fitness. DNC was originally developed for Genetic Algorithms (GAs) to guide the probability of swapping genes in Uniform Crossover (making it, effectively, non-uniform), and was shown to outperform traditional operators substantially. Additionally, we include the Real-Valued Crossover proposed in [9] to investigate the suitability of a floating-point representation of CGP chromosomes. We also analyze whether allowing a variable-length representation for CGP, akin to LGP, helps to improve the program search. Finally, we try to shed light on how different crossover methods influence evolutionary dynamics beyond commonly used learning curves, phenotypic diversity, and the proportion of active nodes. Here we also observe the relationships between the parent-child similarity, the fitness landscape properties of individuals, and the spatio-temporal distribution of crossover and mutation events over parent genotypes categorized by offspring fitness (i.e., whether worse, near-equal, or improved fitness was observed).

We find that (i) DNC does not allow CGP to synthesize better programs than mutation-only CGP, (ii) Cui et al.'s [10] results that CGP nodes are positionally dependent, i.e. the probability of whether a node will be active depends on the node's position, are confirmed, (iii), traditional crossover operators fail because they assume that nodes close to each other are semantically related, which is not true in CGP.

The remainder of the paper is structured as follows: Sect. 2 introduces the established CGP methods and the literature on CGP crossover. Section 3 discusses our experimental methods, whose results are shown in Sect. 4. Finally, in Sect. 5, we discuss the implications of our results and provide future directions for investigation.

2 Background

2.1 CGP

The basis of a Cartesian program is an instruction that is commonly referred to as a *node*. Each node has a structure of $n_m \in N = [o_m \; a_{m,1} \; a_{m,2}]$, which determines the operator and operand nodes, assuming an arity of 2. Note the lack of memory: Each node takes its input directly from the result of the execution of earlier nodes. However, nodes that are neighbors in the genotype (i.e. neighbors in the chromosome) are not necessarily functionally related due to the DAG structure: Each node can connect to any preceding node, regardless of its physical position in the genotype, and some nodes may not contribute to the final output at all (acting as *introns*), referred to as *inactive nodes*. Thus, a program p is defined as $p \in P = \{n_0, n_1, n_2, ..., n_\beta\} \cup [o_0, o_1, o_2, ..., o_n]$, where each output node takes a single operand, and can be an input or program node. CGP programs are necessarily of constant size, determined by parameter β. In CGP, each node is typically concatenated to form a one-dimensional vector. However, for ease of use, each program is represented as a two-dimensional matrix, except during crossover, where the matrix is flattened into a one-dimensional vector and then reshaped into a matrix once the operator is finished.

2.2 Crossover in CGP

Even though they are not commonly used, several existing crossover methods can be applied to CGP, each representing different assumptions about the connectivity in a DAG. Initially, CGP researchers and practitioners adapted GA-like crossover operators, such as n-Point Crossover [31]. Here, different values of n allow a different treatment of gene order and positional dependence in the chromosomes. For a small n, the operator swaps a few consecutive gene segments, allowing neighboring genes to remain together in the offspring. This assumes that genes closer to each other in the genome have some functional relationship and should be passed together to preserve it. However, as n increases, the relationships and positional dependence progressively fade. In an extreme case, such as Uniform Crossover, each gene is treated as having no positional or relational dependence on neighboring genes, discarding any notion of order or connectivity. Because CGP's DAG structure allows each node to connect with any preceding node, crossover operators that assume neighboring genes have functional relationships can disrupt essential dependencies. In contrast, operators that completely disregard positional relationships can amplify these destructive effects, although they introduce greater genetic diversity.

Clegg et al. [9] adapted the Real Valued Crossover for CGP, frequently used in GA literature for solving continuous optimization problems. This crossover essentially performs a randomly-weighted average of the two parent genes. Given that traditional CGP uses an integer-based representation, the authors proposed an algorithm for converting it into a [0–1] range of floating-points. The real-valued CGP crossover was assessed on two symbolic regression problems (Koza2 and Koza3) and was found to outperform the mutation-alone strategy only on one problem. Nevertheless, it was found to achieve faster convergence in earlier iterations, which led the authors to adapt the crossover rate over generations. The adaptive variant turned out to outperform the mutation-alone strategy. However, a later study examined real-valued crossover on three additional classes of problems and found no advantage in using it [39].

In [8], Cai et al. highlight other challenges of traditional CGP representation and its destructive effects on recombination. They argue that a node's location in the CGP chromosome can influence whether it contributes to the final solution. They also note that node position does not correlate with behavior, as identical positions across chromosomes do not imply the same functionality and the same functionality can be observed across different positions. As a result, position-based gene exchanges, typical in traditional crossovers, often disrupt useful substructures. To circumvent this limitation, the authors proposed a more flexible representation where gene location does not influence phenotype, demonstrating superior performance on the even-3 parity problem using the 2-point crossover.

Goldman and Punch [13] observed that the probability of a given node being active varies significantly by position, with a bias toward nodes near the input, which receive more connections due to the DAG structure in CGP, where each node can only connect to preceding nodes to avoid cycles. Two strategies were proposed to mitigate this bias. `DAG` mutates connections to every other node in the genome, as long as the new connection does not create a cycle; `Reorder` shuffles active nodes through the chromosome in a way that phenotype does not change. Both strategies increased the amount of active nodes and achieved faster convergence with a smaller genotype. Although Goldman and Punch [13] studied CGP$(1+\lambda)$ with point mutation, it is reasonable to argue that crossover under-performance can also be attributed to positional bias. To this end, the authors of [10] examined whether the reorder strategy could improve its performance. They compared two standard CGP$(1+\lambda)$ variants against four crossover operators with the reorder strategy on four boolean benchmarks and concluded that crossover with reordering benefits search. In three out of four problems, the crossover was found to outperform CGP$(1+\lambda)$, and Uniform Crossover generally benefits the most.

In 2017, Kalkreuth et al. introduced Subgraph Crossover [21], which exchanges active subgraphs between parents by selecting crossover points within active nodes, focusing recombination on genome segments that impact offspring fitness. Later, Kalkreuth [19] evaluated Subgraph Crossover with large genotypes and small population sizes on symbolic regression and Boolean problems, finding that it generally outperformed mutation-only approaches with faster convergence and better final fitness in regression tasks. He concluded that medium (50) and large (250) population sizes perform best, while larger genotypes offered no significant advantage. In [18], Husa and Kalkreuth presented the *block crossover*, which was inspired by the Subgraph Crossover and the cone-based module creation proposed in [22], which enables CGP to evolve and reuse modules (subcomponents of the program). The method essentially swaps blocks of consecutive active nodes between two parents. The authors assessed the performance of the block, Subgraph, One-Point, and real-valued crossover, and compared them with mutation-only strategies on a subset of problems taken from [21]. Testing block, Subgraph, One-Point, and real-valued crossovers against mutation-only strategies, they found $(1 + \lambda)$ mutation was the best overall, possibly due to the lack of crossover parameter tuning, such as crossover rate.

Preserving gene structures that remain unchanged across generations can be beneficial, as they likely confer a fitness advantage. In biological systems, this concept appears during meiosis, where Homologous Crossover conserves similar sequences while exchanging non-similar parts, promoting genetic diversity [17]. Homologous Crossover has long been explored in GP [6] using both trees [32] and linear representations [12]. Recently, a Homologous Crossover was proposed for CGP in [38], where multiple sequence alignment (MSA) is used to interleave gaps between the sequences so that they achieve the same length and have the maximum possible similarity. After allowing initial generations to stabilize on some meaningful, high-fitness building blocks, MSA identifies intact/conserved sequences without any gaps that appear consistently across multiple genomes. A single representation is then extracted for commonly conserved and non-conserved genes, acting as a footprint of areas of stability and variability in the population's genetic pool. During crossover, two individuals with the highest fitness values that align well with this footprint are selected as parents. In their work, Uniform Crossover is used. The new crossover was explored in the context of neural architecture search (NAS) for image classification, and the results were compared with those found in the literature using human-designed architectures and other types of NAS. The comparison included $CGP(1 + \lambda)$ strategy from a related study [37], where similar CGP hyper-parameters were used. While the accuracy of CGP with the new Homologous Crossover was found to be about 2% above that reported in [37], the evolved network was also found to be slightly more complex. Thus, it is not obvious whether the proposed crossover effectively improves CGP search.

3 Methods

3.1 Problems

We tested each method by evolving models to find eleven univariate functions over 10,000 generations. We chose known optimization problems by virtue of being complex enough to be challenging for the GP to find, as well as to match the literature [20,24]. At each run, we randomly sampled 20 points for simpler problems and 40 for more complex ones from the respective domain. For more details, including functional forms, see Table 1. Also, we assessed the methods on a real-world problem, the Diabetes dataset [11]. A 70–30% train-test split was used and models were trained for 3,000 generations. Note that over time, only training fitness was collected, and testing fitness was collected at the end.

Table 2 shows our notation and basic hyperparameters, and Table 3 shows our hyperparameters for the DNC [35].

Table 1. Problems tested; each model was given a set of random points within the given domain. *: problem too complex to be written out here, see the reference. Mostly reproduced from [24].

Problem	Function	Domain	Points
Koza-1	$x^4 + x^3 + x^2 + x$	$[-1, 1]$	20
Koza-2	$x^5 - 2x^3 + x$	$[-1, 1]$	20
Koza-3	$x^6 - 2x^4 + x^2$	$[-1, 1]$	20
Nguyen-4	$x^6 + x^5 + x^4 + x^3 + x^2 + x$	$[-1, 1]$	20
Nguyen-5	$\sin(x^2)\cos(x) - 1$	$[-1, 1]$	20
Nguyen-6	$\sin(x) + \sin(x + x^2)$	$[-1, 1]$	20
Nguyen-7	$\ln(x+1) + \ln(x^2+1)$	$[0, 2]$	20
Ackley [2]	$-20\exp(-0.2x^2) - \exp(\cos 2\pi x) + 20 + \exp 1$	$[-32.768, 32.768]$	40
Rastrigin [33]	$10 + x^2 - 10\cos 2\pi x$	$[-5.12, 5.12]$	40
Levy [25]	*	$[-10, 10]$	40
Griewank [14]	$x^2/4000 - \cos x + 1$	$[-600, 600]$	40
Diabetes [11]	*	See [11]	442 (70/30% Split)

Table 2. Evolutionary parameters to demonstrate the effects of crossover. Mostly reproduced from [24].

Notation	Crossover	Mutation
CGP(1+4)	None (Canonical)	$\mu = 100\%$ (x4)
CGP-1x(40+40)	One-Point (50%)	$\mu = 2.50\%$
CGP-2x(40+40)	Two-Point (50%)	$\mu = 2.50\%$
CGP-VL1x(40+40)	One-Point | Variable-Length (50%)	$\mu = 2.50\%$
CGP-VL2x(40+40)	Two-Point | Variable Length (50%)	$\mu = 2.50\%$
CGP-Ux(40+40)	Uniform (50%)	$\mu = 2.50\%$
CGP-SGx(40+40)	Subgraph (50%)	$\mu = 2.50\%$
CGP-DNC(40+40)	Uniform Deep Neural Crossover (50%)	$\mu = 2.50\%$
CGP-DNC-1x(40+40)	One-Point Deep Neural Crossover (50%)	$\mu = 2.50\%$
CGP-RV(40+40)	Real-Valued (50%)	$\mu = 2.50\%$

3.2 Deep Neural Crossover (DNC)

Previous research about the crossover in CGP relied upon methods that select parental genes at random. In recent years, conceptually novel evolutionary operators have been proposed that leverage deep learning (DL) capabilities to process

Table 3. Neural Network Hyperparameters for DNC [34]

Hyperparameter	Value
Embedding Dimension	64
Sequence Length	193
Number of Embeddings	75
Running Mean Decay	0.95
Sample Size for Training	820 pairs of parents
Learning Rate	1e−4
Greedy ϵ	0.2
Number of Parents	2

sequence data, like genes, to identify complex relationships. NeuroCrossover [26] utilizes Dual-Aspect Collaborative Transformer architecture [27], coupled with online reinforcement learning (RL), which was shown to be able to optimize the selection of points for order and 2-point crossover applied for routing and packing problem-solving. Specifically, the attention modules learn embedded representations of the parent genomes using what the authors call Cross Information Synergistic Attention (CISA), where each parent embedding is learned separately but can be cross-referenced by the other to find common features. Once the encoder and decoder are trained, the network can then work on finding children with better fitness than the parents, outputting relevant crossover points.

Later, Shem-Tov and Elyasaf extended NeuroCrossover by proposing to learn a probability distribution of exchanging each gene, akin to Uniform Crossover [35]. Their method, called Deep Neural Crossover (DNC), is motivated by the assumptions that distinguish One-Point and Uniform Crossover (previously discussed in Sect. 2.2). However, there are more differences. Although DNC also learns through policy-based online reinforcement learning, a different neural architecture based on an LSTM Pointer Network is employed. The latter uses a set of softmax modules with attention to produce the probability of selecting a gene from each parent. In addition, the authors use what they call an ϵ−greedy method to randomly perform actions with the probability ϵ. Moreover, DNC can operate in a multiple-parent setting, which is different from the two-parent approach in [26]. Diagrams of the encoding and decoding network are shown in Fig. 1.

There are some minor differences between the baseline DNC code and ours. First, where DNC produces a single child from each pairing, we produce two children to match the other forms of crossover we measure. For the same reason, we use a two-parent setting rather than the multi-parent option provided. Second, minor structural alterations are made to accommodate the CGP structures we were already using: we enabled ourselves to plug in our fitness function, mutation operator, and data recording methods. The ANN methods were untouched as DNC is genome-agnostic. In addition, we also test these strategies with *fixed par-*

ents: in short, each starting set of parents in each replicate are identical instead of randomly initialized each time; and we tested different neural network learning rates. In our work, we test both Uniform and One-Point DNC (as in [26]).

3.3 Other Design Details

We also tested the One-Point and Two-Point crossovers with *variable-length individuals*, which uses a mutation operator close to LGP, in that it can add and remove instructions as well as perform point mutation; and crossover between nodes, not genes. All methods except DNC used *forced diversity*, making sure each parent returned by the selection operator has a unique genotype [3,5]. DNC is excluded as we did not want to tamper with the authors' existing selection method, which is standard tournament selection. For selection, the CGP$(1+4)$ strategy uses elite selection, DNC uses the tournament selection as in [35], and the others use tournament selection with elitism following [24]. All trials were run using the MSU High-Powered Computing Cluster [1]. Point mutation is used at a gene level.

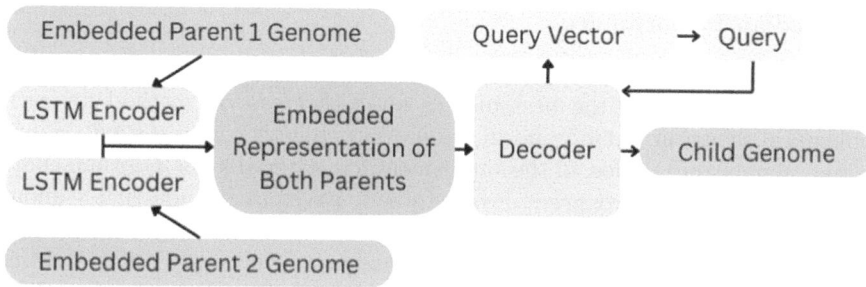

Fig. 1. Encoder and Decoder architecture in DNC. Each parent is converted into an embedding. Embedded child genomes are generated and converted back into a typical genomic representation [35].

In each run and for each generation, we record several metrics in relation to the evolutionary dynamics. These are the fitness and number of active nodes of the best-performing individuals, the change in fitness between the best parent and child, and parent-offspring similarity as stated by [4] (*Similarity*). We also count the amount of deleterious, near-neutral, or beneficial crossover and mutation events at each generation (*Crossover* and *Mutation Impact*); for determining the impact category, the percent difference of the parent and child fitness was measured, and labeled as deleterious if $\%\Delta f > 0.001$, beneficial if $\%\Delta f < -0.001$, and near-neutral otherwise. Finally, we measure the distribution of crossover indices over each generation categorized by impact. These metrics are briefly summarized in Table 4.

For our fitness function, we use the Pearson correlation as suggested in [15, 23], which focuses selection on expressions that are close in shape to the training

Table 4. Summary of Metrics Recorded.

Metric	Description
Best Fitness	Smallest fitness at the end of evolution
Best Fitness per Generation	Smallest fitness during each generation
Semantic Diversity	Standard Deviation of population semantics (fitnesses) in each generation
Instructions	Number of active instructions in the best model after evolution
Program Size per Generation	Number of active nodes in the best program in each generation
Similarity	Similarity score (using alignment [4]) between parents and their best child
Xover Density Distribution	Temporal frequency of indices used as crossover points categorized by impact direction
Mutation Density Distribution	Temporal frequency of indices mutated categorized by impact direction

target, instead of merely requiring that aggregate error is small, as can happen using RMSE simply because the scale of model output is close to targets, despite the quality of fit being poor. To minimize the fitness measure, we use $1 - r^2$, where r is the Pearson Correlation.

4 Results

4.1 Median Performance

For each replicate, we saved the fitness of the best model in the population. Median fitness values of the best models for each crossover method for each problem are shown in Table 5 and in more detail, together with the full Mann-Whitney significance tables, in the supplementary material https://github.com/MarkKocherovsky/cgp_crossover. From Table 5, we can conclude that Canonical CGP is the best performer for most of the problems considered, except being significantly worse at the Levy problem, not significantly different from other results for the Ackley problem, and being insignificantly worse than Subgraph on Nguyen 6. However, One-Point-Variable-Length, Uniform, and One-Point Crossovers show competitive performance on several problems, with Two-Point-Variable-Length and Subgraph also performing well in some cases. The Real-Valued and Deep Neural Crossovers perform significantly worse. On the Diabetes problem, both Variable-Length crossovers perform significantly better than the other methods, and DNC performs significantly worse.

4.2 Temporal Plots

For our assessment of the metrics discussed in Table 4, we have chosen to limit our figures in this paper to Koza 3 and Rastrigin because of the details in each plot. Both problems are difficult to solve and are therefore adequate representatives of the evolutionary dynamics. Full plots are available at https://github.com/MarkKocherovsky/cgp_crossover.

Figure 2a shows the median fitness of the best individuals in the population at each generation. For the paper, we will look at our two example problems starting at generation 100, as the most interesting development occurs between 500—2000 generations into evolution: Canonical CGP, which starts with much

Table 5. Median fitness values of the best models for each crossover method in each problem after 50 runs. The best value is shown in the deepest green and in **bold**, and the worst in the deepest red.

	Koza 1	Koza 2	Koza 3	Nguyen 4	Nguyen 5	Nguyen 6
CGP(1+4)	**6.76E-04**	**5.11E-03**	**9.91E-03**	**8.91E-04**	2.84E-04	5.06E-04
CGP-1x	1.12E-03	1.67E-02	4.18E-02	1.33E-03	1.11E-03	8.76E-04
CGP-DNC	1.70E-03	2.03E-02	1.17E-01	2.02E-03	1.35E-03	9.71E-04
CGP-DNC-1x	1.48E-03	2.73E-02	8.30E-02	1.93E-03	1.51E-03	9.78E-04
CGP-1xVL	1.26E-03	6.60E-03	2.74E-02	1.58E-03	2.36E-04	5.68E-04
CGP-2xVL	1.93E-03	8.94E-03	1.98E-02	2.05E-03	6.78E-04	1.01E-03
CGP-2x	1.12E-03	1.71E-02	4.89E-02	1.30E-03	1.04E-03	8.80E-04
CGP-RVx	3.62E-03	1.00E-01	3.38E-01	9.48E-03	2.77E-03	2.10E-03
CGP-SGx	1.21E-03	2.31E-02	4.16E-02	1.63E-03	**1.09E-04**	8.45E-04
CGP-Ux	9.34E-04	1.19E-02	1.75E-02	1.55E-03	7.55E-04	**4.99E-04**
	Nguyen 7	Ackley_1D	Rastrigin_1D	Levy_1D	Griewank_1D	Diabetes
CGP(1+4)	**8.89E-06**	2.92E-02	**3.61E-01**	1.61E-01	**6.00E-04**	5.09E-01
CGP-1x	1.53E-05	2.78E-02	4.01E-01	**9.78E-02**	6.25E-04	5.22E-01
CGP-DNC	2.59E-05	3.39E-02	4.16E-01	1.91E-01	6.39E-04	5.34E-01
CGP-DNC-1x	4.47E-05	3.38E-02	4.11E-01	1.94E-01	6.45E-04	5.38E-01
CGP-1xVL	1.33E-05	2.70E-02	4.10E-01	1.02E-01	6.31E-04	**4.87E-01**
CGP-2xVL	1.81E-05	2.88E-02	4.24E-01	1.06E-01	6.19E-04	4.92E-01
CGP-2x	1.28E-05	2.85E-02	4.10E-01	1.04E-01	6.30E-04	5.25E-01
CGP-RVx	5.43E-04	3.16E-02	4.41E-01	1.07E-01	6.39E-04	5.24E-01
CGP-SGx	1.02E-04	2.77E-02	4.14E-01	1.06E-01	6.37E-04	5.09E-01
CGP-Ux	1.30E-05	**2.68E-02**	4.01E-01	1.01E-01	6.12E-04	5.18E-01

worse models on average, likely due to its small population size (as it is a (1+4) ES), overtakes all crossover methods rather early in the evolutionary process. Even though each method shows improvement of fitness over time, crossover clearly stunts the rate of improvement. In Fig. 2b, we show the median amount of active nodes for the elite model in each replicate over time. Canonical CGP produces more complex models, which may correspond to its high performance, but some competitive crossovers, namely Uniform Crossover in Koza 3 and Two-Point crossover in Rastrigin, evidently produce less complex models, indicating a complicated solution landscape.

At the population level, we measured *semantic diversity*, defined as the standard deviation of fitnesses in the population. This is shown in Fig. 2c. We can see that the greatest diversity comes from the Variable-Length group, closely followed by Subgraph and Canonical CGP. Though the high diversity of the latter is likely caused by its small population, the former two, as we will see, have high amounts of deleterious events when compared to the other methods, so the higher diversity is reasonable. Figure 2d shows the median *similarity* between

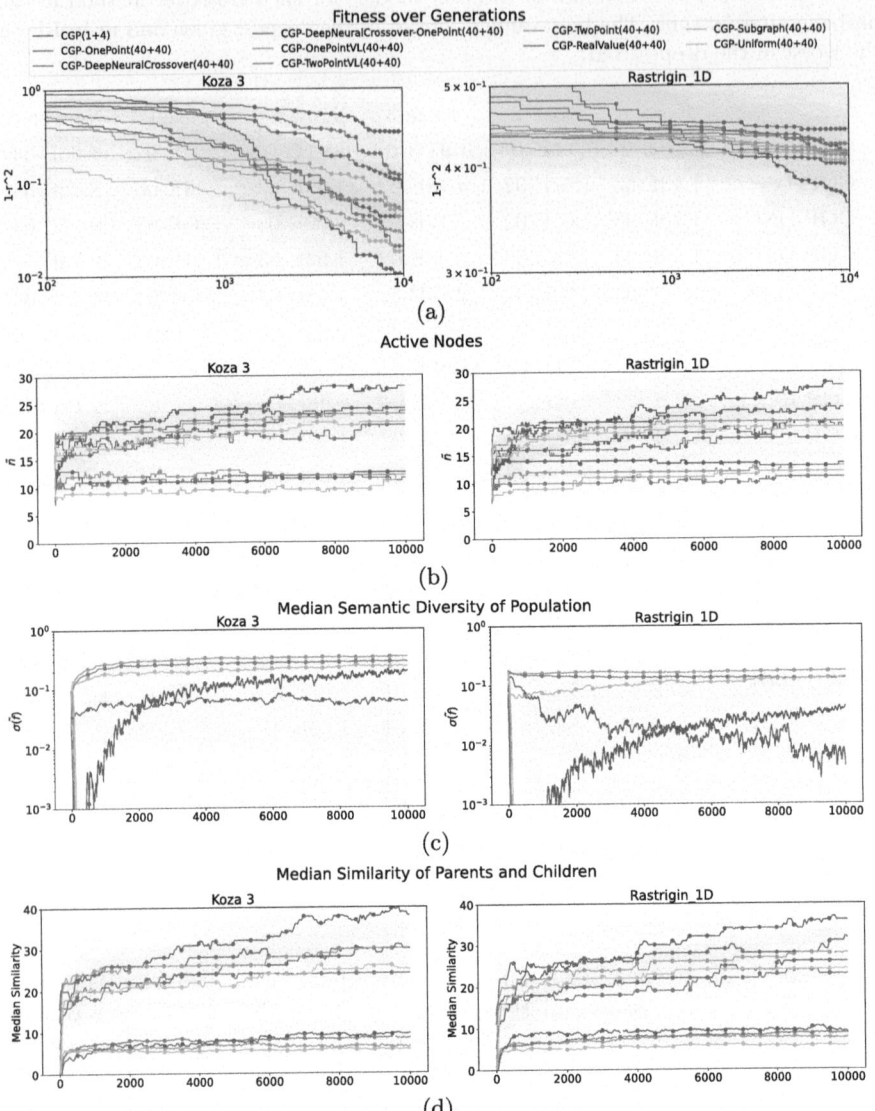

Fig. 2. (**a**): Median fitness of the elite model in the population starting at generation 100, with markers indicating every 500 generations. The error ribbons fill in the area between the first and third quartiles. Full plots are available in the supplementary material. (**b**): Median amount of active nodes in the elite model at each generation. CGP canonical tends to be the most complex by the end of evolution. (**c**): Median Semantic Diversity of the Population at each generation. One-Point-Variable-Length, Subgraph, and Canonical CGP have the most diverse populations. (**d**): Median similarity of the best parent and child in each couple at each generation. Canonical CGP produces the most similar children from each parent, as expected.

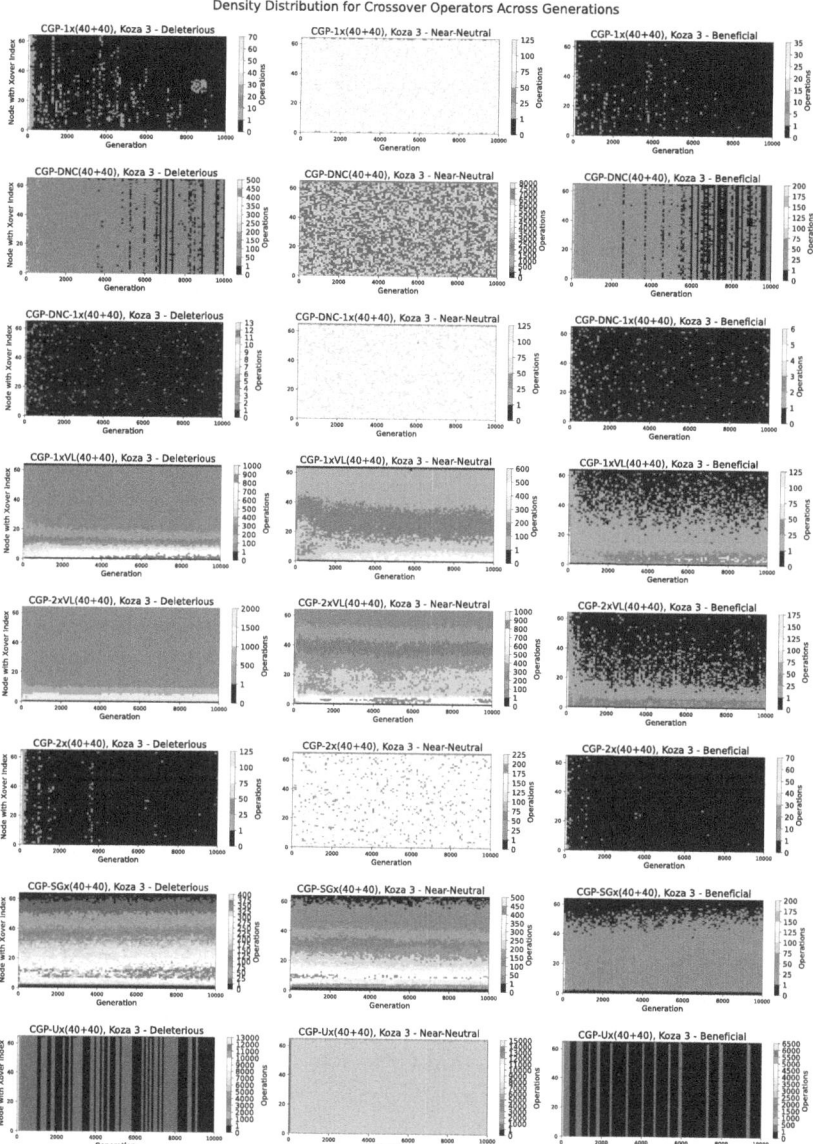

Fig. 3. Density of crossover points over time for Koza 3, grouped by algorithm and effectiveness. The data is condensed into points containing 100 generations and, except for Subgraph and One-Point-Variable-Length, three genes, thus each pixel represents a single node. This decision was made to keep events visible.

the best parent and the best child for each pair of parents, where Canonical CGP, Two-Point methods, and One-Point DNC produce more similar children by the end of evolution.

To check whether there are any patterns in the crossover events themselves, Fig. 3 shows the density of crossover points over time for each crossover method and their effectiveness. The effectiveness of an operation is determined by finding the percent change between the best parent in the couple and the best of that couple's offspring. The operation is then deleterious if $\Delta f < -0.001\%$, beneficial if $\Delta f > 0.001\%$, and near-neutral otherwise, if $-0.001\% < \Delta f < 0.001\%$. First, it is immediately clear that for most cases the overwhelming majority of events are near-neutral, and for the most part, those events are nearly uniformly spread across the genome at all times, which makes sense given the use of a uniform random number generator. It is also obvious that deleterious and beneficial operations are very common at the start of evolution (within the first hundred generations) and from there become increasingly uncommon. Two crossover methods stray from this pattern. Variable-Length methods are more likely to have deleterious than they are to have neutral operations, and operations are more likely closer to the front of the genome. Both observations make sense given that the genome in this case can change size, thus it is more likely to be smaller and more likely to have events at smaller indices, which would make its evolution less stable. Subgraph Crossover, on the other hand, does not have such capabilities: always having a fixed length genome and a randomly selected crossover point (within constraints), the distribution *should* be uniform, yet we see crossover events being much more common closer to the front of the genome. Since crossover points are restricted to active nodes, this seems to confirm Cui et al.'s observations that an instruction closer to the front of the genome is more likely to be active than those at the back [10].

Finally, we collect similar density distributions for mutation operations. However, since we are not testing different mutation operators, the results are relatively uniform across problems and crossover methods: operations are more likely to be deleterious or beneficial if performed near the front of the genome, and more likely to be neutral if performed towards the back.

5 Discussion

Our results indicate that Deep Neural Crossover does not seem to be an effective crossover operator for Cartesian Genetic Programming; DNC and Real-Value Crossover perform significantly worse than Canonical CGP(1+4) and other crossover methods, particularly Two-Point-Variable-Length and Uniform Crossovers. Subgraph Crossover, on the other hand, performs competitively in some scenarios, even when problems are more complex. We posit that the CGP genome does not contain any patterns that can be found with lightweight neural networks. As intimated in Sect. 1, previous literature has largely established constraints on selection and existing crossover methods instead of new ones. We argue that these new constraints assume *both* the semantic equivalence of genes in CGP regardless of position and that nodes closer to each other are likely to be connected. Following [8] and [10], it is most likely that *a node's semantics also rely on its position in the graph*. Even if the phenotype is kept intact, moving a gene from one position to another may have a more dramatic effect on fitness than previously recognized. This might occur because the Cartesian representation lacks a grounding mechanism such as the internal calculation registers in LGP [24].

From here we can posit that the literature operates on each node without considering the context of its connected nodes. A single instruction can be simultaneously dependent on other instructions and itself a dependency for other nodes. *A crossover method that does not take into account the interdependence of nodes is therefore unlikely to succeed.* This may be why Subgraph Crossover is competitive, as it tries to preserve relationships between active nodes. In this vein, we suggest an exploration of the relationship between positional bias and semantics following [10], which used the REORDER operator introduced in Goldman and Punch (2013) [13] to mitigate positional bias in relation to active nodes. We also propose to develop a method that focuses on the semantics (output) of each node rather than simply the genotype and fitness of the phenotype. Positive results have already been published for a semantic *mutation* operator in a $1 + \lambda$ framework [16].

5.1 Summary of Contributions

To summarize, we return to our question from Sect. 1: *why is crossover so destructive in CGP in the first place?* We believe, following [13] and [10], that the problem lies in the assumptions of traditional crossover. Standard n-Point Crossover assumes that genes that are closer to each other are more likely to be positionally related, which in CGP is not the case. We also confirm the findings of Cui *et al.*, namely that the activity of a node is negatively correlated with its position in the genome.

From a theoretical perspective, it would be prudent to run further experiments designed to test the dynamics of positional bias. Regardless of whether this is intuitive, the knowledge of *how* and *why* we observe positional bias is still rather vague at this time. It would also be interesting to put into practice a contextual perspective of CGP nodes and design a crossover operator around the semantics and interconnectivity of individual nodes, .i.e. using a phenotypic perspective on the crossover operator.

Acknowledgments. We would like to thank the Institute for Cyber-Enabled Research at Michigan State University and the John R. Koza Endowment fund administered at MSU.

Disclosure of Interests. The authors declare no conflicts of interest.

References

1. Hardware | institute for cyber-enabled research. https://icer.msu.edu/hpcc/hardware (2024)
2. Ackley, D.: A Connectionist Machine for Genetic Hillclimbing, vol. 28. Springer (2012)
3. Alfaro-Cid, E., Merelo, J., De Vega, F.F., Esparcia-Alcázar, A.I., Sharman, K.: Bloat control operators and diversity in genetic programming: a comparative study. Evol. Comput. **18**(2), 305–332 (2010)
4. Aygün, E., Ecer, D.: Python-alignment (2017). https://github.com/eseraygun/python-alignment
5. Banzhaf, W., Bakurov, I.: On the nature of the phenotype in tree genetic programming. In: Handl, J., Li, X., Wagner, M., Garza-Fabre, M., et al. (eds.) Proceedings of the Genetic and Evolutionary Computation Conference (GECCO 2024), Melbourne, Australia, pp. 868–877. ACM Press (2024)
6. Banzhaf, W., Nordin, P., Keller, R.E., Francone, F.D.: Genetic Programming: An Introduction—On the Automatic Evolution of Computer Programs and its Applications. Morgan Kaufmann Publishers Inc. (1998)
7. Brameier, M., Banzhaf, W.: Linear Genetic Programming, pp. 36–37. Springer (2007)
8. Cai, X., Smith, S.L., Tyrrell, A.M.: Positional independence and recombination in Cartesian Genetic Programming. In: Collet, P., Tomassini, M., Ebner, M., Gustafson, S., et al. (eds.) European Conference on Genetic Programming, pp. 351–360. Springer (2006)
9. Clegg, J., Walker, J.A., Miller, J.F.: A new crossover technique for Cartesian Genetic Programming. In: Thierens, D., Beyer, H.G., Branke, J., et al. (eds.) Proceedings of the 9th Annual Conference on Genetic and Evolutionary Computation, pp. 1580–1587. ACM Press (2007)
10. Cui, H., Margraf, A., Heider, M., Hähner, J.: Towards understanding crossover for Cartesian Genetic Programming. In: van Stein, N., Marcelloni, F., Lam, H.K., Cottrell, M., et al. (eds.) Proceedings of the 15th International Joint Conference on Computational Intelligence (IJCCI 2023), pp. 308–314. SCITEPRESS (2023)
11. Efron, B., Hastie, T., Johnstone, I., Tibshirani, R.: Least angle regression. Ann. Stat. **32**(2), 407–451 (2004)
12. Francone, F.D., Conrads, M., Banzhaf, W., Nordin, P.: Homologous crossover in genetic programming. In: Banzhaf, W., Daida, J.M. (eds.) Proceedings of the 1st Annual Conference on Genetic and Evolutionary Computation, GECCO 1999, vol. 2, pp. 1021-1026. Morgan Kaufmann Publishers Inc., San Francisco (1999)
13. Goldman, B.W., Punch, W.F.: Length bias and search limitations in Cartesian Genetic Programming. In: Silva, S., Esparcia-Alcazar, A.I., Lopez-Ibanez, M., Mostaghim, S., et al. (eds.) Proceedings of the 15th Annual Conference on Genetic and Evolutionary Computation, pp. 933–940 (2013)
14. Griewank, A.O.: Generalized descent for global optimization. J. Optim. Theory Appl. **34**, 11–39 (1981)
15. Haut, N., Banzhaf, W., Punch, B.: Correlation versus RMSE loss functions in symbolic regression tasks. In: Trujillo, L., Winkler, S.M., Silva, S., Banzhaf, W. (eds.) Genetic Programming Theory and Practice XIX, pp. 31–55. Springer (2023)
16. Hodan, D., Mrazek, V., Vasicek, Z.: Semantically-oriented mutation operator in Cartesian Genetic Programming for evolutionary circuit design. In: Coello Coello, C.A., Aguirre, A.H., Uribe, J.C., Fabre, M.G. (eds.) Proceedings of the 2020 Genetic and Evolutionary Computation Conference, pp. 940–948 (2020)

17. Hunter, N.: Meiotic recombination: The Essence of Heredity. Cold Spring Harb. Perspect. Biol. **7** 12 (2015). https://api.semanticscholar.org/CorpusID:33542130
18. Husa, J., Kalkreuth, R.: A comparative study on crossover in Cartesian Genetic Programming. In: Castelli, M., Sekanina, L., Zhang, M., Cagnoni, S., García-Sánchez, P. (eds.) EuroGP 2018. LNCS, vol. 10781, pp. 203–219. Springer, Cham (2018). https://doi.org/10.1007/978-3-319-77553-1_13
19. Kalkreuth, R.: A comprehensive study on subgraph crossover in Cartesian Genetic Programming. In: Merelo Guervós, J.J., Garibaldi, J., Wagner, C., et al. (eds.) Proceedings of the 12th International Joint Conference on Computational Intelligence (IJCCI 2020) - Volume 1: ECTA, pp. 59–70. INSTICC, SciTePress (2020). https://doi.org/10.5220/0010110700590070
20. Kalkreuth, R.: Reconsideration and extension of Cartesian Genetic Programming. Ph.D. thesis, Technical University of Dortmund, Germany (2021)
21. Kalkreuth, R., Rudolph, G., Droschinsky, A.: A new subgraph crossover for cartesian genetic programming. In: McDermott, J., Castelli, M., Sekanina, L., Haasdijk, E., García-Sánchez, P. (eds.) Genetic Programming, pp. 294–310. Springer, Cham (2017)
22. Kaufmann, P., Platzner, M.: Advanced techniques for the creation and propagation of modules in Cartesian Genetic Programming. In: Keijzer, M., Antoniol, G., Congdon, C.B., Deb, K., et al. (eds.) Proceedings of the 10th Annual Conference on Genetic and Evolutionary Computation, GECCO 2008, pp. 1219–1226. Association for Computing Machinery, New York (2008). https://doi.org/10.1145/1389095.1389334
23. Keijzer, M.: Scaled symbolic regression. Genet. Program Evolvable Mach. **5**(3), 259–269 (2004)
24. Kocherovsky, M., Banzhaf, W.: Crossover destructiveness in cartesian versus Linear Genetic Programming. In: Faíña, A., Risi, S., Medvet, E., Stoy, K., et al. (eds.) ALIFE 2024: Proceedings of the 2024 Artificial Life Conference, p. 20. Artificial Life Conference Proceedings, The International Society for Artificial Life (2024). https://doi.org/10.1162/isal_a_00735
25. Laguna, M., Marti, R.: Experimental testing of advanced scatter search designs for global optimization of multimodal functions. J. Glob. Optim. **33**, 235–255 (2005)
26. Liu, H., Zong, Z., Li, Y., Jin, D.: NeuroCrossover: An Intelligent genetic locus selection scheme for genetic algorithm using reinforcement learning. Appl. Soft Comput. **146**, 110680 (2023)
27. Ma, Y., et al.: Learning to iteratively solve routing problems with dual-aspect collaborative transformer. In: Ranzato, M., Beygelzimer, A., Dauphin, Y., Liang, P.S., et al. (eds.) Proceedings of the 35th International Conference on Neural Information Processing Systems, NIPS 2021, Curran Associates Inc., Red Hook (2024)
28. Miller, J.F., et al.: An empirical study of the efficiency of learning Boolean functions using a cartesian genetic programming approach. In: Banzhaf, W., Daida, J.M., Eiben, A.E., Garzon, M.H., Honavar, V. (eds.) Proceedings of the Genetic and Evolutionary Computation Conference, pp. 1135–1142. Morgan Kaufmann (1999)
29. Miller, J.F.: Cartesian genetic programming: Its Status and Future. Genet. Program Evolvable Mach. **21**(1–2), 129–168 (2020)
30. Miller, J.F., Harding, S.L.: Cartesian genetic programming. In: Keijzer, M., Antoniol, G., Bates Congdon, C., Deb, K., et al. (eds.) Proceedings of the 10th Annual Conference Companion on Genetic and Evolutionary Computation, pp. 2701–2726. ACM Press (2008)

31. Oltean, M., Groşan, C., Oltean, M.: Encoding multiple solutions in a Linear Genetic Programming chromosome. In: Bubak, M., Albada, G.D., Sloot, P.M.A., Dongarra, J. (eds.) International Conference on Computational Science, pp. 1281–1288. Springer (2004)
32. Poli, R., Langdon, W.B.: Schema theory for genetic programming with one-point crossover and point mutation. Evol. Comput. **6**(3), 231–252 (1998). https://doi.org/10.1162/evco.1998.6.3.231
33. Rudolph, G.: Globale optimierung mit parallelen Evolutionsstrategien. Ph.D. thesis, Department of Computer Science, University of Dortmund (1990)
34. Shem-Tov, E., Elyasaf, A.: Deep neural crossover: A multi-parent operator that leverages gene correlations. In: Handl, J., Li, X., Wagner, M., Garza-Fabre, M., et al. (eds.) Proceedings of the Genetic and Evolutionary Computation Conference, pp. 1045–1053 (2024)
35. Shem-Tov, E., Sipper, M., Elyasaf, A.: Deep learning-based operators for evolutionary algorithms. In: Winkler, S.M., Banzhaf, W., Hu, T., Lalejini, A. (eds.) Genetic Programming Theory and Practice XXI. Springer (2025)
36. da Silva, J.E., Bernardino, H.S.: Cartesian genetic programming with crossover for designing combinational logic circuits. In: 2018 7th Brazilian Conference on Intelligent Systems (BRACIS), pp. 145–150. IEEE Press (2018)
37. Suganuma, M., Shirakawa, S., Nagao, T.: Designing convolutional neural network architectures using cartesian genetic programming, pp. 185–208. Springer, Singapore (2020). https://doi.org/10.1007/978-981-15-3685-4_7
38. Torabi, A., Sharifi, A., Teshnehlab, M.: Using cartesian genetic programming approach with new crossover technique to design convolutional neural networks. Neural Process. Lett. **55**(5), 5451–5471 (2023)
39. Turner, A.J.: Improving crossover techniques in a genetic program. Master's thesis, Department of Electronics, University of York (2012)
40. Wilson, G., Banzhaf, W.: A comparison of Cartesian Genetic Programming and Linear Genetic Programming. In: O'Neill, M., Vanneschi, L., Gustafson, S., Esparcia Alcázar, A.I., et al. (eds.) Genetic Programming: Proceedings of the 11th European Conference, EuroGP 2008, Naples, Italy, pp. 182–193. Springer (2008)

Population Diversity, Information Theory and Genetic Improvement

William B. Langdon(✉) and David Clark

CREST, Department of Computer Science, UCL,
Gower Street, London WC1E 6BT, UK
W.Langdon@cs.ucl.ac.uk, david.clark@ucl.ac.uk
http://www.cs.ucl.ac.uk/staff/W.Langdon,
http://www.cs.ucl.ac.uk/staff/D.Clark, http://crest.cs.ucl.ac.uk/

Abstract. Compression, e.g. gzip, gives algorithmic information theory (Kolmogorov Complexity) based measures of string population diversity. To boost it we use the GI tool Magpie and select programs of average fitness that contribute most to variety, allowing evolution to automatically tailor triangle.c for production speed. We calculate C source code diversity via approximations to the Normalised Compression Distance on Multisets (NCDm) using both Cohen and Vitanyi's $O(n^2)$ approach and our own, $O(n)$ method, finding the cheaper, $O(n)$, is equally good.

Keywords: Evolutionary computing · EC · genetic programming · GP · SBSE · NCD · Normalised Information Distance · NID · perf · test set diameter

1 Introduction

Diversity plays an important role in optimising finite populations, e.g. in genetic algorithms [1], genetic programming (GP) [2,3] and genetic improvement GI [4–18]. In software engineering the widely used application of diversity is to test sets [20–25], whilst information theory has been applied to software robustness [26] and security [12,27]. The problem of lack of local gradient, or worse deceptive fitness gradients [28,29] or even fitness plateaux [30] is well known in optimisation and evolutionary computing. Lack of local gradient appears to be important in population based genetic improvement, with software engineering benchmarks such as the triangle program (Sect. 3 [31–33]) having search landscapes [34,35] dominated by large plateaus of equal fitness connected by relatively few improving mutations. Although, using normalised compression distance (NCD), we investigate the usefulness of program source code variability as a measure to decide which individuals to discard and which to select for the next generation (Fig. 1), such syntax based population diversity gives mixed results.

Programs are strings but a high quality string diversity measure is a challenging topic. The Rolls Royce measure of string diversity is Vitanyi's Normalised Information Distance [37], which is based on Kolmogorov Complexity.

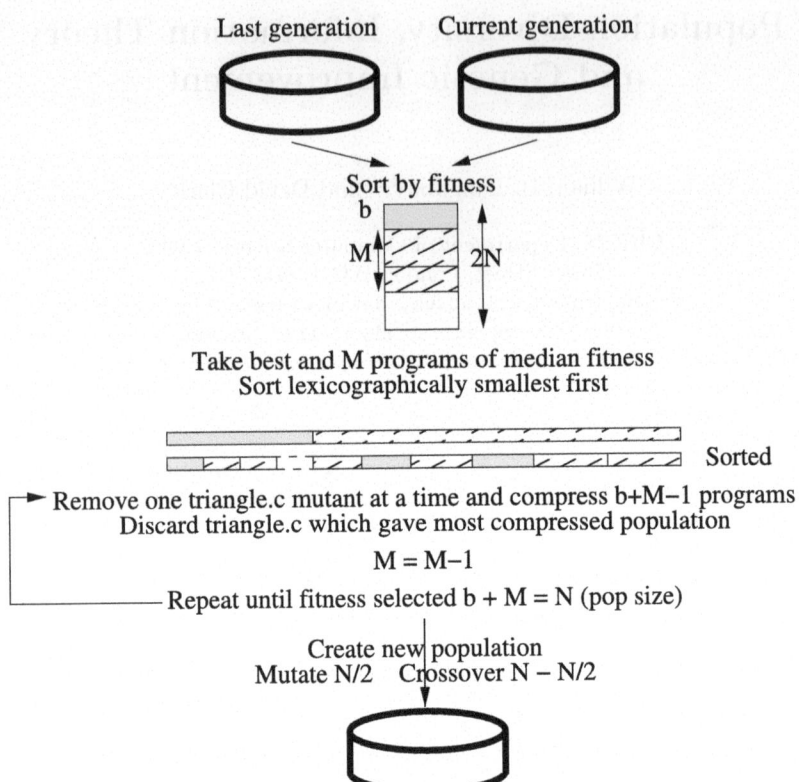

Fig. 1. Incorporating Cohen and Vitani's Normalised Compression Distance on Multisets (NCDm) [36] $O(n^2)$ into genetic algorithms. The GA population can contain duplicates (hence multiset rather than set). Each generation the GA selects from the current and previous generation the best (pink) and those of the average fitness individuals (hatched) which make the population most diverse (i.e. hardest to compress). (Color figure online)

Unfortunately Kolmogorov Complexity is not computable but numerous approximations are available and these offer trade offs between accuracy and efficiency. Commonly used is Cilibrasi and Vitanyi's compression based approximation, the Normalised Compression Distance (NCD) [25,38]. At the other end of the efficiency scale is the well known but more approximate Levenshtein distance [39–41], also satisfying the metric space axioms, and in the middle is dictionary based compression which can be efficient and produce a tight upper Kolmogorov Complexity bound for a known, finite population.

We exploit a variant of NCD known as NCDm, or NCD for multi-sets, that produces a single diversity measure or "diameter" for a multi-set, such as an EC population [21,36]. In addition, we err on the side of efficacy rather than efficiency, using Cohen and Vitanyi's suggested quadratic-in-population-size approximation for NCDm as well as a linear one of our own invention.

When improving the execution speed of the triangle program, we find that these diversity measures are approximately equally useful across our GI runs. Since population size is a confounding variable for diversity, we study a wide range of population sizes from one to 1000. However run time increases quadratically and as no further fitness improvement was found, we limit our experiments to one run per scenario for populations sizes 200–1000.

Section 3 says how we use the popular software engineering triangle program benchmark, whilst Sect. 4 describes our hybrid genetic programming and GI Magpie system, particularly how it incorporates information based diversity into selection. The experiments (Sect. 5) and results (Sect. 6) are followed by a discussion of a 17% improved triangle program (Sect. 7) and our conclusions (Sect. 8). But first the next section briefly describes Cohen and Vitanyi's diversity measure NCDm [36], how Feldt et al. [21] use it to measure test suite diversity and then how we have used their approach within genetic improvement to actively select breeding populations of evolving program source code.

2 Information Theory Applied to Genetic Algorithms

The topic of information theory in genetic algorithms (GAs) and evolutionary computing, e.g. genetic programming [42] and genetic improvement (GI) [43], is vast. We will concentrate upon how we have applied it in population selection in our GI and only note that the approach could be widely used in population based evolutionary computing.

In genetic algorithms the importance of striking the right balance between exploring to find new good regions of the search space and exploiting the good parts already found has long been known [1]. We present (Fig. 1) an information theoretic way of combining fitness based selection and population diversity based on Andrew Cohen and Paul Vitanyi's [36] diversity measure for multisets. Their Normalized Compression Distance (NCD) based multiset distance (NCDm) is very general and has been applied to test set selection [21].

Cohen and Vitanyi's underlying approach is to define the information content of a multiset (which in our case is the population) as using Kolmogorov complexity. The Kolmogorov complexity of a string is the size of the smallest program that can generate the string. However Kolmogorov complexity is not in general computable and so they take their usual NCD approach and approximate it as the length of the compressed string. (Here we will use the size in bytes of the output generated by gzip.) As part of calculating the normalised distance for a collection of strings (technically a multiset, as the collection may contain duplicates) they wish to find the minimum compressed size with all possible orderings and to normalise by dividing by the largest compressed size of the multiset excluding all possible subsets. Since there an exponentially large number of orderings they define an approximation which is still a metric but whose computational complexity is only quadratic in the number of strings $O(n^2)$.

We start with their quadratic algorithm, as used by Robert Feldt et al. [21]'s universal algorithm for measuring the diversity of test suites ("test set diameter"). Our NCD based approach is feasible even for populations of 1000, but as expected it is slow. Therefore we introduce a further, linear time $O(n)$ approximation, which can be orders of magnitude faster (Sect. 6.6) and as effective.

2.1 Normalized Compression Distance (NCD) for Multisets (NCDm)

As there an exponentially large number of orderings, to approximate the smallest compression distance over all possible orderings Cohen and Vitanyi [36] consider only a quadratic number of orderings. To select which ordering, their basic approach is to order the multiset and then concatenate it into a single file which can be compressed (Fig. 1). They order the strings (here members of the population) first by size and then alphabetically. By placing similar strings next to each other, there is a good chance the compression algorithm will perform well and give a small compressed output file. They then in order omit one member of the multiset and compress the new (now shorter) concatenated file. They work through the whole multiset one at a time, to find which string contributed least and discard it. This gives a multiset which is one member smaller. They repeat, again removing the string which has least impact on the compression of the new (smaller) multiset, until only 2 strings are left in the multiset. (Notice the algorithm is described as sequential but parts could be run parallel.) We first follow Feldt et al. [21] and in the next section describe how we use this central part of Cohen and Vitanyi's [36] algorithm as part of parent selection in the evolutionary algorithm.

2.2 Information Based Parent Selection

To incorporate Feldt et al. [21]'s test case selection algorithm into a fitness based evolutionary algorithm with population size N, we start with the current and previous population (both of size N, total size 2 N). From these 2 N we select N to be parents of the next generation. These are sorted by fitness (cf. rank based selection [1,44]). Those better than average (median) fitness are automatically selected. Those of worse than median fitness are automatically discarded. We then apply information theory to chose those individuals of average fitness which will contribute most to the breeding population of parents for the next generation. (The number of programs of average fitness is quite variable, but in these experiments it is typically near 10% of the combined population size.) Like Feldt et al. [21] we apply Cohen and Vitanyi's [36] NCDm to the source code of the programs of average fitness but we do not calculate the distance, we merely run the NCDm algorithm (Fig. 1) until we have reduced the number of files (here C programs) until it plus the number of better than average fitness members of the combined populations is equal to N, the size of the next population. This becomes our breeding population.

This is not in itself an elitist approach. We can choose to make it elitist by passing one or more members of our breeding population unchanged to the next generation. But we choose to create half the children using mutation and the remainder by crossover. The approach could be readily applied to many evolutionary algorithms which use separate populations.

3 Genetic Improvement Triangle.c Benchmark

The software engineering triangle benchmark takes three inputs and returns one of four integer values representing the type of the triangle: scalene, isosceles, equilateral or not a triangle. (Versions of the triangle program seem to go back to 1976 and Fortran [45]. We use our C version[1] and test suite[2] [46].) The important function is 40 lines of C source code (1300 bytes) containing 16 comparisons and 8 return statements. The benchmark's test suite is designed to cover all branches. It spends much of its time checking for errors ("not a triangle", 9 of the 14 tests). In the source code most of these error checks are at the start of the code, with another right at the very end.

In our genetic improvement experiment we suppose that the developers of a real system have taken such a heavy error detection approach and later the customer wants the code to be faster for everyday use. That is, in the triangles example, we assume most of the time the code would be presented with three numbers which are indeed the 3 lengths of the side of a triangle. So in our experiment we start with the original code and tests but now weight the tests so important ones score more in the fitness function (see Table 1).

The **fitness** test harness uses the Linux perf utility's API to measure how many computer instructions the mutated code takes on each of the 14 tests and multiplies it by the weighting for that test. The mutant's fitness is the 14 added together (Table 1). Note we minimise fitness scores. If the mutant gives the wrong answer on any test or there is a run time error, its fitness is so poor it will never be selected to be a parent.

Mutations and crossovers are able to re-arrange the existing C code to get better scores by moving code that deals with lower weighted cases to further from the start, allowing important cases to be dealt with more quickly.

Even in a time sharing network desktop, Linux perf's instruction count proved very stable and gave reliable fitness measurements. In contrast measurements of elapsed time taken during fitness testing are very noisy [47–49].

4 Genetic Programming Based on Magpie

Our genetic programming systems is based on Magpie [50][3]. Magpie is a language independent genetic improvement system written in Python. It has many options. We use only its XML mode. Using srcml (version 1.0.0) we convert the mutable source code into a single triangle.c.xml file. To avoid changes to Magpie, the population selection (Sect. 2.2 and Fig. 1) are done externally. Magpie and GP parameters are given in Table 1. Our GP makes use of Magpie in three ways:

1. Magpie was run with triangle.c.xml to generate a pool of all 2535 possible different XML mutations.

---

[1] https://github.com/wblangdon/triangle/blob/master/jss/triangle.c.
[2] https://github.com/wblangdon/triangle/blob/master/jss/testcases_oracle.txt.
[3] https://github.com/bloa/magpie downloaded 2 October 2023.

2. To create the initial GP population Magpie is run many times to create one random mutant at a time. We reject mutants which do not compile, give runtime errors or fail one or more fitness test. We keep doing this until we have enough credible mutants to fill the initial population (mutant triangle.c mean size 1320.1 ± 41.8 bytes).
3. As our GP is running, Magpie facilities are used to compile, run, test and calculate fitness of each mutant.

4.1 GP Operations: Mutation and Two Point Crossover

The basic Magpie representation is like linear genetic programming [51,52] and consists of a text based list of genes. Therefore it is easy to extract and insert individual genes from and into Magpie genomes.

Mutation: a parent, selected uniformly at random from the breeding population, is copied and the copy mutated by selecting uniformly at random one gene within it and replacing it with one taken at random from the 2535 possible different XML mutations (see item 1 in previous section).

With crossover: two parents are chosen uniformly at random from the breeding population. The first is copied. Two random cut points are chosen uniformly in the copy and in the second parent. The middle part (i.e. between the cut points) of the copy is replaced by genes copied from the middle of the second parent [53, Fig. 2].

Note mutation does not change the number of genes whereas crossover can but on average neither changes the genome's length.

5 Experiments

The GP/Magpie system was run 10 times on populations of 1, 2, 5, 20, 50, 100. Also there were a few runs of 200, 500 and 1000. For each we tried three types of selection (Fig. 1 Sect. 2.2): based on Feldt et al.'s NCDm $O(n^2)$ [21], our linear $O(n)$ approximation to NCDm and finally breaking ties of average fitness at random. The GP representation, fitness and parameters are given in Table 1. The fastest triangle.c mutant on test cases may be found any time up to generation 100.

6 Results

6.1 Speedup

Figure 2 shows the performance of the best in run for all ten repeated runs. As expected there is variation between runs but typically the population needs to contain at least 20 mutated programs for the search to do well. Indeed, although we did a few runs with larger populations (200, 500 and 1000) there seems to be no advantage in increasing it above 100. There is little difference between the three selection algorithms (plotted with +, × or □). Note the new linear

Table 1. Faster triangle.c

Representation	C code converted to XML by srcml. Variable length linear sequence of XML mutations. Mutated XML converted to C code and compiled
Fitness cases	14 test cases, each 3 sides of triangle and expected classification. Test suite designed to cover original C code
	Test suite weighting to favour important outputs: scalene and equilateral (one test each) weight 81, isosceles (three tests) weight 27, not a triangle (nine tests) weight 1, (Sect. 3)
Selection	Fitness is the sum of the number of instructions taken by each test multiplied by its weighting
	fitness = $\sum_{i=1}^{14}$ X86 instructions for test i × weight i
	If mutant fails to compile, fails at run time, exceeds 2 s time out or gives wrong answer on any test its fitness is so bad it will never have children
	1st fitness based rank selection and 2nd contribution to population diversity, see Fig. 1 and Sect. 2.2
Population:	Panmictic, non-elitist, generational, size 1, 2, 5, 10 ⋯ 1000
initial pop	Every triangle.c is mutated exactly once. All compile and run (page 6 item 2.). Initial fitness 7929–12578 (most as unmutated code 9069)
Parameters Magpie	Python version 3.10.1, GGC version 10.2.1, compiler options -O3 -DNDEBUG. Magpie defaults except [search] warmup = 1. XML edits: StmtReplacement StmtInsertion StmtDeletion ComparisonOperatorSetting ArithmeticOperatorSetting NumericSetting RelativeNumericSetting StmtMoving
GP:	50% subtree XML crossover, 50% subtree XML mutation (Sect. 4.1). 100 generations. No size limit

approximation × to estimating population diversity does as well as the quadratic approach inspired by Feldt et al.'s Test Set Diameter $O(n^2)$ [21] + and it is considerably faster (Sect. 6.6). Except for some runs with a population of only one or two, all runs make progress. If we concentrate on runs with a population of 20 or more, the median speed up is 16%.

6.2 Evolution of Performance

The performance of the three types of selection with various population sizes are summarised in Figs. 2 and 3. Figure 3 plots the evolution of the fastest (weighted) triangle.c program in the population at each generation for a typical run at each population size. Typically each run does not converge and the populations contain a range of fitness values. As expected performance depends on population size, with larger populations doing better. Runs with a population containing a single program (which will have been created by crossover) typically make no progress (shown by a horizontal line at the top of Fig. 3).

In the absence of elitism (Sect. 2.2 above), even though the best in the population is guaranteed to be part of the breeding population, they have no guarantee

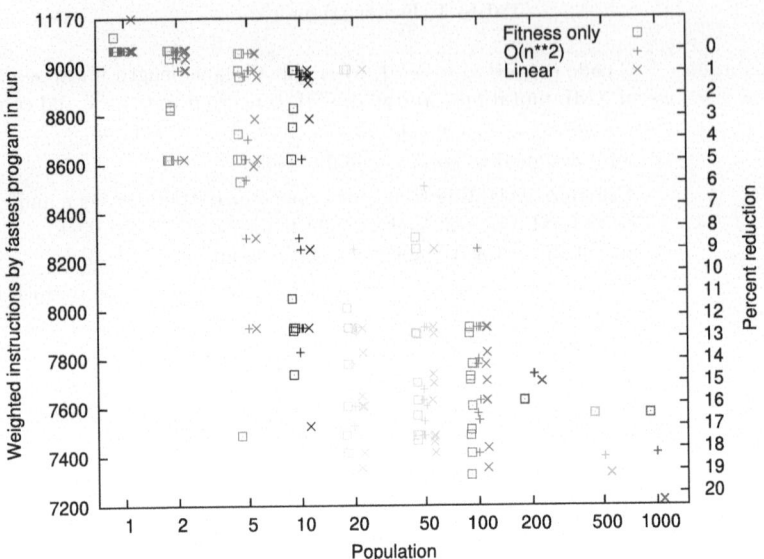

Fig. 2. Fastest triangle.c mutant found in ten runs by generation 100 with pop size 1, 2, 5, ··· up to 1000 (only 1 run 200–1000). Select best from current and previous generations to be parents. (One linear × run with population 1 got stuck at fitness 11 170.) Small horizontal noise added to spread data. Section 6.1.

that they will be selected from it for either way of making genetic changes. And even if selected, their children will be either mutated or created by crossover. Either of which may give a child with worse (or indeed better) fitness. Thus although a downward trend can be seen in Fig. 3, fitness does not usually improve monotonically. Indeed, since we are hoping for diverse populations we should not be disappointed that evolution does not lock into the "best seen so far" fitness value.

6.3 Evolution of Population Diversity

Figure 4 shows the average evolution of information contents of the population in ten runs with a population of 100 for the three selection schemes and Fig. 5 presents a summary by selection scheme and population size (note log scales).

Typically only about 10% of the population have average fitness (where information content is used to break ties, Sect. 2.2). Suggesting, in contrast to typical tournament selection [3, Sect. 2.3], selecting the best of the current and previous generations with 100% (50% crossover + 50% mutation) genetic modification avoids too high a selection pressure and does not drive the population to converge on a single fitness value [54] and instead our GP retains diverse fitness.

As with Fig. 2, Fig. 5 presents a summary of all the runs for each selection type and population size. As typically the population's information content does not tend to rise to a maximum at the end of the run but often falls towards the end,

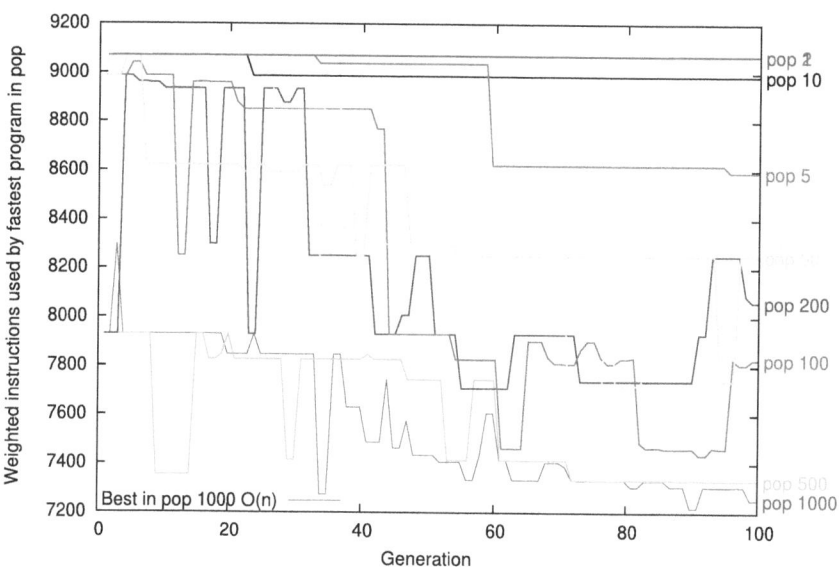

Fig. 3. Evolution of best fitness in typical run using linear $O(n)$ complexity selection. GI populations from 1 to 1000. Runs with $O(n^2)$ and without complexity selection are similar. (Same run colours as Fig. 2.) See Sect. 6.2.

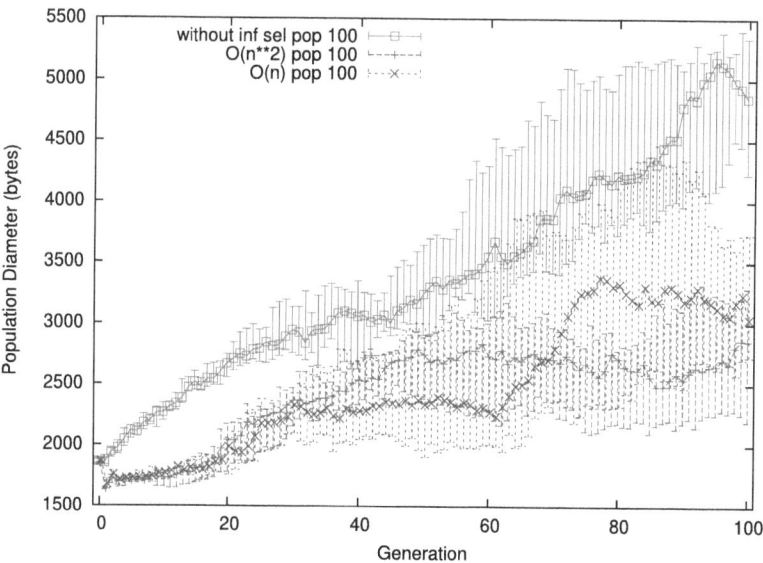

Fig. 4. Evolution of median population diameter for ten runs with the three selection schemes. GI populations of 100. The error bars give the interquartile spread across ten runs. Note change in colour scheme.

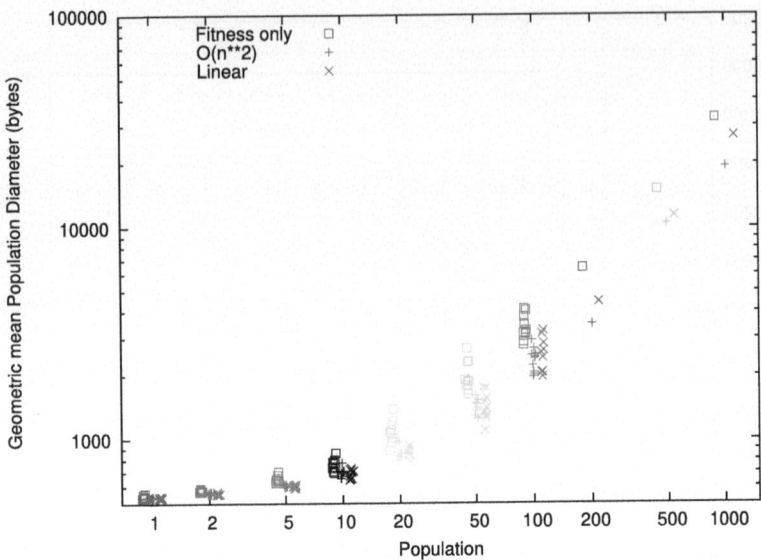

Fig. 5. Average compressed population of triangle.c mutants (see Sect. 6.3) in ten runs up to generation 100 with populations 1, 2, 5, ··· up to 1000 (only 1 run 200–1000). Small horizontal noise added to spread data.

Fig. 5 gives the average information content across each run. As expected, Fig. 5 shows the higher performing larger populations contain more information than the smaller populations of triangle.c mutants. Whilst, in terms of population information content, our linear $O(n)$ × approximation behaves as Cohen and Vitani's $O(n^2)$ NCDm [36] +. Although the fitness only approach □ gives on average less compressible populations ($p = 4\ 10^{-17}$ two-sided non-parametric Mann-Whitney U test), across the 73 runs of each type the median difference is only 13%.

The size of the triangle.c mutants (phenotype) is determined by the mutations applied (genotype). On average both information based selection schemes increase the C source code by about 10% (to 1430 bytes) while with fitness only selection there is more bloat (24%, 1617 bytes). gzip is very good at compressing the populations. For example with populations 1, 2, 5 and 10, it compresses the whole population into less than half the space of the original program. Even with the larger populations (e.g. 100) gzip gives average compression ratios of 46–74.

6.4 Evolution of Genome Size

To illustrate the evolution of the number of genes, Fig. 6 shows the growth of the average genome size for ten runs with a population of 100 triangle.c programs. It plots 10 runs with $O(n^2)$ (solid lines), 10 runs with our linear information based selection (dashed lines) and 10 runs with fitness only selection, where fitness ties are broken randomly (dotted lines), showing the median number of genes

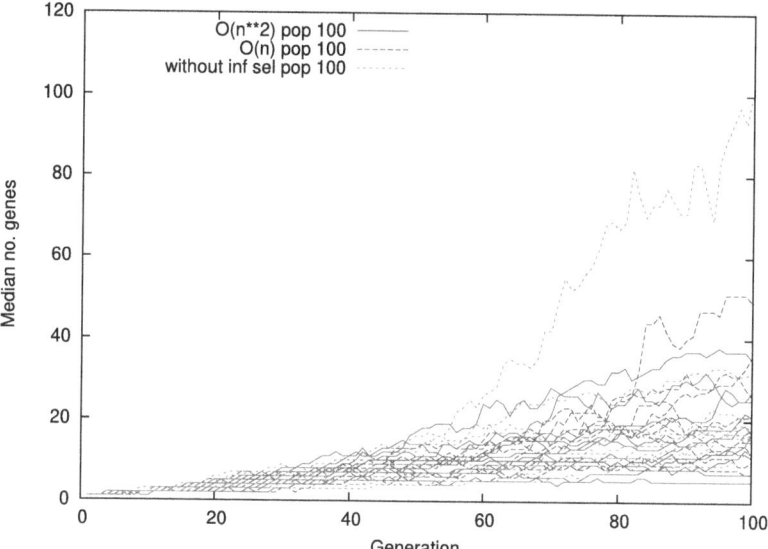

Fig. 6. Evolution of genome size for ten runs with the three selection schemes. GI populations of 100. (Same colour scheme as Fig. 4.)

rising from 1 initially to 17 ± 14 in generation 100. (The instances of triangle.c similarly grow from 1320 on average to 2000 ± 400 bytes by generation 100.)

Although there are fluctuations between generations, an upward trend, known as bloat [55], can be seen. Therefore Fig. 7 gives genome statistics for the end of each run with each of the population sizes and each of the three selection schemes. Figure 7 shows there is considerable variation between independent runs (note log scales). However there is a trend for larger populations (which tend to contain fitter programs) to contain more genes (i.e. more Magpie mutations of triangle.c, see also Fig. 8). In runs with the same population size, all three approaches (+ × □) tend to have on average a similar number of mutations.

6.5 Size of Best Solutions

Figure 8 shows all the best in run mutant's performance (y-axis) and their number of mutations (x-axis). Figure 8 shows a clear trend for faster (weighted) programs to have had more changes. However a few mutated triangle.c with only one, two or three changes do very well. The three selection schemes do approximately as well as each other.

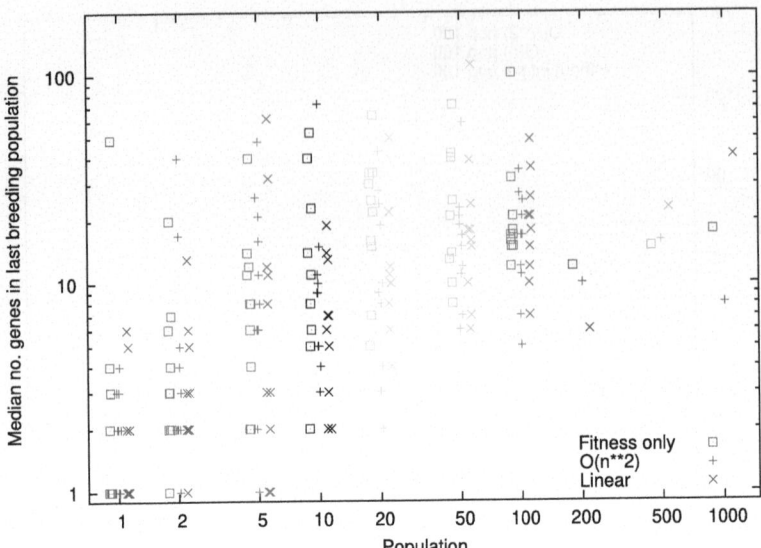

Fig. 7. Average number of genes in ten runs at generation 100 showing little difference in the 3 selection schemes. Small horizontal noise added to spread data (same run colours as Fig. 2). See Sect. 6.4.

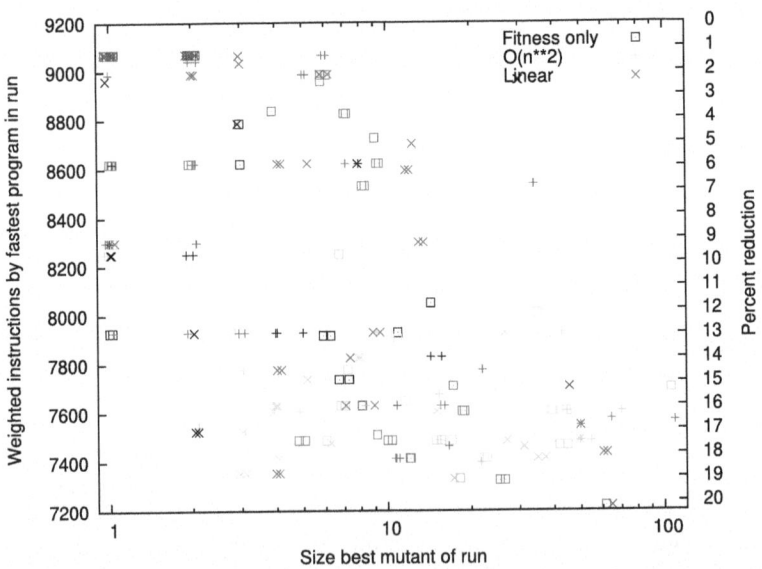

Fig. 8. Mutant gnome size v. fitness. Size and fitness of fastest triangle.c mutant found by generation 100 with populations 1, 2, 5, ··· up to 1000. Small horizontal noise added to spread data (same run colours as Fig. 2). See Sect. 6.5.

```
@@ -14,9 +14,7 @@
       int triang ;

-      if( side1 <= 0 || side2 <= 0 || side3 <= 0){
-          return 4;
-      }
+
       triang = 0;

@@ -27,6 +25,9 @@
          triang = triang + 2;
      }
      if(side2 == side3){
+         if( side1 <= 0 || side2 <= 0 || side3 <= 0){
+             return 4;
+         }
          triang = triang + 3;
      }

@@ -45,6 +46,10 @@
         return 3;
      }
      else if ( triang == 1 && side1 + side2 > side3) {
+         if(side1 + side2 <= side3 ||
+ side2 + side3 <= side1 || side1 + side3 <= side2){
+             return 4;
+         }
          return 2;
      }
      else if (triang == 2 && side1 + side3 > side2){
```

Fig. 9. Example Magpie changes to triangle.c which reduces its (weighted) instruction count from 9069 to 7544 (17% improvement). Pink code removed, green inserted. (Initially 1300 bytes, mutant 1438 bytes.) See Sect. 7 (Color figure online).

6.6 Time to Do Selection

With fitness only selection and our linear $O(n)$ approximation, selection typically takes < 1 s per generation on an otherwise unladen multi-core 32 GB 3.6 GHz Intel i7-4790 desktop. Naturally the $O(n^2)$ algorithm [36] scales badly with population size and in the worst case the time to select the parents with the largest population (1000) reaches almost two hours.

7 Discussion: Example Small High Fitness Mutant

Figure 9 shows a high scoring Magpie mutation as a C source code patch. The example is from run 6 with a population of 50 using $O(n^2)$ selection (we have deliberately chosen a small example to make explaining it easier). The mutated triangle.c is now larger (and so more difficult to compress). It was discovered in generation 15. As the run continued, evolution found similar mutations with identical scores containing the same genes (some repeated) plus others giving a still larger C source code. The Magpie genome for this mutation contains two genes StmtInsertion | StmtMoving. Both StmtInsertion and StmtMoving mutations have two components: the XML stmt level code to be inserted or moved and the XML location where it is to be placed.

StmtMoving is perhaps the easiest to explain, it moves the compound if (XML stmt 1, C sources lines 17–19, shown with pink shading in Fig. 9) and inserts it at XML _inter_block 19 (before line 30 in triangle.c, central green shading in Fig. 9). This means two of the three highly weighted tests for isosceles (return 2) and the even higher weight scalene (return 1) (see Table 1) do not incur the cost of checking for non-positive lengths, which correspond to not a valid triangle (return 4). However it also means risking missing some error conditions, which are not in the test suite.

The first mutation StmtInsertion copies another compound if from XML stmt 11 (lines 34–36 of triangle.c) and inserts it at XML _inter_block 30 (before line 48 in triangle.c, last green shaded region in Fig. 9). In the benchmark's test suite there are three tests with invalid zeros as input, notice that the now duplicated compound if detects them all. Thus, if because the initial tests for zero side length have been removed, execution reaches line 48, the duplicated code will still correctly detect the bad inputs and return 4. So the mutated compiled (with -O3) code gets a higher score by classifying important test cases earlier.

Notice how Magpie's operation on XML, effectively at the compiler AST level, means the mutations can easily operate with compound statements covering multiple lines and (except in a few odd ball cases) the mutant remains valid C code. (Even the indentation is correct.)

8 Conclusion

We have demonstrated how evolutionary algorithms can use information theory based population diversity alongside fitness selection. In the case of genetic improvement we used program source code, whereas Genetic Programming might use trees or instructions and Genetic Algorithms would use bit strings. Being compression based Cohen and Vitanyi's Normalised Information Distance (NID) in its single measure for a multiset (population) form (NCDm) would work with GP and GAs as well. We also invented a much faster linear version of Cohen and Vitanyi's quadratic approximation.

Apart from Pareto multi-objective combinations of our information and traditional fitness measures or systematically investigating more and weaker approximations to NID and NCDm in the hopes of a Goldilocks trade off, perhaps the next thing to consider is other ways to exploit this diversity measure. Population size is a confounding random variable for diversity so rather than theory that predicts convergence to optimal on the basis of population size [56] we could predict convergence on the basis of NCDm diversity diameter.

Acknowledgements. I am grateful for the assistance of Aymeric Blot, Dan Hoffman and Dan Blackwell.

Example C code in https://github.com/wblangdon/linux_perf_api etc.

References

1. Goldberg, D.E.: Genetic Algorithms in Search Optimization and Machine Learning. Addison-Wesley, Boston (1989)
2. Koza, J.R.: Genetic Programming: On the Programming of Computers By Means of Natural Selection. MIT Press, Cambridge (1992). https://mitpress.mit.edu/9780262527910/genetic-programming/
3. Poli, R., Langdon, W.B., McPhee, N.F.: A field guide to genetic programming (2008). Published via http://lulu.com and freely available at http://www.gp-field-guide.org.uk. (With contributions by J.R. Koza)
4. Langdon, W.B.: Genetic improvement of programs. In: Matousek, R. (ed.) 18th International Conference on Soft Computing, MENDEL 2012, Brno University of Technology. Brno 27–29 Jun 2012. http://www.cs.ucl.ac.uk/staff/W.Langdon/ftp/papers/Langdon_2012_mendel.pdf. invited keynote
5. Langdon, W.B., Harman, M.: Optimising existing software with genetic programming. IEEE Trans. Evol. Comput. **19**(1), 118–135 (2015). https://doi.org/10.1109/TEVC.2013.2281544
6. Petke, J., Langdon, W.B., Harman, M.: Applying genetic improvement to MiniSAT. In: Ruhe, G., Zhang, Y. (eds.) SSBSE 2013. LNCS, vol. 8084, pp. 257–262. Springer, Heidelberg (2013). https://doi.org/10.1007/978-3-642-39742-4_21
7. Petke, J.: Genetic improvement of software: a comprehensive survey. IEEE Trans. Evol. Comput. **22**(3), 415–432 (2018). https://doi.org/10.1109/TEVC.2017.2693219
8. Langdon, W.B., Harman, M.: Genetically improved CUDA C++ software. In: Nicolau, M., Krawiec, K., Heywood, M.I., Castelli, M., García-Sánchez, P., Merelo, J.J., Rivas Santos, V.M., Sim, K. (eds.) EuroGP 2014. LNCS, vol. 8599, pp. 87–99. Springer, Heidelberg (2014). https://doi.org/10.1007/978-3-662-44303-3_8
9. Langdon, W.B., Lam, B., Modat, M., Petke, J., Harman, M.: Genetic improvement of GPU software. Genet. Program Evolvable Mach. **18**(1), 5–44 (2016). https://doi.org/10.1007/s10710-016-9273-9
10. Petke, J., Harman, M., Langdon, W.B., Weimer, W.: Specialising software for different downstream applications using genetic improvement and code transplantation. IEEE Trans. Softw. Eng. **44**(6), 574–594 (2018). https://doi.org/10.1109/TSE.2017.2702606
11. Blot, A., Petke, J.: Empirical comparison of search heuristics for genetic improvement of software. IEEE Trans. Evol. Comput. **25**(5), 1001–1011 (2021). https://doi.org/10.1109/TEVC.2021.3070271
12. Mesecan, I., et al.: HyperGI: automated detection and repair of information flow leakage. In: Khalajzadeh, H., Schneider, J.G. (eds.) The 36th IEEE/ACM International Conference on Automated Software Engineering, New Ideas and Emerging Results track, ASE NIER 2021, pp. 1358–1362. Melbourne (2021). https://doi.org/10.1109/ASE51524.2021.9678758
13. Brownlee, A., et al.: Enhancing genetic improvement mutations using large language models. In: Arcaini, P., Yue, T., Fredericks, E. (eds.) SSBSE 2023. LNCS, vol. 14415, pp. 153–159. Springer, Cham (2023). https://doi.org/10.1007/978-3-031-48796-5_13

14. Pinna, G., et al.: Enhancing large language models-based code generation by leveraging genetic improvement. In: Giacobini, M., Xue, B., Manzoni, L. (eds.) EuroGP 2024. LNCS, vol. 14631, pp. 108–124. Springer, Cham (2024). https://doi.org/10.1007/978-3-031-56957-9_7
15. Nemeth, Z., Rainford, P.F., Porter, B.: Phenotypic species definitions for genetic improvement of source code. In: Faina, A., et al. (eds.) ALIFE 2024: Proceedings of the 2024 Artificial Life Conference, pp. 530–539. The International Society for Artificial Life, MIT Press, Cambridge (2024). https://doi.org/10.1162/isal_a_00795
16. Guizzo, G., et al.: Speeding up genetic improvement via regression test selection. ACM Trans. Softw. Eng. Methodol. **33**(8), 1–31 (2024). https://doi.org/10.1145/3680466
17. Brownlee, A.E.I., et al.: Large language model based mutations in genetic improvement. Autom. Softw. Eng. 15 (2025). https://doi.org/10.1007/s10515-024-00473-6. Special Issue on Advances in Search-Based Software
18. Blot, A., Petke, J.: A comprehensive survey of benchmarks for improvement of software's non-functional properties. ACM Comput. Surv. (2025). https://discovery.ucl.ac.uk/id/eprint/10203326/1/main.pdf. in press
19. Harman, M., Jones, B.F.: Search based software engineering. Inf. Softw. Technol. **43**(14), 833–839 (2001). https://doi.org/10.1016/S0950-5849(01)00189-6
20. Clark, D., Feldt, R., Poulding, S.M., Shin Yoo: Information transformation: an underpinning theory for software engineering. In: Bertolino, A., Canfora, G., Elbaum, S.G. (eds.) 37th IEEE/ACM International Conference on Software Engineering, ICSE 2015, Florence, 16–24 May 2015, vol. 2, pp. 599–602. IEEE Computer Society (2015). https://doi.org/10.1109/ICSE.2015.202
21. Feldt, R., Poulding, S.M., Clark, D., Shin Yoo: Test set diameter: quantifying the diversity of sets of test cases. In: IEEE International Conference on Software Testing, Verification and Validation, pp. 223–233. ICST Chicago (2016). https://doi.org/10.1109/ICST.2016.33
22. Chen, T.Y., Kuo, F.C., Merkel, R.G., Tse, T.H.: Adaptive random testing: the art of test case diversity. J. Syst. Softw. **83**(1), 60–66 (2010). https://doi.org/10.1016/J.JSS.2009.02.022
23. Arcuri, A., Briand, L.C.: Adaptive random testing: an illusion of effectiveness? In: Dwyer, M.B., Tip, F. (eds.) Proceedings of the 20th International Symposium on Software Testing and Analysis, ISSTA 2011, pp. 265–275. ACM, Toronto (2011). https://doi.org/10.1145/2001420.2001452
24. Anand, S., et al.: An orchestrated survey of methodologies for automated software test case generation. J. Syst. Softw. **86**(8), 1978–2001 (2013). https://doi.org/10.1016/j.jss.2013.02.061
25. Elgendy, I.T., Hierons, R.M., McMinn, P.: Evaluating string distance metrics for reducing automatically generated test suites. In: Lonetti, F., et al. (eds.) Proceedings of the 5th ACM/IEEE International Conference on Automation of Software Test (AST 2024), pp. 171–181. Lisbon (2024). https://doi.org/10.1145/3644032.3644455
26. Petke, J., Clark, D., Langdon, W.B.: Software robustness: a survey, a theory, and some prospects. In: Avgeriou, P., Dongmei Zhang (eds.) ESEC/FSE 2021, Ideas, Visions and Reflections, pp. 1475–1478. ACM, Athens (2021). https://doi.org/10.1145/3468264.3473133
27. Kosorukov, I., et al.: Mining for mutation operators for reduction of information flow control violations. In: IEEE/ACM International Conference on Automated Software Engineering, The New Ideas and Emerging Results (ASE-NIER 2024), Sacramento (2024). https://doi.org/10.1145/3691620.3695308

28. Goldberg, D.E.: Genetic algorithms and Walsh functions: part II, deception and its analysis. Complex Syst. **3**(2), 153–171 (1989). https://www.complex-systems.com/abstracts/v03_i02_a03/
29. Grefenstette, J.J.: Deception considered harmful. In: Whitley, L.D. (ed.) Foundations of Genetic Algorithms, vol. 2, pp. 75–91. Morgan Kaufmann, Vail (1992). https://doi.org/10.1016/B978-0-08-094832-4.50011-8
30. Ochoa, G., Veerapen, N.: Mapping the global structure of TSP fitness landscapes. J. Heuristics **24**(3), 265–294 (2018). https://doi.org/10.1007/S10732-017-9334-0
31. Langdon, W.B., Veerapen, N., Ochoa, G.: Visualising the search landscape of the triangle program. In: Castelli, M., McDermott, J., Sekanina, L. (eds.) EuroGP 2017. LNCS, vol. 10196, pp. 96–113. Springer, Cham (2017). https://doi.org/10.1007/978-3-319-55696-3_7
32. Veerapen, N., Daolio, F., Ochoa, G.: Modelling genetic improvement landscapes with local optima networks. In: Petke, J., White, D.R., Langdon, W.B., Weimer, W. (eds.) GI-2017, pp. 1543–1548. ACM, Berlin (2017). https://doi.org/10.1145/3067695.3082518. best presentation prize
33. Veerapen, N., Ochoa, G.: Visualising the global structure of search landscapes: genetic improvement as a case study. Genet. Program Evolvable Mach. **19**(3), 317–349 (2018). https://doi.org/10.1007/s10710-018-9328-1
34. Petke, J., et al.: A survey of genetic improvement search spaces. In: Alexander, B., Haraldsson, S.O., Wagner, M., Woodward, J.R. (eds.) 7th edition of GI @ GECCO 2019, pp. 1715–1721. ACM, Prague(2019). https://doi.org/10.1145/3319619.3326870
35. Langdon, W.B., Bruce, B.R.: The gem5 C++ glibc heap fitness landscape. In: Blot, A., Nowack, V., Rainford, P.F., Krauss, O. (eds.) 14th International Workshop on Genetic Improvement @ICSE 2025, Ottawa (2025). http://www.cs.ucl.ac.uk/staff/W.Langdon/ftp/papers/langdon_2025_GI.pdf. forthcoming
36. Cohen, A.R., Vitanyi, P.: Normalized compression distance of multisets with applications. IEEE Trans. Pattern Anal. Mach. Intell. **37**(8), 1602–1614 (2015). https://doi.org/10.1109/TPAMI.2014.2375175
37. Vitanyi, P., Balbach, F.J., Cilibrasi, R.L., Li, M.: Normalized information distance. In: Emmert-Streib, F., Dehmer, M. (eds.) Information Theory and Statistical Learning, vol. 3, pp. 45–82. Springer, Cham (2009). https://doi.org/10.1007/978-0-387-84816-7_3
38. Cilibrasi, R., Vitanyi, P.: Clustering by compression. IEEE Trans. Inf. Theory **51**(4), 1523–1545 (2005). https://doi.org/10.1109/TIT.2005.844059
39. Sapna, P.G., Mohanty, H.: Automated test scenario selection based on Levenshtein distance. In: Janowski, T., Mohanty, H. (eds.) 6th Distributed Computing and Internet Technology (ICDCIT 2010). LNCS, vol. 5966, pp. 255–266. Springer, Cham (2010). https://doi.org/10.1007/978-3-642-11659-9_28
40. Sakal, J., Fieldsend, J., Keedwell, E.: Genotype diversity measures for escaping plateau regions in university course timetabling. In: Thomson, S.L., et al. (eds.) Workshop on Landscape-Aware Heuristic Search (LAHS 2022), pp. 2090–2098. GECCO 2023, Association for Computing Machinery, Lisbon (2023). https://doi.org/10.1145/3583133.3596334
41. Elgendy, I.T., Hierons, R.M., McMinn, P.: A survey of the metrics, uses, and subjects of diversity-based techniques in software testing. ArXiv (2023). https://arxiv.org/abs/2311.09714
42. Johnson, C.G., Woodward, J.R.: Information theory, fitness, and sampling semantics. In: Johnson, C., Krawiec, K., Moraglio, A., O'Neill,

M. (eds.) Semantic Methods in Genetic Programming. Ljubljana (2014). https://citeseerx.ist.psu.edu/document?repid=rep1&type=pdf&doi=1bdff27d8e4dbc6321bef2aab06feb13f642b977. workshop at Parallel Problem Solving from Nature 2014 conference
43. Haraldsson, S.O., Woodward, J.R., Brownlee, A.E.I.: The use of automatic test data generation for genetic improvement in a live system. In: Galeotti, J.P., Petke, J. (eds.) Search-Based Software Testing, pp. 28–31. IEEE/ACM, Buenos Aires (2017). https://doi.org/10.1109/SBST.2017.10
44. Blickle, T., Thiele, L.: A comparison of selection schemes used in evolutionary algorithms. Evol. Comput. **4**(4), 361–394 (1996). https://doi.org/10.1162/evco.1996.4.4.361
45. Ramamoorthy, C.V., Ho, S.B., Chen, W.T.: On the automated generation of program test data. IEEE Trans. Softw. Eng. **2**(4), 293–300 (1976). https://doi.org/10.1109/TSE.1976.233835
46. Langdon, W.B., Harman, M., Jia, Y.: Efficient multi-objective higher order mutation testing with genetic programming. J. Syst. Softw. **83**(12), 2416–2430 (2010). https://doi.org/10.1016/j.jss.2010.07.027
47. Blot, A., Petke, J.: Comparing genetic programming approaches for non-functional genetic improvement case study: improvement of MiniSAT's running time. In: Hu, T., Lourenco, N., Medvet, E. (eds.) EuroGP 2020. LNCS, vol. 12101, pp. 68–83. Springer Verlag, Seville, Spain (2020). https://doi.org/10.1007/978-3-030-44094-7_5
48. Blot, A., Petke, J.: Using genetic improvement to optimise optimisation algorithm implementations. In: Hadj-Hamou, K. (ed.) 23ème congrès annuel de la Société Française de Recherche Opérationnelle et d'Aide à la Décision, ROADEF 2022. INSA Lyon, Villeurbanne - Lyon (2022). https://hal.archives-ouvertes.fr/hal-03595447
49. Langdon, W.B., Clark, D.: Deep imperative mutations have less impact. Autom. Softw. Eng. **32**(6), 1–39 (2025). https://doi.org/10.1007/s10515-024-00475-4
50. Blot, A., Petke, J.: MAGPIE: machine automated general performance improvement via evolution of software. arXiv (2022). https://doi.org/10.48550/arxiv.2208.02811
51. Banzhaf, W., Nordin, P., Keller, R.E., Francone, F.D.: Genetic Programming – An Introduction; On the Automatic Evolution of Computer Programs and its Applications. Morgan Kaufmann, San Francisco (1998). https://www.amazon.co.uk/Genetic-Programming-Introduction-Artificial-Intelligence/dp/155860510X
52. Brameier, M., Banzhaf, W.: Linear Genetic Programming. No. XVI in Genetic and Evolutionary Computation. Springer, Cham (2007). https://doi.org/10.1007/978-0-387-31030-5
53. Langdon, W.B., Banzhaf, W.: Repeated sequences in linear genetic programming genomes. Complex Syst. **15**(4), 285–306 (2005). http://www.cs.ucl.ac.uk/staff/W.Langdon/ftp/papers/wbl_repeat_linear.pdf
54. Langdon, W.B.: Genetic programming convergence. Genet. Program Evolvable Mach. **23**(1), 71–104 (2021). https://doi.org/10.1007/s10710-021-09405-9
55. Langdon, W.B., Poli, R.: Fitness causes bloat. In: Chawdhry, P.K., Roy, R., Pant, R.K. (eds.) Soft Computing in Engineering Design and Manufacturing, pp. 13–22. Springer, Cham (1997). https://doi.org/10.1007/978-1-4471-0427-8_2
56. Schmitt, L.M.: Theory of genetic algorithms. Theor. Comput. Sci. **259**(1–2), 1–61 (2001). https://doi.org/10.1016/S0304-3975(00)00406-0

Introducing Crossover in SLIM-GSGP

Gloria Pietropolli[1(✉)], Davide Farinati[2(✉)], Luca Manzoni[1], Mauro Castelli[2], Sara Silva[3], and Leonardo Vanneschi[2]

[1] Department of Mathematics, Informatics, and Geosciences, University of Trieste, Trieste, Italy
{gloria.pietropolli,lmanzoni}@units.it, lvanneschi@novaims.unl.pt
[2] NOVA Information Management School (NOVA IMS), Universidade Nova de Lisboa, Lisboa, Portugal
{dfarinati,mcastelli}@novaims.unl.pt
[3] LASIGE, Faculty of Sciences, University of Lisbon, Lisbon, Portugal
sgsilva@fc.ul.pt

Abstract. The Semantic Learning algorithm based on Inflate and deflate Mutations (SLIM-GSGP, or simply SLIM) is a variant of Geometric Semantic Genetic Programming (GSGP) designed to generate compact and interpretable models while maintaining the beneficial characteristic of GSGP of inducing an error surface without local optima. To date, no crossover operator has been defined for SLIM and the existing SLIM framework relies solely on two mutation operators: inflate and deflate mutation. This paper introduces two novel crossover operators for SLIM: Swap Crossover (XOSw) and Donor Crossover (XODn). These crossovers capitalize on SLIM's linked-list representation to facilitate genetic exchange while controlling program size. Experimental results on five symbolic regression problems demonstrate that the new crossover operators often improve fitness and reduce model size when compared to standard SLIM and to GSGP. Our findings establish these operators as solid improvements of traditional GSGP crossover.

Keywords: SLIM · Geometric Semantic Genetic Programming · Geometric Semantic Crossover · Inflate Mutation · Deflate Mutation

1 Introduction

In traditional Genetic Programming (GP) [17], genetic operators act on the syntax of the individuals without any knowledge of how they affect semantics, often leading to drastic and unpredictable effects during the evolution [24]. To

---

G. Pietropolli and D. Farinati—These authors contributed equally to this work.

The original version of the chapter has been revised. The Chapter 7 author affiliation has been corrected. A correction to this chapter can be found at https://doi.org/10.1007/978-3-031-89991-1_16

Supplementary Information The online version contains supplementary material available at https://doi.org/10.1007/978-3-031-89991-1_7.

address this problem, some researchers have focused on studying semantic-aware operators [32,39], both acting directly [20,26,27] and indirectly [3,4,7,18,29].

Notably, Geometric Semantic Genetic Programming (GSGP), introduced in [26], is a GP variant where traditional crossover and mutation are replaced by Geometric Semantic Crossover (GSC) and Geometric Semantic Mutation (GSM), collectively known as Geometric Semantic Operators (GSOs). GSOs, although acting on the syntax of the programs, have known effects on their semantics. The main strength of GSOs lies in their ability to induce a unimodal fitness landscape on any supervised learning problem, which facilitates the search for good solutions, compared to standard GP [9]. Since its inception, various improvements have been made to GSGP [5,13,28,33–35,38], leading to several achievements across a range of applications [10,21,42]. Most improvements have focused on refining GSOs, with a particular emphasis on GSC [2,12,16,37]. However, a remaining limitation of GSGP is that, by design, GSOs always generate offspring larger than their parents, causing program sizes to grow rapidly. To avoid building and storing massive programs, which would make GSGP impractical for real-world applications, implementations were introduced [8,25,38] that store only the semantic vectors of newly created individuals, along with all individuals of the initial population and the random programs generated across generations. While this solution made GSGP fast enough to be used in practice, it is a workaround to the problem, since it does not limit code growth. Although other efforts have been made to solve this problem [22], no major breakthroughs were achieved. Moreover, despite its theoretical advantages, GSC in GSGP shows limited effectiveness, as it cannot generate individuals outside the convex hull of the current population.

Recently, the Semantic Learning algorithm based on Inflate and deflate Mutation (SLIM), a new variant of GSGP [36,40] (also referred to as SLIM-GSGP), was shown to preserve the ability to induce a unimodal error surface while evolving much smaller models than GSGP and standard GP. SLIM evolves individuals that can be represented as linked lists of expressions and uses two different mutation operators: Inflate Geometric Semantic Mutation (IGSM), analogous to the GSM of GSGP, which generates larger offspring by appending one or more items at the end of the list, and Deflate Geometric Semantic Mutation (DGSM), which produces smaller offspring by deleting one of these items from the list.

To date, no crossover has been defined for SLIM. This is not surprising, since mutation is a more effective operator than crossover for traditional GSGP [9,27]. However, improving the search ability of SLIM with a semantic-aware crossover is an open possibility. Leveraging the SLIM linked-list implementation, we can define crossovers that address both limitations of GSC, enabling offspring to extend beyond the convex hull of the population and avoiding unnecessary increases in offspring size. To answer this call, we introduce two different crossover operators for SLIM, providing their mathematical formulations and integrating them in different variants of SLIM, with and without using the deflate mutation. We study their effects on fitness and size of the resulting models on a set of real-world problems.

2 Semantic Learning Algorithm Based on Inflate and Deflate Mutation

This section describes how SLIM-GSGP differs from traditional GSGP. More specifically, after a brief introduction to GSGP (Sect. 2.1), it presents the two mutation operators, IGSM and DGSM, it introduces the different studied SLIM variants and it discusses the linked-list representation of the individuals (Sect. 2.2).

2.1 Geometric Semantic Genetic Programming

In GP, the term *semantics* refers to the vector of the output values of a program on a set of observations. GSGP [26] is a variant of GP that replaces traditional syntax-based genetic operators with GSOs. These operators have a known effect on the semantics of the individuals, inducing geometric properties in the semantic space. Specifically, given two parent functions $T_1, T_2 : \mathbb{R}^n \to \mathbb{R}$, the GSC produces an offspring function defined as: $T_{XO} = T_1 \cdot T_R + (1 - T_R) \cdot T_2$, where T_R is a random real function with outputs in $[0, 1]$. Similarly, the GSM generates a mutated function from a parent $T : \mathbb{R}^n \to \mathbb{R}$ as: $T_M = T + ms \cdot (T_{R1} - T_{R2})$ where T_{R1} and T_{R2} are random real functions with outputs in $[0, 1]$ and ms is a parameter called *mutation step*. GSC corresponds to a geometric crossover in the semantic space. In other words, it produces offspring whose semantics lie on the segment connecting the semantics of the two parents. As a consequence, the offspring cannot be worse than the worst of its parents on the training set. GSM, on the other hand, corresponds to a ball mutation in the semantic space, generating offspring within a sphere of radius ms centered in the semantics of the parent. One of the principal advantages of GSGP is that these operators induce a unimodal error surface, simplifying the search process by eliminating local optima. The interested reader is referred to [41] for a detailed explanation of the reason why GSOs induce a unimodal error surface. However, these operators also have a significant limitation: by design, they always produce offspring larger than their parents, leading to a fast growth in the size of the individuals during the evolution.

2.2 SLIM-GSGP

SLIM-GSGP [36,40] is an extension of GSGP that, while maintaining the property of inducing a unimodal error surface, generates smaller models. As the name suggests, it incorporates two types of mutation: the IGSM which, as traditional GSM, generates an offspring larger than its parent; and DGSM, which produces offspring that are smaller than their parents.

The IGSM shares the same definition of GSM. The definition and validity of DGSM are based on two key observations: first, GSM can be rewritten as $\text{GSM}(T) = T + ms \cdot (T_{R1} - T_{R2}) = T - ms \cdot (T_{R2} - T_{R1})$; second, the random trees T_{R1} and T_{R2} are interchangeable, as they are independently sampled from the same distribution. Thus, the GSM can be written as $\text{GSM}(T) = T - ms \cdot (T_{R1} - T_{R2})$. To provide some intuition of why this reasoning can be used to generate offspring that are smaller than their parent, consider for instance an individual T to which GSM is applied, say, three consecutive times. The result of this operation is:

$$T_M = \text{GSM}^3(T) = T + ms \cdot (T_{R1} - T_{R2}) + ms \cdot (T_{R3} - T_{R4}) + ms \cdot (T_{R5} - T_{R6})$$

If we now apply again GSM, but this time using subtraction instead of addition, the generated individual is:

$$T_{M'} = T + ms \cdot (T_{R1} - T_{R2}) + ms \cdot (T_{R3} - T_{R4}) + ms \cdot (T_{R5} - T_{R6}) - ms \cdot (T_{R7} - T_{R8})$$

Before proceeding, it is important to recall that a widely adopted approach in the literature is to reuse random trees instead of generating new ones [38]. For example, we could use T_{R3} and T_{R4} instead of T_{R7} and T_{R8}, which would result in the following individual:

$$T' = T + ms \cdot (T_{R1} - T_{R2}) \underbrace{+ ms \cdot (T_{R3} - T_{R4})}_{} + ms \cdot (T_{R5} - T_{R6}) \underbrace{- ms \cdot (T_{R3} - T_{R4})}_{}$$

After this simplification, the individual has a smaller genotype. Using this idea, DGSM simply works by removing one of the terms that IGSM had previously added to an individual, effectively performing a geometric semantic mutation. Both IGSM and DGSM perturb the semantics of the individual within the same range, specifically $[-ms, ms]$, where ms is the mutation step parameter. This allows us to reduce the size of the individuals while maintaining the semantic properties of the search.

The IGSM and DGSM are the only operators defined so far within the SLIM framework. To date, no crossover operator has been introduced for SLIM.

Different SLIM Variants. In [40], the authors considered three different methods (SIG2, ABS, SIG1) for defining a function that returns values within the range $[-ms, ms]$, and two different approaches for perturbing an individual to achieve ball mutation in the semantic space.

We provide the definition for two of the three functions that return values within $[-ms, ms]$, specifically those used in this paper, as follows:

1. Using two random trees T_{R1} and T_{R2}, each passed through the sigmoid function S (which maps outputs to $[0, 1]$) as follows:

$$\text{SIG2} = ms \cdot (S(T_{R1}) - S(T_{R2})),$$

which is the same as traditional GSM.
2. Using a single random tree T_R passed through the sigmoid function S:

$$\text{SIG1} = ms \cdot (2 \cdot S(T_R) - 1).$$

On the other hand, a small perturbation of a numeric value can be achieved either by adding a small positive or negative quantity close to zero, or by multiplying by a factor slightly smaller or larger than one. Accordingly, there are two methods for perturbing an individual within the SLIM framework:

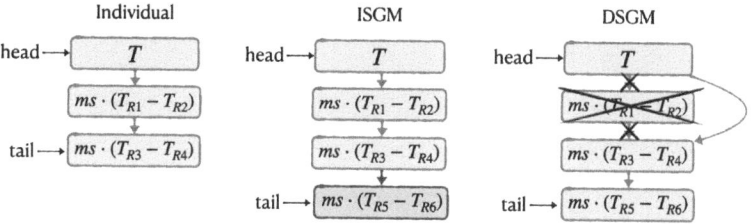

Fig. 1. An example of a SLIM+SIG2 individual, with a visual representation of the effect of the IGSM and DGSM on its genotype.

1. IGSM adds the function, while DGSM subtracts it.
2. IGSM multiplies by (1 + function), while DGSM divides by it.

By combining these two methods of perturbing an individual with the three functions defined for mutation, we obtain six SLIM variants. These are denoted as SLIM+ or SLIM∗, depending on whether they use addition or multiplication, followed by SIG2, ABS, or SIG1, based on the function applied. For instance, the variant called SLIM+SIG2 uses the sum of a small positive expression close to zero to perform ball mutation and the SIG2 function to define a perturbation within $[-ms, ms]$.

Linked-List Implementation of SLIM. In SLIM, it is useful to represent the individuals as linked lists, where each item contains a subprogram. In this way, an IGSM can be implemented by appending one or more items at the end of the list, while DGSM can be implemented by deleting one of these items from the list. To evaluate an individual, it is sufficient to accumulate either the sum (for the SLIM+ variants) or the product (for the SLIM* variants) of the terms of the list. A visual representation of the linked-list implementation and the application of the IGSM and DGSM is shown in Fig. 1 for the specific case of the SLIM+SIG2 variant.

Thus, using a simple notation, a SLIM individual can be represented as:

$$\mathcal{T}_j = (T_j, \Delta T_1^{(j)}, \ldots, \Delta T_n^{(j)}),$$

where T_j is the head of the list (the initial individual before any mutation), n is the number of elements that belong to the linked list, excluding the head, and each $\Delta T_i^{(j)}$ denotes the i-th block belonging to individual j. To minimize memory occupation, these blocks can consist exclusively of the single tree used for mutation in the ABS and SIG1 cases, or they can represent the difference between two random trees in the SIG2 case.

3 Crossover in SLIM

In SLIM, the linked-list representation not only facilitates the implementation of the deflate operator but also allows for new approaches to recombine genetic

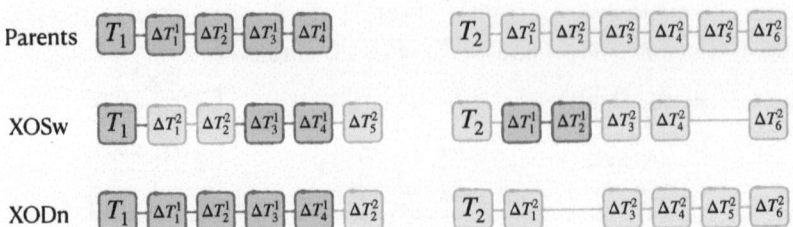

Fig. 2. An intuitive representation of the XOSw and XODn crossovers.

material within the population. Standard GSGP faces well-recognized limitations: (1) Even more than GSM, GSC generates offspring larger than the parents, leading to exponential growth in population size [26], and (2) GSC shows restricted search ability as, despite its theoretical advantages, it cannot generate new individuals outside the convex hull of the current population [9,27]. To address these issues, we propose leveraging the linked-list structure to define new crossover methods aimed at overcoming both limitations of GSC. For (1), we will design these crossover methods to control individual size, avoiding the rapid growth characteristic of GSC. For (2), we will move beyond standard geometricity, meaning these crossovers will not generate offspring to the segment between the parents, as GSC. However we can still define crossover operators with a clear geometrical meaning.

Firstly, we define Swap Crossover (XOSw), which swaps items of the linked list between parents, allowing recombination of genetic material while keeping offspring the same size as the parents. Next, we introduce Donor Geometric Crossover (XODn), where one parent (donor) donates a block to the other parent (receiver). This crossover mimics the SLIM mutation operators, essentially performing IGSM on the receiver (which gains a block) and DGSM on the donor (which loses a block). Unlike mutation, however, this operator uses genetic material already present in the population rather than a random block. A visual intuition of these crossovers is provided in Fig. 2.

As discussed in Sect. 2, SLIM can be implemented in two ways: additive (the SLIM+ variants) and multiplicative (the SLIM* variants). While the crossovers we introduce modify blocks in the same way in both forms, the subsequent evaluation of the offspring differs. Thus, we provide two variants for each crossover: one for the additive case (denoted by **A**) and one for the multiplicative case (denoted by **M**).

Swap Crossover (**XOSw**): Given two parent functions $T_1 = (T_1, \Delta T_1^{(1)}, \ldots, \Delta T_{n_1}^{(1)})$ and $T_2 = (T_2, \Delta T_1^{(2)}, \ldots, \Delta T_{n_2}^{(2)})$, where $n_1 < n_2$

without loss of generality, XOSw produces two offspring functions:

A. $\mathcal{T}_{\text{XOSw}_1} = \text{sel}(T_1, T_2) + \sum_{i=1}^{n_1} \text{sel}(\Delta T_i^{(1)}, \Delta T_i^{(2)}) + \sum_{i=n_1+1}^{n_2} \text{sel}(\varnothing, \Delta T_i^{(2)}),$
$\mathcal{T}_{\text{XOSw}_2} = \overline{\text{sel}}(T_1, T_2) + \sum_{i=1}^{n_1} \overline{\text{sel}}(\Delta T_i^{(1)}, \Delta T_i^{(2)}) + \sum_{i=n_1+1}^{n_2} \overline{\text{sel}}(\varnothing, \Delta T_i^{(2)})$

M. $\mathcal{T}_{\text{XOSw}_1} = \text{sel}(T_1, T_2) \cdot \prod_{i=1}^{n_1} \text{sel}(\Delta T_i^{(1)}, \Delta T_i^{(2)}) \cdot \prod_{i=n_1+1}^{n_2} \text{sel}(1, \Delta T_i^{(2)}),$
$\mathcal{T}_{\text{XOSw}_2} = \overline{\text{sel}}(T_1, T_2) \cdot \prod_{i=1}^{n_1} \overline{\text{sel}}(\Delta T_i^{(1)}, \Delta T_i^{(2)}) \cdot \prod_{i=n_1+1}^{n_2} \overline{\text{sel}}(1, \Delta T_i^{(2)})$

where $\text{sel}(a, b)$ selects a or b with equal probability, with $\varnothing$ indicating no additive term and 1 serving as the neutral multiplicative element. The operator $\overline{\text{sel}}(a, b)$ returns the complement of $\text{sel}(a, b)$, such that if $\text{sel}(a, b) = a$, then $\overline{\text{sel}}(a, b) = b$, and vice versa.

This crossover generates two distinct individuals. For the first offspring, each block is randomly selected from the corresponding position in either parent with equal probability. The second offspring is then constructed by taking all blocks not selected by the first; for each position, if the first offspring inherits a block from one parent, the second offspring inherits the corresponding block from the other parent. When parents differ in length, any extra blocks from the longer parent (for $i = n_1 + 1, \ldots, n_2$) may or not be assigned to the first offspring at random. The second offspring then inherits any remaining unassigned blocks, ensuring distinct contributions without overlap. As shown in Fig. 2, the two offspring contain blocks randomly selected from each parent. Being the second parent longer than the first, the two extra blocks are randomly assigned between the offspring: specifically, ΔT_5^2 goes to the first, while ΔT_6^2 goes to the second.

When defining a crossover in SLIM, we need to ensure that offspring blocks remain within the range $[-ms, ms]$, as this is appropriate for applying DGSM in the next evolutionary step. This crossover generates two offspring that maintain block values within this range, as they are composed of preexisting blocks, that already had this property.

The geometrical properties of XOSw are quite peculiar. In fact, while the crossover itself is not a geometric crossover, it *can* be defined (under some additional assumptions) as the crossover between two different individuals that derive from the original parents. In fact, let us define two trees, $\mathcal{T}_{\min}$ and $\mathcal{T}_{\max}$, representing the element-by-element minimum (resp., maximum) of each of the components of the trees $\mathcal{T}_1$ and $\mathcal{T}_2$. These trees are a (not necessarily strict) upper bound and lower bound to what can be achieved by crossover, and each individual resulting from the crossover between $\mathcal{T}_1$ and $\mathcal{T}_2$ is in the segment joining $\mathcal{T}_{\min}$ and $\mathcal{T}_{\max}$. Hence, while the crossover is not geometric in the classical sense, it is geometric on an *extension* of the original parents' semantic vectors, in a way similar to line crossovers. Thus, an "escape" from the convex hull of the population via crossover is possible, but in a tightly controlled way. A formal investigation of this fact, while interesting, is outside the scope of this paper.

A key property of this crossover is that the offspring match the size of their parents, ensuring that applying this operator does not lead to an overall growth in individual size; the total number of blocks between the two parents is maintained in the two offspring.

Donor Crossover **(XODn):** Given two parent functions $\mathcal{T}_1 = (T_1, \Delta T_1^{(1)}, \ldots, \Delta T_{n_1}^{(1)})$ and $\mathcal{T}_2 = (T_2, \Delta T_1^{(2)}, \ldots, \Delta T_{n_2}^{(2)})$, with $\mathcal{T}_1$ as the donor and $\mathcal{T}_2$ as the receiver, XODn produces two offspring:

A. $\mathcal{T}_{\text{XODn}_1} = T_1^{(1)} + \sum_{i \neq j} \Delta T_i^{(1)}, \quad \mathcal{T}_{\text{XODn}_2} = T_1^{(2)} + \Delta T_j^{(1)} + \sum_{i=1}^{n_2} \Delta T_i^{(2)}$

M. $\mathcal{T}_{\text{XODn}_1} = T_1^{(1)} \cdot \prod_{i \neq j} \Delta T_i^{(1)}, \quad \mathcal{T}_{\text{XODn}_2} = T_1^{(2)} \cdot \Delta T_j^{(1)} \cdot \prod_{i=1}^{n_2} \Delta T_i^{(2)}$

where j represents the donated block index.

XODn leverages the SLIM linked-list structure by transferring a block from a donor to a receiver. An example can be found in Fig. 2, where the first parent is randomly chosen as the receiver and the second as the donor. A block is then randomly selected from the donor (in this example ΔT_2^2) and appended to the receiver.

Intuitively, it acts as DGSM on the donor and IGSM on the receiver, generating offspring within a radius ms around the parents. This approach preserves the geometricity property, thus maintaining the GSGP trait of inducing a unimodal error surface, while also enabling genetic material exchange without the GSC limitation of confining offspring within the initial solution space. Unlike mutation, which introduces a random block, XODn transfers a block from an existing individual, justifying its classification as a crossover due to the genetic material exchange. Moreover, offspring remain close in size to the parents (within one block more or less), preventing size growth.

4 Experimental Study

We now examine the impact of both XOSw and XODn crossovers on the fitness and size of the evolved individuals. We design two experimental setups: one where the crossover is added to the regular SLIM framework and one where the crossover is added to a modified version of SLIM from which the deflate mutation is removed.

We perform the experiments using two configurations—(SLIM+2SIG) (additive) and (SLIM*1SIG) (multiplicative)—as they were shown to be the best performing variants in [36]. Notice that, in SLIM+2SIG, the second setup (that uses no deflate) is equivalent to running GSGP replacing its traditional GSC by the new crossover, since the IGSM in SLIM+2SIG is equivalent to the standard GSM in GSGP. We will compare the results with two baselines: the standard SLIM (that uses only the two mutations IGSM and DGSM) and the standard GSGP (that uses GSC and GSM=IGSM).

4.1 Dataset and Experimental Settings

The test problems in this study are five widely recognized symbolic regression tasks frequently used as benchmarks in the GP literature (e.g., [9,19,31,36]): Istanbul (7 features, 536 samples [1]), Airfoil (5 features, 1052 samples [6]), Yacht (6 features, 307 samples [30]), Concrete Strength (8 features, 1030 samples [11]), and Concrete Slump (9 features, 102 samples [44]). Further details can be found in [23,43]. We performed Monte Carlo cross-validation [14] with 30 random data partitions into training (70%) and test (30%) sets. All the experiments were performed 30 times, each using one of the 30 partitions.

Regarding parameter settings, trees are initialized with a maximum depth of 6 using the ramped half-and-half initialization. Selection uses tournaments of size 2, with elitism that ensures the survival of the best individual from one generation to the next. The function set includes the four binary arithmetic operators (+, −, *, and /), with/protected as in [17]. The terminal set comprises only the variables of the problem, without any constants. The population size is set to 100 individuals, allowed to evolve for 1000 generations. All the methods use the same number of fitness evaluations, ensuring a fair comparison. Fitness is assessed by the Root Mean Squared Error (RMSE) between predicted and expected outputs.

For GSGP, we use standard probabilities of 0.8 for crossover and 0.2 for mutation. For the SLIM algorithm, the probability of applying DGSM is always 0.7. At every mutation event, the mutation step is randomly generated according to a uniform distribution in the range [0, 1]. Regarding XODn and XOSw, the probabilities are also set according to each problem. Then, the IGSM probability is equal to 1 minus the crossover probability. When XOSw is applied, a probability of 0.5 is used across all datasets but Istanbul, which is set to 0.8. Meanwhile, when using XODn a probability of 0.8 is used for both SLIM*1SIG and SLIM+2SIG on the Yacht and Istanbul datasets; 0.5 for both variants on Airfoil; and 0.5 for SLIM*1SIG and 0.8 for SLIM+2SIG on the Strength and Slump datasets.

The code, for the complete reproducibility of the proposed experiments, is available at slim.

5 Results

Figure 3 shows the distribution of test fitness of the best individual at the final generation over 30 runs, while Fig. 4 shows its evolution across generations, median of the 30 runs. Figure 5 shows the distribution of the size of the best individual at the final generation over 30 runs.

Each row corresponds to a different problem. The left column shows the results for SLIM*1SIG, and the right one for SLIM+2SIG. Both compare the standard version (black) with the variants using the two crossovers introduced in Sect. 3. For each variant, we present results without DGSM, shown with dashed contours in the boxplot and dashed lines in the line plot, and with DGSM, drawn with solid contours and solid lines. We also include a boxplot in Fig. 3 and a line plot in Fig. 4 with the results of GSGP (in dashed contours, since it also does not include DGSM). In the boxplots, a G and/or S under or above the boxes means a statistically significant difference (according to the Mann-Whitney U test with $\alpha = 0.01$) when comparing to GSGP (G) or the respective SLIM (S). A character under/above a box means the respective variant is better/worse than the baseline (GSGP or SLIM). The supplementary material contains tables with the p-values of all pairwise comparisons.

5.1 Test Fitness

These plots reveal the effectiveness of the crossover operators. With the addition of crossover—whether DGSM is included or not—the test fitness in the last generation (Fig. 3) is always statistically equal to or better than traditional SLIM, except for a single case (top right plot, XODn without DGSM) where SLIM is significantly better. Looking at the evolution of test fitness along the generations (Fig. 4) we realize that this exception is one of the few cases where overfitting occurred, precisely with most crossover variants but not with SLIM. Had the evolution been stopped earlier, there would be no exception to the successful performance of the crossover operators when compared with SLIM. Regarding the comparison with GSGP, the crossover variants achieve statistically equal or better test fitness in 28 out of 40 comparisons (whereas SLIM only achieves this in 4 out of 10 comparisons). Looking again at the evolution of test fitness we realize that in some cases (e.g., on the Istanbul dataset) early stopping could have benefited the SLIM variants when compared to GSGP.

5.2 Model Size

The improvement of test fitness brought by the crossover operators often comes at the cost of model size (Fig. 5), which is generally equal to or significantly larger than in traditional SLIM, with the notable exception of XOSw with DGSM that only once (in 40 comparisons) produced models significantly larger than SLIM. Regarding GSGP, the distribution of model size is not shown because it exceeds the plot limits, except on the Istanbul dataset. The median size of GSGP individuals is 2628 for the Istanbul dataset and reaches as high as 10^{60} for Airfoil, 10^{68} for Yacht, 10^{70} for Strength, and 10^{86} for Slump. All SLIM variants produce significantly smaller models than GSGP, except for a few cases on the Istanbul dataset, where the differences are not significant. This difference substantially impacts computational efficiency and, even more importantly, model interpretability.

5.3 To Deflate or Not To Deflate?

As expected, variants that apply crossover with DGSM consistently produce smaller individuals compared to those without DGSM (Fig. 5), but the use of DGSM tends to decrease the performance in terms of fitness (Fig. 3). In XOSw fitness becomes significantly worse in 4/10 cases, while in XODn this happens in 2/10 cases. We hypothesize that this happens because crossover arranges blocks in useful combinations and then DGSM may discard valuable components of these combinations, diminishing their usefulness. However, an interesting observation emerges from the fitness evolution plots (Fig. 4). In the SLIM*1SIG Istanbul plot, the crossover variants without DGSM tend to overfit, as illustrated

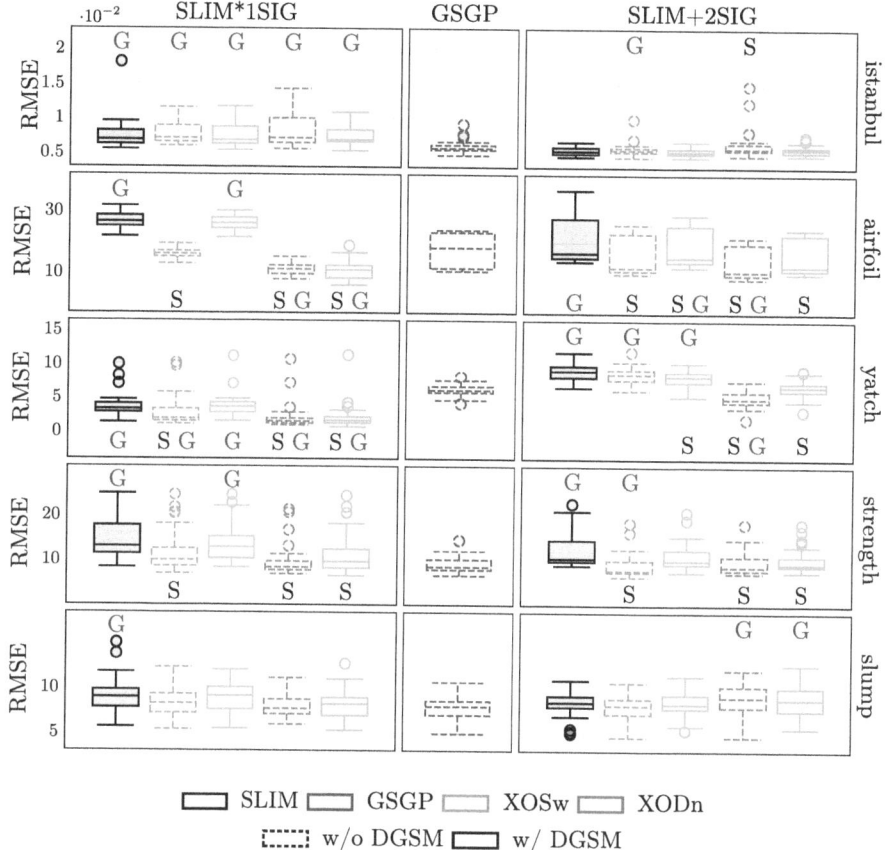

Fig. 3. Test fitness distribution of the best individual across 30 runs. Dashed and solid box contours indicate versions without and with DGSM, respectively. A G/S below or above a box indicates a statistically significant difference (Mann-Whitney U test with $\alpha = 0.01$) from GSGP (G) or SLIM (S), with the position indicating whether the variant is better (under) or worse (above).

by an increase in test fitness over time. A similar trend is observed in standard SLIM. However, when DGSM is included alongside crossover, overfitting appears to be reduced, likely due to the ability of DGSM to simplify the structure of the individuals.

5.4 Swap Crossover or Donor Crossover

XOSw performs well both with and without DGSM: it is never surpassed by standard SLIM and is equal to or significantly better than GSGP in 12/20 cases (half with DGSM and half without DGSM). Let us focus on XOSw with DGSM. Notably, in [36,40] the authors show that SLIM models are either smaller or of comparable size to standard GP models and that this compactness allows

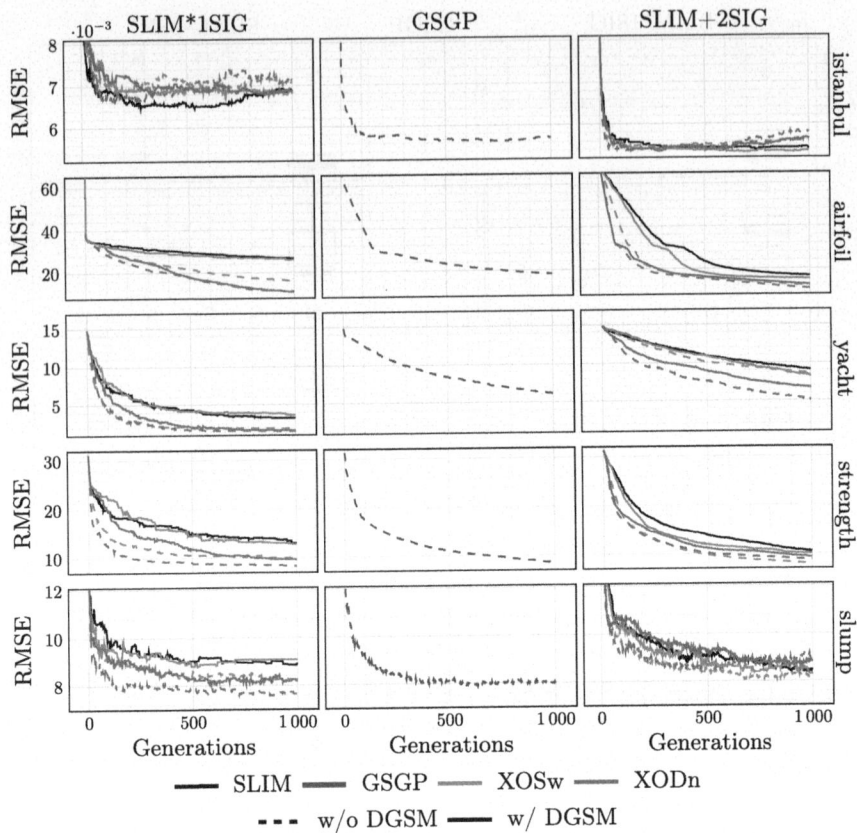

Fig. 4. Evolution of test fitness for the best individual in SLIM across 30 runs. Dashed lines represent versions without DGSM, while solid lines indicate versions with DGSM.

for human interpretation. XOSw with DGSM maintains (or even improves) this compactness while matching (or significantly outperforming) the fitness of standard SLIM in every case. This means that the exchange of genetic material is very beneficial when coupled with SLIM's deflate operator.

XODn produces larger models than both standard SLIM and XOSw. When compared to standard SLIM, the larger models of XODn with DGSM suggest that adding blocks via genetic material exchange from a donor—rather than from random trees—creates individuals with higher fitness, which increases their chances of survival and, over time, leads to growth in population size. In standard SLIM, when the DGSM operator randomly removes a block from an individual, the information in that block is lost. In contrast, with XODn, a block removed from one parent is always transferred to the other, allowing the information to survive across generations if it proves beneficial. These observations highlight the benefit of exchanging genetic material in evolution, which is confirmed by the fact that XODn often outperforms others in fitness: it is surpassed

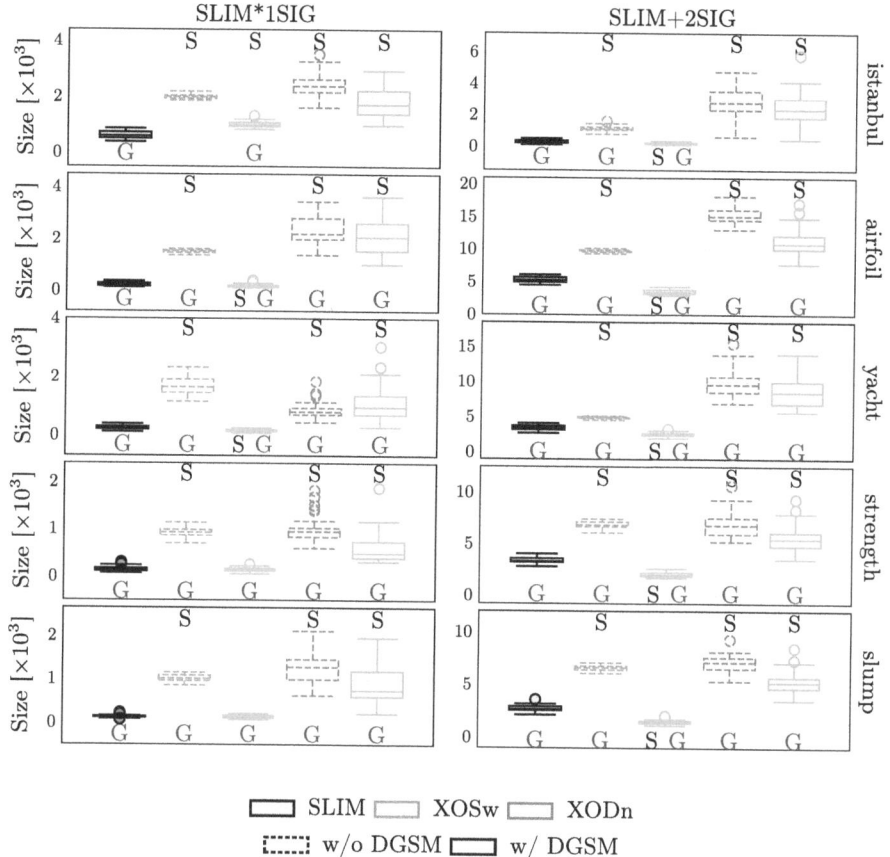

Fig. 5. Size distribution of the best individual in SLIM across 30 runs, following the codification detailed in Fig. 3.

by standard SLIM in only one case, and it is equal to or significantly better than GSGP in 16/20 cases (half with DGSM and half without DGSM), also achieving better results early in the evolution (Fig. 4).

5.5 SLIM*1SIG or SLIM+2SIG

Comparing the results between SLIM*1SIG and SLIM+2SIG, we observe that the best-performing alternative is generally XODn in SLIM*1SIG, which often outperforms GSGP. For maintaining smaller individual sizes, XOSw combined with DGSM in SLIM*1SIG proves to be the most effective.

In SLIM+2SIG, individuals tend to be larger than in SLIM*1SIG variants, as SLIM*1SIG mutations add only one extra tree, compared to two in SLIM+2SIG. Let us recall that SLIM+2SIG algorithms without DGSM essentially function as GSGP variants with an alternative crossover replacing GSC, since the IGSM in SLIM+2SIG is equivalent to the standard GSM in GSGP. Thus, the dashed results on the SLIM+2SIG plot represent GSGP variants with these alternative crossovers, enabled by the linked-list structure in both SLIM and GSGP. When comparing GSC with XODn or XOSw, we see not only a significant reduction in model size but also statistically equal or better performance than GSGP in most cases. This supports findings in [9,27] on the limited effectiveness of GSC in GSGP and suggests that the crossovers here defined offer a strong alternative, addressing both the limited search ability and size-growth issues inherent to GSC.

6 Conclusion

We introduced two novel crossover operators within the Semantic Learning algorithm based on Inflate and deflate Mutations (SLIM), designed to overcome the main limitations of traditional GSGP, namely the vast growth in program size and the restricted search ability of traditional geometric semantic crossover (GSC). Capitalizing on SLIM's linked-list representation, we proposed the Swap Crossover (XOSw) and the Donor Crossover (XODn), which facilitate effective genetic material exchange while maintaining controlled model sizes. XOSw works by exchanging linked-list components between two parents, while XODn transfers one block from a parent (donor) to the other parent (receiver). Our experimental results across five symbolic regression problems reveal that these crossovers often enhance the predictive ability of SLIM, achieving statistically significant improvements over traditional SLIM in terms of fitness on unseen data and model size, especially when combined with deflate mutation (DGSM). These findings establish XOSw and XODn as robust improvements of standard GSC.

Future work will focus on several fronts to further refine and expand the capabilities of SLIM. First, exploring adaptive crossover mechanisms [15] could allow SLIM to dynamically adjust the choice and application of crossover operators based on evolving population characteristics. Additionally, applying SLIM in a broader array of real-world domains will enable further evaluation of the versatility and effectiveness of these crossovers.

Acknowledgments. This work was supported by national funds through FCT (Fundação para a Ciência e a Tecnologia), under the project - UIDB/04152/2020 (DOI: 10.54499/UIDB/04152/2020) - Centro de Investigação em Gestão de Informação (MagIC)/NOVA IMS),(https://doi.org/10.54499/UIDB/04152/2020) the project 2024.07277. IACDC, and the LASIGE Research Unit, ref. UID/000408/2025.

References

1. Akbilgic, O., Bozdogan, H., Balaban, M.E.: A novel hybrid RBF neural networks model as a forecaster. Stat. Comput. **24**, 365–375 (2014)
2. Albinati, J., Pappa, G.L., Otero, F.E., Oliveira, L.O.V.: The effect of distinct geometric semantic crossover operators in regression problems. In: Genetic Programming: 18th European Conference, EuroGP 2015, Copenhagen, Denmark, 8–10 Apr 2015, Proceedings 18, pp. 3–15. Springer (2015)
3. Beadle, L., Johnson, C.G.: Semantically driven crossover in genetic programming. In: 2008 IEEE Congress on Evolutionary Computation (IEEE World Congress on Computational Intelligence), pp. 111–116. IEEE (2008)
4. Beadle, L., Johnson, C.G.: Semantically driven mutation in genetic programming. In: 2009 IEEE Congress on Evolutionary Computation, pp. 1336–1342. IEEE (2009)
5. Bonin, L., Rovito, L., De Lorenzo, A., Manzoni, L.: Cellular geometric semantic genetic programming. Genet. Program Evolvable Mach. **25**(1), 8 (2024)
6. Brooks, T.F., Pope, D.S., Marcolini, M.A.: Airfoil self-noise and prediction. Technical report (1989)
7. Bryant, R.E.: Graph-based algorithms for Boolean function manipulation. Comput. IEEE Trans. **100**(8), 677–691 (1986)
8. Castelli, M., Manzoni, L.: Gsgp-c++ 2.0: a geometric semantic genetic programming framework. SoftwareX **10**, 100313 (2019). https://doi.org/10.1016/j.softx.2019.100313, https://www.sciencedirect.com/science/article/pii/S2352711019301736
9. Castelli, M., Manzoni, L., Gonçalves, I., Vanneschi, L., Trujillo, L., Silva, S.: An analysis of geometric semantic crossover: a computational geometry approach. In: IJCCI (ECTA), pp. 201–208 (2016)
10. Castelli, M., Trujillo, L., Vanneschi, L.: Energy consumption forecasting using semantic-based genetic programming with local search optimizer. Comput. Intell. Neurosci. **2015**(1), 971908 (2015)
11. Castelli, M., Vanneschi, L., Silva, S.: Prediction of high performance concrete strength using genetic programming with geometric semantic genetic operators. Expert Syst. Appl. **40**(17), 6856–6862 (2013)
12. Chen, Q., Zhang, M., Xue, B.: New geometric semantic operators in genetic programming: perpendicular crossover and random segment mutation. In: Proceedings of the Genetic and Evolutionary Computation Conference Companion, pp. 223–224 (2017)
13. Chen, X., Ong, Y.S., Lim, M.H., Tan, K.C.: A multi-facet survey on memetic computation. IEEE Trans. Evol. Comput. **15**(5), 591–607 (2011)
14. Coombes, K., Baggerly, K., Morris, J., Dubitzky, W., Granzow, M., Berrar, D.: Fundamentals of Data Mining in Genomics and Proteomics, pp. 79–99. Kluwer, Boston (2007)
15. Ferreira, J., Castelli, M., Manzoni, L., Pietropolli, G.: A self-adaptive approach to exploit topological properties of different gas' crossover operators. In: European Conference on Genetic Programming (Part of EvoStar), pp. 3–18. Springer (2023)
16. Hara, A., Kushida, J.I., Tanemura, R., Takahama, T.: Deterministic crossover based on target semantics in geometric semantic genetic programming. In: 2016 5th IIAI International Congress on Advanced Applied Informatics (IIAI-AAI), pp. 197–202. IEEE (2016)

17. Koza, J.R.: Genetic programming as a means for programming computers by natural selection. Stat. Comput. **4**, 87–112 (1994)
18. Krawiec, K.: Medial crossovers for genetic programming. In: Genetic Programming: 15th European Conference, EuroGP 2012, Málaga, Spain, 11–13 Apr 2012. Proceedings 15, pp. 61–72. Springer (2012)
19. La Cava, W., et al.: Contemporary symbolic regression methods and their relative performance. arXiv preprint arXiv:2107.14351 (2021)
20. Mambrini, A., Manzoni, L., Moraglio, A.: Theory-laden design of mutation-based geometric semantic genetic programming for learning classification trees. In: 2013 IEEE Congress on Evolutionary Computation, pp. 416–423. IEEE (2013)
21. Marchetti, F., Pietropolli, G., Camerota Verdù, F.J., Castelli, M., Minisci, E.: Control law automatic design through parametrized genetic programming with adjoint state method gradient evaluation (2023). Available at SSRN 4490005
22. Martins, J.F.B.S., Oliveira, L.O.V.B., Miranda, L.F., Casadei, F., Pappa, G.L.: Solving the exponential growth of symbolic regression trees in geometric semantic genetic programming. In: Proceedings of the Genetic and Evolutionary Computation Conference, GECCO 2018, pp. 1151–1158. Association for Computing Machinery, New York (2018). https://doi.org/10.1145/3205455.3205593
23. McDermott, J., et al.: Genetic programming needs better benchmarks. In: Proceedings of the 14th Annual Conference on Genetic and Evolutionary Computation, pp. 791–798 (2012)
24. McPhee, N.F., Ohs, B., Hutchison, T.: Semantic building blocks in genetic programming. In: Genetic Programming: 11th European Conference, EuroGP 2008, Naples, Italy, 26–28 Mar 2008, Proceedings 11, pp. 134–145. Springer (2008)
25. Moraglio, A.: An efficient implementation of GSGP using higher-order functions and memoization. In: Johnson, C., Krawiec, K., Moraglio, A., O'Neill, M. (eds.) Semantic Methods in Genetic Programming, Ljubljana, Slovenia 13 Sep 2014. http://www.cs.put.poznan.pl/kkrawiec/smgp2014/uploads/Site/Moraglio2.pdf. workshop at Parallel Problem Solving from Nature 2014 conference
26. Moraglio, A., Krawiec, K., Johnson, C.G.: Geometric semantic genetic programming. In: Coello, C., Cutello, V., Deb, K., Forrest, S., Nicosia, G., Pavone, M. (eds.) PPSN 2012. LNCS, vol. 7491, pp. 21–31. Springer, Heidelberg (2012). https://doi.org/10.1007/978-3-642-32937-1_3
27. Moraglio, A., Mambrini, A.: Runtime analysis of mutation-based geometric semantic genetic programming for basis functions regression. In: Proceedings of the 15th Annual Conference on Genetic and Evolutionary Computation, pp. 989–996 (2013)
28. Nadizar, G., Sakallioglu, B., Garrow, F., Silva, S., Vanneschi, L.: Geometric semantic GP with linear scaling: Darwinian versus Lamarckian evolution. Genet. Program Evolvable Mach. **25**(2), 1–24 (2024)
29. Nguyen, Q.U., Nguyen, X.H., O'Neill, M.: Semantic aware crossover for genetic programming: the case for real-valued function regression. In: Genetic Programming: 12th European Conference, EuroGP 2009 Tübingen, Germany, 15–17 Apr 2009 Proceedings 12, pp. 292–302. Springer (2009)
30. Ortigosa, I., Lopez, R., Garcia, J.: A neural networks approach to residuary resistance of sailing yachts prediction. In: Proceedings of the International Conference on Marine Engineering Marine, vol. 2007, p. 250 (2007)
31. Orzechowski, P., La Cava, W., Moore, J.H.: Where are we now? A large benchmark study of recent symbolic regression methods. In: Proceedings of the Genetic and Evolutionary Computation Conference, pp. 1183–1190 (2018)
32. Pawlak, T.P., Wieloch, B., Krawiec, K.: Review and comparative analysis of geometric semantic crossovers. Genet. Program Evolvable Mach. **16**, 351–386 (2015)

33. Pietropolli, G., Manzoni, L., Paoletti, A., Castelli, M.: Combining geometric semantic GP with gradient-descent optimization. In: European Conference on Genetic Programming (Part of EvoStar), pp. 19–33. Springer (2022)
34. Pietropolli, G., Manzoni, L., Paoletti, A., Castelli, M.: On the hybridization of geometric semantic GP with gradient-based optimizers. Genet. Program Evolvable Mach. **24**(2), 16 (2023)
35. Trujillo, L., et al.: Local search is underused in genetic programming. In: Genetic Programming Theory and Practice XIV, pp. 119–137 (2018)
36. Vanneschi, L.: SLIM_GSGP: the non-bloating geometric semantic genetic programming. In: Giacobini, M., Xue, B., Manzoni, L. (eds.) Genetic Programming, pp. 125–141. Springer, Cham (2024)
37. Vanneschi, L., et al.: PSXO: population-wide semantic crossover. In: Proceedings of the Genetic and Evolutionary Computation Conference Companion, pp. 257–258 (2017)
38. Vanneschi, L., Castelli, M., Manzoni, L., Silva, S.: A new implementation of geometric semantic GP and its application to problems in pharmacokinetics. In: Genetic Programming: 16th European Conference, EuroGP 2013, Vienna, Austria, 3–5 Apr 2013, Proceedings 16, pp. 205–216. Springer (2013)
39. Vanneschi, L., Castelli, M., Silva, S.: A survey of semantic methods in genetic programming. Genet. Program Evolvable Mach. **15**(2), 195–214 (2014). https://doi.org/10.1007/s10710-013-9210-0
40. Vanneschi, L., Farinati, D., Rasteiro, D., Rosenfeld, L., Pietropolli, G., Silva, S.: Exploring non-bloating geometric semantic genetic programming. In: Genetic Programming Theory and Practice. (to appear)
41. Vanneschi, L., Silva, S.: Lectures on Intelligent Systems. Springer, Cham (2023)
42. Vanneschi, L., Silva, S., Castelli, M., Manzoni, L.: Geometric semantic genetic programming for real life applications. In: Genetic Programming Theory and Practice XI, pp. 191–209 (2014)
43. White, D.R., et al.: Better GP benchmarks: community survey results and proposals. Genet. Program Evolvable Mach. **14**, 3–29 (2013)
44. Yeh, I.C.: Simulation of concrete slump using neural networks. Proc. Inst. Civ. Eng. Constr. Mater **162**(1), 11–18 (2009)

Exploring the Integration of Cellular Structures in Genetic Programming-Based Methods

Luigi Rovito[1](✉), Lorenzo Bonin[1], Davide Farinati[2],
Leonardo Vanneschi[2], Luca Manzoni[1], Andrea De Lorenzo[1],
and Gloria Pietropolli[1]

[1] University of Trieste, 34127 Trieste, Italy
luigi.rovito@phd.units.it
[2] NOVA Information Management School (NOVA IMS), Universidade Nova de Lisboa, Campus de Campolide, 1070-312 Lisbon, Portugal

Abstract. The introduction of a Cellular Automata (CA)-like structure on the population of Evolutionary Algorithms (EAs) has been verified to be a method to improve solutions quality. However, the study of CA-like structures for Genetic Programming (GP) has been, so far, limited. In this work, we focus on the effect of introducing these structures on Geometric Semantic variants of GP, focusing on the well-known Geometric Semantic GP (GSGP) and its recently introduced variant SLIM-GSGP, which emphasizes producing smaller and more interpretable individuals. Here we provide guidance on how CA-like structures can impact the quality and size of the solutions for GSGP and SLIM-GSGP, giving a clear understanding of the trade-offs involved in applying these methods.

Keywords: Evolutionary Computation · Evolutionary Algorithms · Genetic Programming · Geometric Semantic Genetic Programming · Cellular Automata · Symbolic Regression

1 Introduction

Cellular Automata (CA) [19,59,69] is classical nature-inspired computing model consisting of a n-dimensional grid of cells, each in one of a finite set of possible states. Each cell is confined within a neighborhood, and at each discrete time step, cell states are all synchronously updated based on a uniform local rule.

The intrinsic properties of CAs make this model a promising substrate for evolution in Evolutionary Algorithms (EAs), as modeling the population as a toroidal structure restricts interactions to individuals within the same neighborhoods [3,4,68]. This structure can offer several advantages, depending on the

---

Supplementary Information The online version contains supplementary material available at https://doi.org/10.1007/978-3-031-89991-1_8.

algorithm. By constraining interactions, it helps prevent local-optima stagnation and encourages the population to explore diverse areas of the search space, often leading to the discovery of better solutions [3,10,28,52,58,66,90].

While combining CA with EAs [40,62] is a promising approach and despite Genetic Programming (GP) is an effective algorithm for Symbolic Regression (SR), its integration with spatial structures like CA is still underexplored. Existing studies on CA-inspired GP largely focus on problems other than SR [26,73,74] or on parallel implementations aimed primarily at improving computational performance [27]. Moreover, these works have not been evaluated using modern, well-established SR benchmarks [40,86]. In addition, these applications have not focused on semantic-based variations of GP, such as Geometric Semantic Genetic Programming (GSGP) [50,77], until very recently, when a cellular structure was introduced to slow the spread of dominant individuals and mitigate premature convergence in GSGP [10]. Despite this, the issue of exponential tree growth in GSGP remains. To counteract those issues, a promising approach is one of Semantic Learning Algorithm Based on Inflate and Deflate Mutation (SLIM) [78,81]. In this recent GSGP variant, a modified Geometric Semantic Mutation (GSM) operator is employed to generate smaller offspring.

Motivated by the findings in [10], we investigate integrating cellular structures within different GP methods. We begin with a comprehensive literature review that unifies prior work on GP methods and cellular structures (Sect. 2). Thereafter, we perform a comprehensive experimental campaign focusing on two key metrics: model performance, to assess quality and model size. Smaller models, as shown in [78,81], lead to more interpretable formulae and facilitate analysis with explainability techniques. Additionally, compact models are faster in computation, require less memory, and are easier to deploy and monitor in real-world applications.

In our experiments, we integrate cellular structures with GP, GSGP, and SLIM, chosen for their distinct strengths and limitations across the two metrics of interest. To the best of our knowledge, this is the first analysis to compare these popular GP methods in terms of accuracy and model size with the addition of cellular structures-known to influence evolutionary dynamics. Our goal is to provide a definitive understanding of how cellular toroidal grids impact both the population and outcomes of key GP algorithms and to clarify the trade-offs between accuracy and model size achievable by combining GP methods with CA-inspired structures.

2 Related Works

Previous research has explored the introduction of spatial structures in EAs [3,4,68]. This strategy limits interactions to smaller subsets of the population, called neighborhoods, typically organized in a grid. In Genetic Algorithms (GAs) [30], spatial structuring is implemented through a Cellular Genetic Algorithm (cGA) [3,4,68] with studies [3,68] analyzing how neighborhood configurations impact the performance. Alba et al. [3] introduced a dynamic cGA adjusting exploration-exploitation during evolution. For multi-objective optimization,

Murata et al. [52] developed C-MOGA, a GA with cellular structuring for local selection. Later, Nebro et al. [58] presented MOCell, inspired by traditional cGA. Mariot et al. [45] applied GA and GP to design orthogonal Latin squares using CAs. However, other EAs also used a cellular structure to manage the spread of solutions within the population [2,71,90]. Hence, it is generally possible to apply a cellular structure to the population regardless of the employed algorithm.

GP [36] is an EA that, by evolving computer programs, showed its flexibility and effectiveness in tackling several problems [1,9,21,33,38,39], especially because the learning process and the solution representation potentially enable the discovery of interpretable models [12,25,32,42,49,54–56,83,84]. Folino et al. [27] introduced a scalable parallel implementation of GP using a cellular structure, outperforming both traditional GP and the island model [46], where evolutionary runs occur on separate population subsets. They also proposed a parallel cellular-based implementation of GP to address classification problems [26]. Later, Takac proposed a cellular-based GP method for classification [74] as well as data mining [73]. Additionally, studies by [85] and [23] demonstrated that combining spatial population structure with local elitist replacement effectively reduces bloat (unnecessary growth in tree size) in GP, while maintaining performance.

To preserve biodiversity, speciation can be applied by dividing the population into distinct niches of individuals with comparable structural traits, forcing crossover only between individuals of the same niche. Della Cioppa et al. [22] introduced an adaptive species discovery strategy to address the limitations of traditional niching methods, which often require prior knowledge of the fitness landscape. Building on the NEAT algorithm [72], Trujillo et al. [76] applied speciation to control program growth in GP through neat-GP, which promotes complexity only when necessary. Juarez et al. [34] improved neat-GP by incorporating a local search operator. Cussat et al. [20] introduced a network distance metric to speciate populations, promoting diversity and maintaining smaller individuals. Martins et al. [48] combined GAs with speciation and grid pattern recognition to reduce investment risks and boost profits. Wickman et al. [87] applied speciation in Reinforcement Learning (RL) to evolve diverse policies, while Pietropolli et al. [66] proposed using substrates with barriers to slow genetic spread and enhance diversity.

Several studies have examined the effects of selection pressure and sampling strategies both in vanilla EAs [29,88] and cellular-based EAs [28,70]. In cellular-based EAs, spatial structure plays a critical role in controlling takeover time, i.e., the number of generations needed for a dominant individual to spread across the population. The higher the takeover time, the higher the diversity [28].

GSGP [50,77], which outperformed GP in various tasks [17,18,40,50,62,79,80,82], is affected by several drawbacks mainly regarding the presence of local optima where evolution may stagnate and the premature convergence towards dominating individuals (low takeover time) [63]. To this end, Bonin et al. [10] implemented Cellular Geometric Semantic Genetic Programming (cGSGP), which imposes a cellular toroidal structure to the GSGP population, showing that this variant can outperform GSGP and improve the diversity in the early stage of the evolution by increasing the takeover time.

Even though these efforts mitigate premature convergence, GSGP still suffers from exponential growth of the trees across the generations. Motivated by the large research literature regarding the enhancement of this technique [14,24,53,57,64,65] and the availability of efficient implementations [15,75], many research works tried to control bloating and reduce the trees exponential size [15,16,24,35,47,64,79]. However, these methods still rely on Geometric Semantic Operators (GSOs) that produce offspring that are larger than their parents [7].

Recently, Vanneschi et al. [78] presented the SLIM, a variant of GSGP that employs a specialized GSM that produces offspring smaller than their parents without affecting performances, thus revolutionizing the way GSGP and its solutions are created and employed.

The variety of GP and GSGP variants and the existence of CA-inspired methods makes it challenging to understand the impact of combining these techniques. In this paper, we explore the integration of a cellular toroidal structure over the population of the main GP algorithms, i.e., GSGP and SLIM. We take these tree GP algorithms (on their standard version for a clear comparison) as they have different strengths and limitations, and we perform an analysis on different metrics to provide, once for all, a clear understanding of their trade-off levels along with the impact of the toroidal grid.

3 Methods

In this section, we briefly describe the two variants [50,81] of GP [36,41] we will investigate alongside vanilla GP. We also describe the cellular-based selection strategy adopted in [10].

3.1 Geometric Semantic Genetic Programming

GP individuals are typically represented as computer programs, visualized as trees, that map inputs to outputs. Given an individual P and inputs $X = \{x_1, x_2, \ldots, x_n\}$, the predictions $P(X) = \{P(x_1), P(x_2), \ldots, P(x_n)\}$ define its semantics. Traditional GP operators modify the syntactic structure, but their effect on semantics is unpredictable, often complicating the search process [60].

Moraglio et al. [50] proposed replacing traditional operators with Geometric Semantic Crossover (GSC) and GSM, which reflects structural changes in an individual's semantics, creating a unimodal fitness landscape. Recent studies [13] found that using only GSM performs similarly or better than combining both operators.

In GSM, given a parent function $T : \mathbb{R}^d \to \mathbb{R}$, mutation is defined as $T_m = T + ms \cdot (T_{R1} - T_{R2})$, where ms is the mutation step, and T_{R1} and T_{R2} are random programs with a sigmoid function to restrict values to $[0, 1]$. This mutation produces offspring whose semantics lie within a sphere of radius

ms around T. While GSO creates a unimodal fitness landscape, it also increases individual size, causing rapid solution growth and making GSGP a black-box model [15].

3.2 Semantic Learning Algorithm Based on Inflate and Deflate Mutation

Recently, Vanneschi et al. [81] introduced a *deflate* GSM that generates smaller offspring. The definition and validity of Deflate Geometric Semantic Mutation (DGSM) are based on two key observations: first, GSM can be rewritten as $\text{GSM}(T) = T + ms \cdot (T_{R1} - T_{R2}) = T - ms \cdot (T_{R2} - T_{R1})$; second, the random trees T_{R1} and T_{R2} are interchangeable, as they are independently sampled from the same distribution. Thus, the GSM can be written as $\text{GSM}(T) = T - ms \cdot (T_{R1} - T_{R2})$. For instance, let T be an individual to which GSM is applied three consecutive times. We obtain:

$$T_M = \text{GSM}^3(T) = T + ms \cdot (T_{R1} - T_{R2}) + ms \cdot (T_{R3} - T_{R4}) + ms \cdot (T_{R5} - T_{R6})$$

If we apply GSM again by using subtraction, we obtain:

$$T_{M'} = T + ms \cdot (T_{R1} - T_{R2}) + ms \cdot (T_{R3} - T_{R4}) + ms \cdot (T_{R5} - T_{R6}) - ms \cdot (T_{R7} - T_{R8})$$

We recall that reusing random trees is widely adopted [79]. For example, we could use T_{R3} and T_{R4} instead of T_{R7} and T_{R8}:

$$T' = T + ms \cdot (T_{R1} - T_{R2}) \cancel{+ ms \cdot (T_{R3} - T_{R4})} + ms \cdot (T_{R5} - T_{R6}) \cancel{- ms \cdot (T_{R3} - T_{R4})}$$

After this simplification, the individual has a smaller genotype. This alternating inflate and deflate approach forms the basis of SLIM. A hyper-parameter controls the likelihood of applying deflate (if zero, SLIM resembles GSGP)

3.3 New Ways of Defining a Geometric Semantic Mutation

In recent works [7,81], the authors introduced three methods to define mutation values within $[-ms, ms]$ and two strategies for perturbing an individual. Given that T_R, T_{R1}, and T_{R2} are random programs with arbitrary output and let S be the sigmoid function, the functions are: (i) $\text{SIG2} = ms \cdot (S(T_{R1}) - S(T_{R2}))$; (ii) $\text{ABS} = ms \cdot (1 - \frac{2}{1+|T_R|})$; (iii) $\text{SIG1} = ms \cdot (2 \cdot S(T_R) - 1)$.

To introduce minor perturbations, function values can either be added/subtracted (SLIM$^+$) or scaled by factors around one (SLIM*) w.r.t the parent tree.

3.4 Cellular Methods

In CA-inspired algorithms, a spatial structure is imposed. As in [10], individuals are arranged according to a toroidal grid, with each individual belonging to a specific cell, randomly chosen at initialization, which belongs to a specific neighborhood.

The neighborhood of radius r for a given individual is defined by a hypercube with side $2r + 1$, corresponding to the Moore neighborhood centered on that individual's cell. Each individual belongs to its own neighborhood. The toroidal configuration ensures all cells have complete neighborhoods. The n-dimensional toroidal grid with radius $r \in \mathbb{N}$ is denoted as $\mathcal{T}_r^n$.

We employ this structure along with the selection strategy in [10]. Especially, given an individual T_i in a cell i in the grid, and its neighborhood $\mathcal{N}_i$, T_i is replaced by a new tree derived from others in $\mathcal{N}_i$. Selection is limited to each cell neighborhood: for each cell i, the two individuals chosen for crossover are selected from within the same neighborhood $\mathcal{N}_i$. If no crossover is performed, T_i is replaced with another (eventually mutated) individual in $\mathcal{N}_i$.

This method (TRS_p) is a tournament-based selection within local neighborhoods, where the selection pressure is defined by a value $p \in (0, 1]$. A subset $\mathcal{S}$ of $\mathcal{N}_i$ is selected by iterating across the elements of $\mathcal{N}_i$ and inserting them into $\mathcal{S}$ with probability p. Then, a rank-based selection is applied on $\mathcal{S}$: if no crossover is performed, then the new individual in cell i will be the best one in $\mathcal{S}$, while, if crossover is performed, then it will be between the two best individuals in $\mathcal{S}$. Finally, the new individual is eventually mutated.

4 Experimental Methodology

We set a population size of 100 that is evolved for 1000 generations with elitism enforced and tournament selection with a tournament size of 4 in case of standard methods ($\mathcal{T}^0$). We use a 10×10 bi-dimensional toroidal grid in the case of cellular methods. We test cellular methods with $p = 1$ and radius between 2 ($\mathcal{T}_2^2$) and 3 ($\mathcal{T}_3^2$).

Trees are initialized with ramped half-and-half [36,41,67], with initial maximum depth set to 6. The function set is composed by $+, -, \times, \div^*$ where $*$ indicates that the division is protected (if the denominator is zero, then one is returned). The terminal set is composed of the variables of the problem. No ephemeral constants are explicitly added to the terminal set, as in [78,81].

GP performs sub-tree crossover with probability 0.8 and sub-tree mutation with probability 0.2, GSGP only performs GSM with probability 1.0. In SLIM, mutation occurs with a probability of 1.0, as the standard implementation does not include crossover, with inflate and deflate probabilities set to 0.3 and 0.7, respectively, following the original paper [78]. The mutation step ms is uniformly sampled at random in $[0, 1]$ for each mutation event in both GSGP and SLIM.

Regarding the specific SLIM algorithms, we test $\mathrm{SLIM}^+_{\mathrm{ABS}}$, $\mathrm{SLIM}^+_{\mathrm{SIG1}}$, $\mathrm{SLIM}^+_{\mathrm{SIG2}}$, $\mathrm{SLIM}^*_{\mathrm{ABS}}$, $\mathrm{SLIM}^*_{\mathrm{SIG1}}$, and $\mathrm{SLIM}^*_{\mathrm{SIG2}}$, which are, currently, all the SLIM versions available.

We adopt six datasets that are commonly used in GP tasks [14,79,82,86]: Airfoil (ARF) [11] with 1502 records and 5 variables, Concrete (CNC) [17] with 1029 records and 8 variables, Slump (SLM) [89] with 102 records and 9 variables, Yacht (YCH) [61] with 307 records and 6 variables, Parkinson (PRK) [18] with 5875 records and 18 variables, and QSAR Aquatic Toxicity (QSR) [8] with

546 records and 8 variables. Each dataset is randomly partitioned in 30 train-test splits with ratio 7:3. For each method, dataset, and combination of hyper-parameters, we perform 30 repetitions (one for each split).

In our experiments, we analyze three key measures: (i) the Root Mean Squared Error (RMSE) [6], which evaluates the qualitative performance of the models; (ii) the size ($\log_{10}(\ell)$) of the models (the 10-base logarithm of the number of nodes Nodes), which serves as a proxy for the interpretability, as outlined in [78], and both memory occupation and computational complexity; (iii) the Global Moran's I (I) [43,51], which is a spatial autocorrelation measure [5] that represents a measure of diversity for neighborhood-based populations.

I assesses the similarity (or dissimilarity) of values located in neighboring locations. Provided that an individual is represented by its semantics, we have:

$$I = \frac{M}{W} \frac{\sum_{i=1}^{M} \sum_{j=1}^{M} \mathbf{w}_{ij} (\mathbf{x}^i - \bar{\mathbf{x}})(\mathbf{x}^j - \bar{\mathbf{x}})}{\sum_{i=1}^{M}(\mathbf{x}^i - \bar{\mathbf{x}})^2}$$

where M is the population size, $\mathbf{w}_{ij} \in \mathbb{R}$ is the value located at the i-th row and the j-th column of the matrix $\mathbf{w} \in \mathbb{R}^{M \times M}$, $\mathbf{w}_{ii} = 0$ for $1 \leq i \leq M$, $W = \sum_{i=1}^{M} \sum_{j=1}^{M} \mathbf{w}_{ij}$, $\mathbf{x}^i$ is the i-th individual in the population as a semantic vector, and $\bar{\mathbf{x}}$ is the mean of all the semantic vectors of the individuals in the population. In our case, $\mathbf{w}_{ij} = 1$ if $i \neq j$ and the j-th individual belongs to the neighborhood of the i-th individual (in a non-cellular method the entire population is a single neighborhood), otherwise $\mathbf{w}_{ij} = 0$.

I ranges in the interval $[-1, 1]$, where values close to 1 indicate positive spatial autocorrelation (similar semantic content is spatially clustered in the population) and values close to -1 indicate negative autocorrelation. Values close to 0 depict a spatial pattern not different from a random one. On the other hand, a strictly positive value of I indicates that each neighborhood tends to be a cluster containing similar individuals that are, at the same time, dissimilar from individuals in other neighborhoods, meaning that the population is spatially arranged in different search space areas and, hence, diversity is better enforced by the toroidal grid.

For the sake of simplicity, we use "*algorithm*" to refer to the different GP variants (GSGP and SLIM), while we use "*method*" to refer to the different variants of cellular structure, including the non-cellular standard structure of the non-cellular baseline. Our Python code cslim is publicly available online.[1]

5 Results and Discussion

In this section, we present the results of our experimental campaign and discuss the main findings.

---

[1] https://github.com/lurovi/cslim

5.1 Statistical Analysis and Convergence Rates

Table 1. Table with the median of RMSE on the test set and $\log_{10}(\ell)$ of the best individual for all the different methods.

		RMSE						$\log_{10}(\ell)$					
		ARF	CNC	SLM	YCH	PRK	QSR	ARF	CNC	SLM	YCH	PRK	QSR
GP	$\mathcal{T}^0$	22.67	9.54	5.28	2.80*	10.41*	1.45	3.04	2.90	**2.55**	2.70	2.65	2.83
	$\mathcal{T}_2^2$	25.86	10.27*	5.56	3.55	10.72	1.47	2.91	2.73*	2.56	2.50	**2.28***	2.39*
	$\mathcal{T}_3^2$	25.18	11.93	6.33	4.08	10.67	1.50	**2.78**	**2.66***	2.58	2.47	2.45*	2.33*
GSGP	$\mathcal{T}^0$	15.20	7.59	3.88	4.84	9.92	1.34	4.26*	4.08*	4.03*	4.19*	3.89*	3.88*
	$\mathcal{T}_2^2$	11.30*	6.96*	**3.85**	3.97*	9.75*	**1.27***	4.32	4.13	4.06	4.24	3.94	4.03
	$\mathcal{T}_3^2$	**10.15***	**6.83***	3.97	**3.74***	**9.69***	1.29*	4.33	4.15	4.08	4.26	3.96	4.06
SLIM$_{SIG2}^+$	$\mathcal{T}^0$	19.74	9.14	4.59	7.34	10.22	**1.27**	3.94*	3.75*	3.63*	3.79*	3.60*	3.13*
	$\mathcal{T}_2^2$	15.53*	7.66*	4.31	6.22*	9.98*	1.28	4.24	4.02	3.90	4.11	3.91	3.57
	$\mathcal{T}_3^2$	**14.53***	**7.58***	**4.22***	**5.90***	**9.93***	1.28	4.30	4.09	3.96	4.19	3.99	3.69
SLIM$_{SIG2}^*$	$\mathcal{T}^0$	20.72	14.09	**6.76**	2.71	10.05	1.36	**2.90***	**2.99***	**2.83***	**3.10***	**2.92***	**2.70***
	$\mathcal{T}_2^2$	14.24*	12.35	8.15	**1.84***	**9.86***	**1.34**	3.55	3.52	3.47	3.52	3.48	3.17
	$\mathcal{T}_3^2$	**13.96***	13.73	7.00	2.13*	**9.84***	1.36	3.70	3.62	3.59	3.69	3.62	3.35
SLIM$_{SIG1}^+$	$\mathcal{T}^0$	17.38	8.69	4.61	6.82	10.22	1.32	**3.76***	**3.51***	**3.34***	**3.57***	**3.43***	**2.73***
	$\mathcal{T}_2^2$	14.57*	7.72*	**4.21***	5.78*	10.10*	**1.30**	4.04	3.78	3.62	3.89	3.77	3.27
	$\mathcal{T}_3^2$	**13.41***	**7.61***	4.31	**5.40***	**10.06***	1.31	4.12	3.87	3.73	3.97	3.86	3.42
SLIM$_{SIG1}^*$	$\mathcal{T}^0$	21.82	13.47	7.61	2.63	10.41	1.43	**2.84***	**2.60***	**2.47***	**2.76***	**2.90***	**2.42***
	$\mathcal{T}_2^2$	13.26*	10.74*	6.01	**2.15***	10.25	1.40	3.58	3.16	3.04	3.19	3.50	2.95
	$\mathcal{T}_3^2$	**12.54***	13.24	**5.96***	2.05*	10.30	1.41	3.80	3.36	3.26	3.36	3.64	3.14
SLIM$_{ABS}^+$	$\mathcal{T}^0$	19.25	9.32	4.96	6.37	10.19	1.32	**3.74***	**3.58***	**3.46***	**3.68***	**3.29***	**2.74***
	$\mathcal{T}_2^2$	15.75*	8.13*	**3.86***	5.23*	10.08*	**1.29**	4.05	3.83	3.70	3.93	3.55	3.18
	$\mathcal{T}_3^2$	**14.94***	**7.91***	3.87*	**4.89***	**10.04***	1.29	4.13	3.91	3.78	4.00	3.62	3.30
SLIM$_{ABS}^*$	$\mathcal{T}^0$	28.78	9.84	6.28	2.62	10.40	1.43	**2.39***	**2.53***	**2.37***	**2.81***	**2.29***	**2.35***
	$\mathcal{T}_2^2$	23.72*	11.76	6.61	**2.17***	10.35	**1.38**	2.96	2.89	2.78	3.09	2.74	2.85
	$\mathcal{T}_3^2$	**22.73***	19.37	5.98	2.21*	10.34	1.40	3.17	3.01	2.93	3.20	2.90	3.02

Our initial analysis evaluates the impact of incorporating a cellular structure on RMSE and $\log_{10}(\ell)$, with results detailed in Table 1. The term $\mathcal{T}^0$ refers to the non-cellular standard version, while $\mathcal{T}_2^2$ and $\mathcal{T}_3^2$ indicate, respectively, cellular methods with radius 2 and 3. Statistical comparisons are conducted using the Wilcoxon-Mann-Whitney test [44] ($\alpha = 0.05$), with p-values adjusted by the Holm-Bonferroni correction [31] (these tests are performed after a preliminary Kruskal-Wallis test [37]). For each dataset and algorithm, the statistically superior method among $\mathcal{T}^0$, $\mathcal{T}_2^2$, and $\mathcal{T}_3^2$ is marked with a blue asterisk (*), and the one with lowest median value is indicated in bold. Additionally, cellular methods that outperform their non-cellular baseline are marked with a black asterisk (*). We further analyze performance and size by presenting the median test fitness evolution of the best individual over 30 runs in Fig. 1 and the median size evolution of the best individual in Fig. 2.[2]

Table 1 shows that cellular methods generally achieve lower errors for GSGP and SLIM$^+$, but tend to increase solution size. Figure 1 confirms that this improved performance persists throughout the evolutionary process. Exceptions include SLM and QSR, where cellular and non-cellular methods yield similar results, even though cellular methods converge faster. Another exception is YCH, where SLIM* outperforms SLIM$^+$. Regarding SLIM*, cellular methods produce lower errors less frequently than for SLIM$^+$. Nevertheless, for both GSGP and SLIM variants, they are never statistically significantly worse than their non-cellular counterparts. This property is a first important take-home message: embedding GSGP and SLIM in a cellular substrate never leads to statistically worse results.

SLIM* variants often produce smaller solutions, frequently even smaller than those from GP, aligning with findings in [78,81]. Overall, cellular methods tend to increase tree size, except in GP, where they result in smaller solutions. In particular, the longer the radius, the greater the size increment (which often correlates with more accurate models). In Fig. 2, we show the trend of the size of the best-so-far individual during the evolution, aggregating values from all datasets and repetitions.[3] The plot confirms that the trend of the size we analyzed in Table 1 holds for each generation.

5.2 Diversity

Following the intuition of [10], we employ I to analyze diversity in cellular-based methods. In Fig. 3, we track the trend of I computed on the whole population across the generations, aggregating values from all datasets and repetitions.[4] The plot shows that for GSGP and SLIM$^+$ the cellular structure effectively preserves diversity during the first half of the evolution. This suggests that the grid preserves the spatial organization of the population, slowing the spread of dominant solutions by confining them to their local neighborhoods. However, this trend holds in the first phase of the evolution, then dominant solutions spread across all the neighborhoods, and eventually I converges to zero.

Building on the previous analysis, we conclude that cellular methods can outperform their non-cellular counterparts when I-based semantic diversity is

---

[2] The showed lineplots depict the median of the examined measure across all repetitions, with shaded area denoting the interquartile range.

[3] Preliminary analysis indicated similar trends across datasets, so $\log_{10}(\ell)$ values are aggregated. The trend of $\log_{10}(\ell)$ for each dataset separately is shown in the supplementary materials.

[4] Preliminary analysis indicated similar trends across datasets, so I values are aggregated. The trend of I for each dataset separately is shown in the supplementary materials.

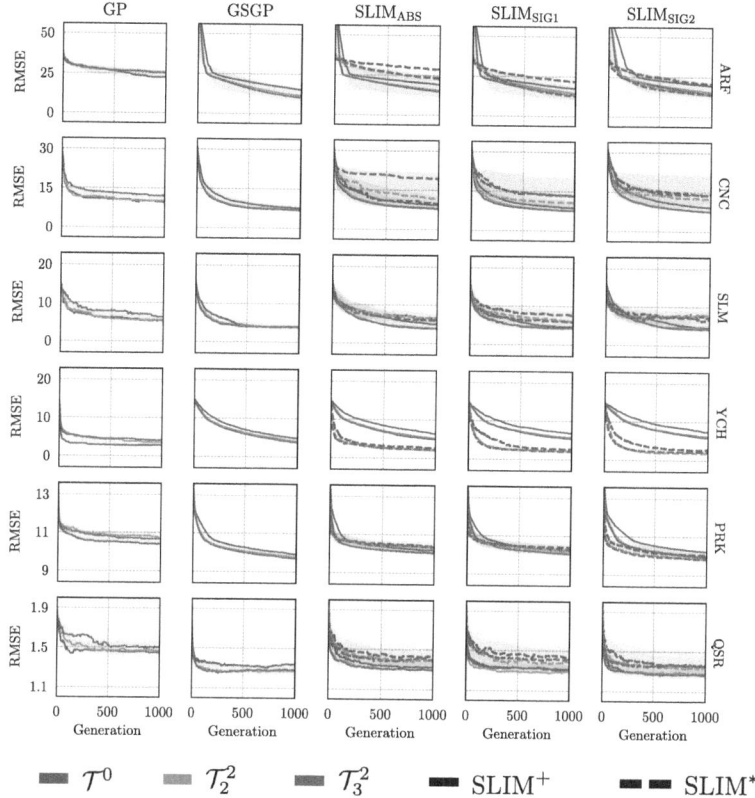

Fig. 1. Evolution of test fitness for the best individual across 30 runs.

preserved during part of the optimization process. On the other hand, cellular algorithms where I remains near zero (indicating a random spatial pattern in the population) generally fail to achieve better solutions, despite the cellular structuring. Notably, GP shows a constant trend for I across all methods, indicating that, even with a cellular structure, the spatial disposition of individuals resembles a random pattern.

To analyze spatial distribution and diversity in detail, we use heatmaps to visualize training fitness across individuals. In Fig. 4, we present the heatmaps for a single repetition of ARF for different algorithms and methods.[5]

These plots confirm that for GP and its cellular variant, the toroidal grid induces low intra-neighborhood similarity in training fitness. Individuals are randomly distributed, with good and poor solutions often placed close together, lacking spatially coherent patterns. For GSGP and its cellular variant, dominant individuals gradually spread through the population over generations. In

---

[5] Despite we show a single example, we provide additional heatmaps for the other datasets and for $SLIM^*_{SIG1}$ and $SLIM^+_{SIG1}$ in the supplementary materials.

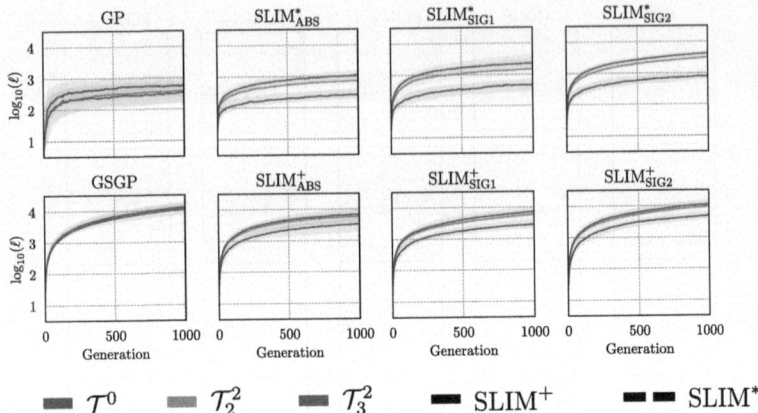

Fig. 2. Evolution of the size of the best individual across 30 runs. Each line represents the median across all datasets and repetitions for each algorithm.

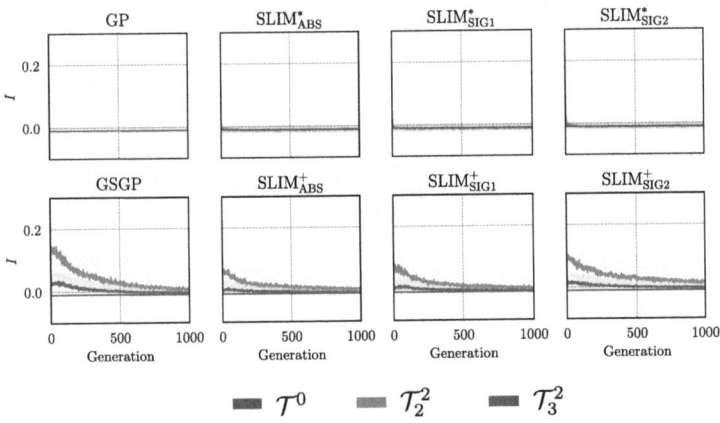

Fig. 3. Trend of the Global Moran's I calculated on the population of each algorithm and method. Each line represents the median across all datasets and repetitions for each algorithm.

the early generations of cGSGP, a distinct spatial clustering emerges, forcing the optimization to focus on different (good and bad) areas of the search space and slowing interactions with dominant solutions. This clustering enables the cellular variant to construct a solution with a lower error than that found by GSGP by the end of the process.

The behavior observed in Fig. 3 and Fig. 4 can be explained by examining how algorithms transform individuals to balance exploration and exploitation. Diversity, according to I, is better preserved in GSGP and SLIM$^+$, as these algorithms generate new solutions by adding components to existing individuals. In contrast, SLIM* generates new solutions by continuously multiplying individuals

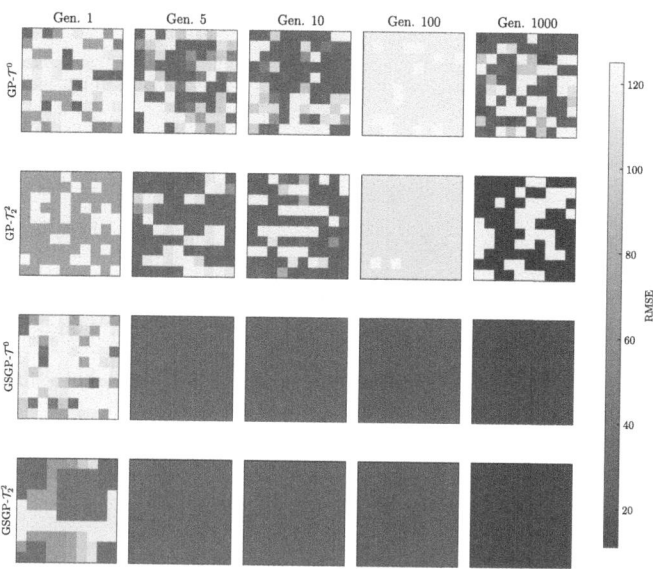

Fig. 4. Heatmaps of the individuals training fitness in a random repetition of GP and GSGP (and their cellular variants with radius 2) with ARF dataset.

with new components. This approach risks quickly invalidating solutions, as the size of the new component-whether very small or very large-can have a significant impact. Similarly, crossover in GP can drastically alter the behavior of individuals by swapping code segments between parents, leading to significant changes in structure. To sum up, disruptive operations quickly erase spatial patterns in the population, whereas conservative operations build on previous populations, thus preserving diversity and spatial patterns, which in cellular structures are often correlated since a toroidal grid may force interactions between individuals to happen only within neighborhoods and preserve diversity among different population areas. However, this holds if individuals within a neighborhood do not change too much when mutation is applied to them, otherwise, random patterns can be quickly formed despite the enforcement of a spatial arrangement.

5.3 Algorithms Pareto Front

We summarize our main findings and provide a global comparison among all the algorithms by representing them in Pareto fronts highlighting the trade-offs between model error and model size (Fig. 5). In the following plot, we aggregate the results from all datasets and repetitions for each algorithm and method.[6]

This figure indicates that cGSGP achieves the best performance, while $\text{SLIM}^*_{\text{ABS}}$ produces the smallest solutions (even smaller than GP itself). As

---

[6] Supplementary materials contain a version of this plot for each dataset separately.

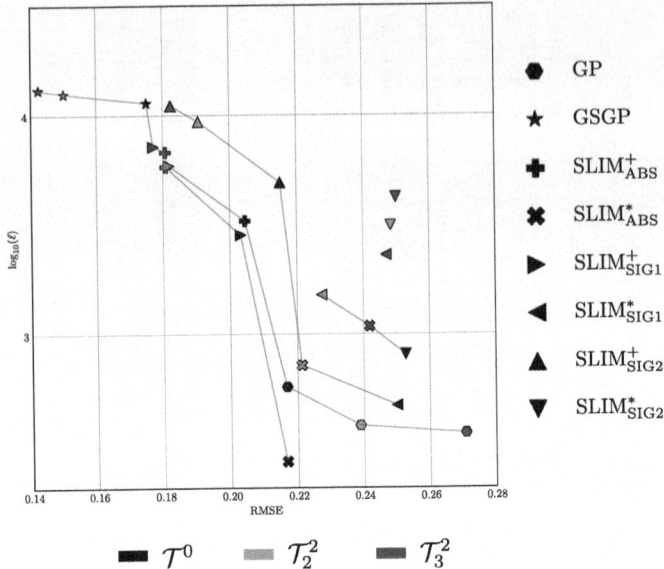

Fig. 5. Pareto fronts of the tested methods. Each Pareto front represents a set of non-dominated solutions w.r.t. to both error and size. Each method is identified by the median across all datasets and repetitions. RMSE is scaled in $[0, 1]$ by using the maximum RMSE discovered among the best solutions from all the experiments.

noted in the previous analysis, GP is the only algorithm where the cellular variant yields smaller solutions than the non-cellular one, though this comes with a slight reduction in performance quality. In general, SLIM* leads to smaller solutions while SLIM$^+$ leads to more accurate models, especially when a cellular structure is employed. This overall outlines the possible trade-offs between error and size when choosing a GP algorithm for a SR problem. cGSGP and SLIM* offer the best performance and interpretability, respectively, while the cellular SLIM variants fall in-between, providing varying levels of trade-off between these two quality criteria. In some variants, cellular structure may lead to worst models as regards both accuracy and size, as in SLIM$^*_{ABS}$, perhaps because of the disruptive effect of the multiplication combined with mutation. A general rule of thumb can be that cellular methods are worth employing with algorithms that do not contain disruptive operations when creating new solutions, i.e., genetic operators lead to offspring that are similar to their parents.

6 Conclusion and Future Work

In this paper, we explore the effects of the integration of CA-inspired spatial structures on popular GP methods that are characterized by different strengths and limitations as regards both the qualitative performance of the solutions and their size. Taking inspiration from [10], we apply a cellular-based toroidal grid to the population of three GP algorithms, namely, standard tree-based GP [36], GSGP [50], and SLIM [78,81], the latter being a recent and innovative variant of GSGP in which the problem of exponential tree growth is addressed by using genetic operators that lead to smaller offspring. In particular, with our study, we assess the effects of cellular structures on SLIM for the first time. Our comparative analysis highlights the different levels of trade-off that these methods exhibit both when executed in their standard version and when integrated with a cellular-based spatial structure. Especially, cellular helps in boosting the qualitative performance of GSGP and SLIM$^+$ at the cost of increasing the model size. On the other hand, SLIM* provides the smallest models, with cellular that occasionally improves its performance, even though it increases the size, but not as much as cellular SLIM$^+$. We additionally assess the diversity preservation mechanisms examined via Global Moran's I, in which we see that diversity is better preserved in the early stage of the evolution when cellular is applied to GSGP and SLIM$^+$ (the algorithms that benefit more from a toroidal grid as regards model accuracy). As a take-home message, we hypothesize that cellular can be beneficial for algorithms that do not perform mutations that can quickly alter the individuals' behavior. Future research works should focus on enlarging the spectrum of the algorithms tested with cellular to other variants of GP that account for the syntactical structure of the individuals to evaluate the impact of the toroidal grid also on particular methods that do not directly explore the semantic space.

Acknowledgements. This work was supported by national funds through FCT (Fundação para a Ciência e a Tecnologia), under the project - UIDB/04152/2020 - Centro de Investigação em Gestão de Informação (MagIC)/NOVA IMS (https://doi.org/10.54499/UIDB/04152/2020).

This research is partially supported by the PRIN 2022 PNRR project "Cellular Automata Synthesis for Cryptography Applications (CASCA)" (P2022MPFRT) financed by the European Union - Next Generation EU.

References

1. Ahvanooey, M.T., Li, Q., Wu, M., Wang, S.: A survey of genetic programming and its applications. KSII Trans. Internet Inf. Syst. (TIIS) **13**(4), 1765–1794 (2019)
2. Al-Betar, M.A., Khader, A.T., Awadallah, M.A., Alawan, M.H., Zaqaibeh, B.: Cellular harmony search for optimization problems. J. Appl. Math. **2013** (2013)
3. Alba, E., Dorronsoro, B.: The exploration/exploitation tradeoff in dynamic cellular genetic algorithms. IEEE Trans. Evol. Comput. **9**(2), 126–142 (2005)
4. Alba, E., Dorronsoro, B.: Introduction to Cellular Genetic Algorithms, pp. 3–20. Springer, Boston (2008). https://doi.org/10.1007/978-0-387-77610-1_1
5. Anselin, L., Rey, S.: Properties of tests for spatial dependence in linear regression models. Geogr. Anal. **23**(2), 112–131 (1991)
6. Armstrong, J.S., Collopy, F.: Error measures for generalizing about forecasting methods: empirical comparisons. Int. J. Forecast. **8**(1), 69–80 (1992)

7. Bakurov, I., et al.: Geometric semantic genetic programming with normalized and standardized random programs. Genet. Program. Evolvable Mach. **25**(1) (2024). https://doi.org/10.1007/s10710-024-09479-1
8. Ballabio, D., Cassotti, M., Consonni, V., Todeschini, R.: QSAR aquatic toxicity. UCI Mach. Learn. Repository (2014). https://doi.org/10.24432/C5SG7H
9. Banzhaf, W., Koza, J., Ryan, C., Spector, L., Jacob, C.: Genetic programming. IEEE Intell. Syst. Appl. **15**(3), 74–84 (2000). https://doi.org/10.1109/5254.846288
10. Bonin, L., Rovito, L., De Lorenzo, A., Manzoni, L.: Cellular geometric semantic genetic programming. Genet. Program Evolvable Mach. **25**(1), 8 (2024)
11. Brooks, T.F., Pope, D.S., Marcolini, M.A.: Airfoil self-noise and prediction. Technical report (1989)
12. Brotto Rebuli, K., Giacobini, M., Silva, S., Vanneschi, L.: A comparison of structural complexity metrics for explainable genetic programming. In: Proceedings of the Companion Conference on Genetic and Evolutionary Computation, pp. 539–542 (2023)
13. Castelli, M., Manzoni, L., Gonçalves, I., Vanneschi, L., Trujillo, L., Silva, S.: An analysis of geometric semantic crossover: a computational geometry approach. pp. 201–208 (2016). https://doi.org/10.5220/0006056402010208
14. Castelli, M., Manzoni, L., Vanneschi, L., Silva, S., Popovič, A.: Self-tuning geometric semantic genetic programming. Genet. Program Evolvable Mach. **17**, 55–74 (2016)
15. Castelli, M., Silva, S., Vanneschi, L.: A c++ framework for geometric semantic genetic programming. Genet. Program Evolvable Mach. **16**, 73–81 (2015)
16. Castelli, M., Vanneschi, L., Popovič, A.: Controlling individuals growth in semantic genetic programming through elitist replacement. Comput. Intell. Neurosci. **2016**, 42–42 (2016)
17. Castelli, M., Vanneschi, L., Silva, S.: Prediction of high performance concrete strength using genetic programming with geometric semantic genetic operators. Expert Syst. Appl. **40**(17), 6856–6862 (2013)
18. Castelli, M., Vanneschi, L., Silva, S.: Prediction of the unified Parkinson's disease rating scale assessment using a genetic programming system with geometric semantic genetic operators. Expert Syst. Appl. **41**(10), 4608–4616 (2014)
19. Codd, E.F.: Cellular Automata. Academic Press, Cambridge (1968)
20. Cussat-Blanc, S., Harrington, K., Pollack, J.: Gene regulatory network evolution through augmenting topologies. IEEE Trans. Evol. Comput. **19**(6), 823–837 (2015)
21. Dabhi, V.K., Chaudhary, S.: Empirical modeling using genetic programming: a survey of issues and approaches. Nat. Comput. **14**, 303–330 (2015)
22. Della Cioppa, A., Marcelli, A., Napoli, P.: Speciation in evolutionary algorithms: adaptive species discovery. In: Proceedings of the 13th Annual Conference on Genetic and Evolutionary Computation, pp. 1053–1060 (2011)
23. Dick, G., Whigham, P.A.: Controlling bloat through parsimonious elitist replacement and spatial structure. In: European Conference on Genetic Programming, pp. 13–24. Springer (2013)
24. Farinati, D., Bakurov, I., Vanneschi, L.: A study of dynamic populations in geometric semantic genetic programming. Inf. Sci. **648**, 119513 (2023). https://doi.org/10.1016/j.ins.2023.119513, https://www.sciencedirect.com/science/article/pii/S0020025523010988
25. Ferreira, L.A., Guimarães, F.G., Silva, R.: Applying genetic programming to improve interpretability in machine learning models. In: 2020 IEEE Congress on Evolutionary Computation (CEC), pp. 1–8 (2020). https://doi.org/10.1109/CEC48606.2020.9185620

26. Folino, G., Pizzuti, C., Spezzano, G.: A cellular genetic programming approach to classification. In: GECCO, pp. 1015–1020 (1999)
27. Folino, G., Pizzuti, C., Spezzano, G.: A scalable cellular implementation of parallel genetic programming. IEEE Trans. Evol. Comput. **7**(1), 37–53 (2003)
28. Giacobini, M., Tomassini, M., Tettamanzi, A.G., Alba, E.: Selection intensity in cellular evolutionary algorithms for regular lattices. IEEE Trans. Evol. Comput. **9**(5), 489–505 (2005)
29. Goldberg, D.E., Deb, K.: A comparative analysis of selection schemes used in genetic algorithms. In: Foundations of Genetic Algorithms, vol. 1, pp. 69–93. Elsevier (1991)
30. Holland, J.H.: Genetic algorithms and the optimal allocation of trials. SIAM J. Comput. **2**(2), 88–105 (1973)
31. Holm, S.: A simple sequentially rejective multiple test procedure. Scand. J. Stat. 65–70 (1979)
32. Hu, T.: Genetic Programming for Interpretable and Explainable Machine Learning, pp. 81–90. Springer, Singapore (2023). https://doi.org/10.1007/978-981-19-8460-0_4
33. Huynh, Q., Singh, H., Ray, T., Oyama, A.: Improved genetic programming for symbolic regression: case studies on practical applications. In: 2022 IEEE Symposium Series on Computational Intelligence (SSCI), pp. 1135–1142 (2022). https://doi.org/10.1109/SSCI51031.2022.10022279
34. Juárez-Smith, P., Trujillo, L., García-Valdez, M., Fernández de Vega, F., Chávez, F.: Local search in speciation-based bloat control for genetic programming. Genet. Program Evolvable Mach. **20**(3), 351–384 (2019). https://doi.org/10.1007/s10710-019-09351-7
35. Koga, D., Ohnishi, K.: Non-generational geometric semantic genetic programming. In: 2021 IEEE Symposium Series on Computational Intelligence (SSCI), pp. 1–7. IEEE (2021)
36. Koza, J.R.: Genetic programming as a means for programming computers by natural selection. Stat. Comput. **4**(2), 87–112 (1994)
37. Kruskal, W.H., Wallis, W.A.: Use of ranks in one-criterion variance analysis. J. Am. Stat. Assoc. **47**(260), 583–621 (1952)
38. La Cava, W., Danai, K., Spector, L.: Inference of compact nonlinear dynamic models by epigenetic local search. Eng. Appl. Artif. Intell. **55**, 292–306 (2016)
39. La Cava, W., Danai, K., Spector, L., Fleming, P., Wright, A., Lackner, M.: Automatic identification of wind turbine models using evolutionary multiobjective optimization. Renew. Energy **87**, 892–902 (2016)
40. La Cava, W., et al.: Contemporary symbolic regression methods and their relative performance. In: Thirty-Fifth Conference on Neural Information Processing Systems Datasets and Benchmarks Track (2021)
41. Langdon, W.B., Poli, R., McPhee, N.F., Koza, J.R.: Genetic programming: an introduction and tutorial, with a survey of techniques and applications. In: Computational Intelligence: A Compendium, pp. 927–1028 (2008)
42. Lensen, A., Xue, B., Zhang, M.: Genetic programming for evolving a front of interpretable models for data visualization. IEEE Trans. Cybern. **51**(11), 5468–5482 (2021). https://doi.org/10.1109/TCYB.2020.2970198
43. Li, H., Calder, C.A., Cressie, N.: Beyond Moran's i: testing for spatial dependence based on the spatial autoregressive model. Geogr. Anal. **39**(4), 357–375 (2007)
44. Mann, H.B., Whitney, D.R.: On a test of whether one of two random variables is stochastically larger than the other. Ann. Math. Stat. **18**(1), 50–60 (1947)

45. Mariot, L., Picek, S., Jakobovic, D., Leporati, A.: Evolutionary algorithms for the design of orthogonal Latin squares based on cellular automata. In: Proceedings of the Genetic and Evolutionary Computation Conference, GECCO 2017, pp. 306–313. Association for Computing Machinery, New York (2017). https://doi.org/10.1145/3071178.3071284
46. Martin, W.N.: Island (migration) models: evolutionary algorithms based on punctuated equilibria. In: Handbook of Evolutionary Computation (1997)
47. Martins, J.F.B., Oliveira, L.O.V., Miranda, L.F., Casadei, F., Pappa, G.L.: Solving the exponential growth of symbolic regression trees in geometric semantic genetic programming. In: Proceedings of the Genetic and Evolutionary Computation Conference, pp. 1151–1158 (2018)
48. Martins, T.M., Neves, R.F.: Applying genetic algorithms with speciation for optimization of grid template pattern detection in financial markets. Expert Syst. Appl. **147**, 113191 (2020)
49. Mei, Y., Chen, Q., Lensen, A., Xue, B., Zhang, M.: Explainable artificial intelligence by genetic programming: a survey. IEEE Trans. Evol. Comput. **27**(3), 621–641 (2023). https://doi.org/10.1109/TEVC.2022.3225509
50. Moraglio, A., Krawiec, K., Johnson, C.G.: Geometric semantic genetic programming. In: Coello, C., Cutello, V., Deb, K., Forrest, S., Nicosia, G., Pavone, M. (eds.) PPSN 2012. LNCS, vol. 7491, pp. 21–31. Springer, Heidelberg (2012). https://doi.org/10.1007/978-3-642-32937-1_3
51. Moran, P.A.: Notes on continuous stochastic phenomena. Biometrika **37**(1/2), 17–23 (1950)
52. Murata, T., Gen, M.: Cellular genetic algorithm for multi-objective optimization. In: Proceedings of the 4th Asian Fuzzy System Symposium, pp. 538–542. Citeseer (2002)
53. Nadizar, G., Garrow, F., Sakallioglu, B., Canonne, L., Silva, S., Vanneschi, L.: An investigation of geometric semantic GP with linear scaling. In: Proceedings of the Genetic and Evolutionary Computation Conference, GECCO 2023, pp. 1165–1174. Association for Computing Machinery, New York (2023). https://doi.org/10.1145/3583131.3590418
54. Nadizar, G., Medvet, E., Wilson, D.: Searching for a diversity of interpretable graph control policies. In: Proceedings of the Genetic and Evolutionary Computation Conference, GECCO 2024, pp. 933–941. Association for Computing Machinery, New York (2024). https://doi.org/10.1145/3638529.3653987
55. Nadizar, G., Medvet, E., Wilson, D.G.: Naturally interpretable control policies via graph-based genetic programming. In: Giacobini, M., Xue, B., Manzoni, L. (eds.) Genetic Programming. LNCS, pp. 73–89. Springer Nature Switzerland, Cham (2024)
56. Nadizar, G., Rovito, L., De Lorenzo, A., Medvet, E., Virgolin, M.: An analysis of the ingredients for learning interpretable symbolic regression models with human-in-the-loop and genetic programming. ACM Trans. Evol. Learn. Optim. **4**(1) (2024). https://doi.org/10.1145/3643688
57. Nadizar, G., Sakallioglu, B., Garrow, F., Silva, S., Vanneschi, L.: Geometric semantic GP with linear scaling: Darwinian versus Lamarckian evolution. Genet. Program Evolvable Mach. **25**(2), 1–24 (2024)
58. Nebro, A.J., Durillo, J.J., Luna, F., Dorronsoro, B., Alba, E.: Mocell: a cellular genetic algorithm for multiobjective optimization. Int. J. Intell. Syst. **24**(7), 726–746 (2009)
59. Neumann, J.V.: Theory of self-reproducing automata. Math. Comput. **21**, 745 (1966)

60. Nguyen, Q.U.: Examining Semantic Diversity and Semantic Locality of Operators in Genetic Programming. Ph.D. thesis, University College Dublin, Ireland (2011). http://ncra.ucd.ie/papers/Thesis_Uy_Corrected.pdf
61. Ortigosa, I., Lopez, R., Garcia, J.: A neural networks approach to residuary resistance of sailing yachts prediction. In: Proceedings of the International Conference on Marine Engineering Marine, vol. 2007, p. 250 (2007)
62. Orzechowski, P., La Cava, W., Moore, J.H.: Where are we now? A large benchmark study of recent symbolic regression methods. In: Proceedings of the Genetic and Evolutionary Computation Conference, pp. 1183–1190 (2018)
63. Pawlak, T.P., Wieloch, B., Krawiec, K.: Review and comparative analysis of geometric semantic crossovers. Genet. Program Evolvable Mach. **16**, 351–386 (2015)
64. Pietropolli, G., Manzoni, L., Paoletti, A., Castelli, M.: Combining geometric semantic GP with gradient-descent optimization. In: Medvet, E., Pappa, G., Xue, B. (eds.) Genetic Programming, pp. 19–33. Springer, Cham (2022)
65. Pietropolli, G., Manzoni, L., Paoletti, A., Castelli, M.: On the hybridization of geometric semantic GP with gradient-based optimizers. Genet. Program Evolvable Mach. **24**(2), 16 (2023)
66. Pietropolli, G., Nichele, S., Medvet, E.: The role of the substrate in ca-based evolutionary algorithms. In: Proceedings of the Genetic and Evolutionary Computation Conference, GECCO 2024, pp. 768–777. Association for Computing Machinery, New York (2024). https://doi.org/10.1145/3638529.3654112
67. Poli, R., Langdon, W.B., McPhee, N.F., Koza, J.R.: Genetic programming: an introductory tutorial and a survey of techniques and applications. Univ. Essex School Computer Science and Electronic Engineering Technical report No. CES-475, pp. 1–112 (2007)
68. Salto, C., Alba, E.: Cellular genetic algorithms: understanding the behavior of using neighborhoods. Appl. Artif. Intell. **33**(10), 863–880 (2019)
69. Sarkar, P.: A brief history of cellular automata. ACM Comput. Surv. (CSUR) **32**(1), 80–107 (2000)
70. Sarma, J., De Jong, K.: An analysis of the effects of neighborhood size and shape on local selection algorithms. In: Voigt, H.-M., Ebeling, W., Rechenberg, I., Schwefel, H.-P. (eds.) PPSN 1996. LNCS, vol. 1141, pp. 236–244. Springer, Heidelberg (1996). https://doi.org/10.1007/3-540-61723-X_988
71. Shi, Y., Liu, H., Gao, L., Zhang, G.: Cellular particle swarm optimization. Inf. Sci. **181**(20), 4460–4493 (2011)
72. Stanley, K.O., Miikkulainen, R.: Evolving neural networks through augmenting topologies. Evol. Comput. **10**(2), 99–127 (2002)
73. Takac, A.: Application of cellular genetic programming in data mining. In: Proceedings of Conference Knowledge, Citeseer (2004)
74. Takac, A.: Cellular genetic programming algorithm applied to classification task. Neural Netw. World **14**, 435–452 (2004)
75. Trujillo, L., Contreras, J., Hernandez, D.E., Castelli, M., Tapia, J.J.: GSGP-CUDA-a CUDA framework for geometric semantic genetic programming. SoftwareX **18**, 101085 (2022)
76. Trujillo, L., Muñoz, L., Galván-López, E., Silva, S.: neat genetic programming: controlling bloat naturally. Inf. Sci. **333**, 21–43 (2016)
77. Vanneschi, L.: An introduction to geometric semantic genetic programming. In: NEO 2015: Results of the Numerical and Evolutionary Optimization Workshop NEO 2015 held at 23–25 Sep 2015 in Tijuana, Mexico, pp. 3–42. Springer (2016)

78. Vanneschi, L.: SLIM_GSGP: The non-bloating geometric semantic genetic programming. In: European Conference on Genetic Programming (Part of EvoStar), pp. 125–141. Springer (2024)
79. Vanneschi, L., Castelli, M., Manzoni, L., Silva, S.: A new implementation of geometric semantic gp and its application to problems in pharmacokinetics. In: Genetic Programming: 16th European Conference, EuroGP 2013, Vienna, Austria, 3–5 Apr 2013. Proceedings 16, pp. 205–216. Springer (2013)
80. Vanneschi, L., Castelli, M., Silva, S.: A survey of semantic methods in genetic programming. Genet. Program Evolvable Mach. **15**(2), 195–214 (2014). https://doi.org/10.1007/s10710-013-9210-0
81. Vanneschi, L., Farinati, D., Rasteiro, D., Rosenfeld, L., Pietropolli, G., Silva, S.: Exploring non-bloating geometric semantic genetic programming. In: Genetic Programming Theory and Practice. (to appear)
82. Vanneschi, L., Silva, S., Castelli, M., Manzoni, L.: Geometric semantic genetic programming for real life applications. In: Genetic Programming Theory and Practice xi, pp. 191–209 (2014)
83. Virgolin, M., Alderliesten, T., Witteveen, C., Bosman, P.A.: Improving model-based genetic programming for symbolic regression of small expressions. Evol. Comput. **29**(2), 211–237 (2021)
84. Virgolin, M., De Lorenzo, A., Randone, F., Medvet, E., Wahde, M.: Model learning with personalized interpretability estimation (ml-pie). In: Proceedings of the Genetic and Evolutionary Computation Conference Companion, GECCO 2021, pp. 1355–1364. Association for Computing Machinery, New York (2021). https://doi.org/10.1145/3449726.3463166
85. Whigham, P.A., Dick, G.: Implicitly controlling bloat in genetic programming. IEEE Trans. Evol. Comput. **14**(2), 173–190 (2009)
86. White, D.R., McDermott, J., Castelli, M., Manzoni, L., Goldman, B.W., Kronberger, G., Jaśkowski, W., O'Reilly, U.M., Luke, S.: Better GP benchmarks: community survey results and proposals. Genet. Program Evolvable Mach. **14**, 3–29 (2013)
87. Wickman, R., Poudel, B., Villarreal, T.M., Zhang, X., Li, W.: Efficient quality-diversity optimization through diverse quality species. In: Proceedings of the Companion Conference on Genetic and Evolutionary Computation, pp. 699–702 (2023)
88. Xie, H., Zhang, M.: Impacts of sampling strategies in tournament selection for genetic programming. Soft. Comput. **16**, 615–633 (2012)
89. Yeh, I.C.: Simulation of concrete slump using neural networks. Proc. Inst. Civ. Eng. Constr. Mater. **162**(1), 11–18 (2009)
90. Zhang, M., Tian, N., Palade, V., Ji, Z., Wang, Y.: Cellular artificial bee colony algorithm with gaussian distribution. Inf. Sci. **462**, 374–401 (2018)

Ant-Based Metaheuristics Struggle to Solve the Cartesian Genetic Programming Learning Task

Julian Trautwein[ID], Michael Heider[✉][ID], Henning Cui[ID], and Jörg Hähner[ID]

University of Augsburg, 86159 Augsburg, Germany
{julian.trautwein,michael.heider}@uni-a.de

Abstract. Ant-based metaheuristics have successfully been applied to a variety of different graph-based problems. However, for Cartesian Genetic Programming (CGP) only the impact of Max-Min Ant Systems has been tested. In this work, we try to fill this gap by applying four different popular ant-based metaheuristics as the optimizer (and therefore training algorithm) of CGP. The idea of combining CGP with ant-based metaheuristics is not novel but older works' experimental design may not meet today's standard. To compare these metaheuristics to the Evolution Strategies (ESs) commonly used in CGP, we benchmark against a standard CGP variant that uses a simplistic $(1 + 4)$-ES, mutation, and no crossover. Additionally, we include $(\mu + \lambda)$-ES and (μ, λ)-ES in our experiments.

We analyse the performance on datasets from the symbolic regression, regression, and classification domains. By tuning and evaluating various configurations, we can not affirm a significant improvement by using ant-based methods with CGP as we encounter premature convergence— even with those ant-based metaheuristics that were originally proposed to overcome such problems. Despite our results being of negative nature, this work still gives important and interesting insights into the training of CGP models. The key contributions of our work are thus a more thorough benchmarking of these optimizers than has been done before. This should clear up doubts about the capabilities of ant-based metaheuristics in CGP. Furthermore, we include a roadmap on how they can be addressed to solve this complex optimization problem from the model building domain of machine learning.

Keywords: Cartesian Genetic Programming · Evolution Strategies · Evolutionary Algorithm · Ant Colony Optimization · Ant-based Metaheuristic

1 Introduction

Cartesian Genetic Programming (CGP) is a form of *Genetic Programming* (GP) developed by Miller in 1999 [26]. CGP—in contrast to GP—is represented by a *feed-forward, directed, and acyclic* graph instead of a *tree based representation*. This makes it easy to be applied to graph-based applications like *neural architecture search* [32] or *image processing* [25].

CGP often omits *crossovers*, which are an archetypical operator of genetic algorithms, even though there have been experiments where including crossover shows an increase in fitness [9]. As a result, only selection and mutation operators induce changes to optimise a graph for a given learning task. Metaheuristics or other learning paradigms are typically not considered in the context of CGP. Since CGP is represented by said graph, it can be viewed as a pathfinding problem, for which ant system–related algorithms are a natural solution. In this work, we apply four different variants of ant-based metaheuristics and analyse their impact on CGP.

We start by reintroducing the core principles of CGP in Sect. 2 to serve as an easy entrance to the reader. In Sect. 5 the different ant-based metaheuristics used in this work are presented, which is followed by a summary of previous work on ant-based CGP in Sect. 4. After that, we give a description of the implementation of our ant-based metaheuristics into CGP (Sect. 3). Then, the performance of all used metaheuristics is analysed in Sect. 6. At last, Sect. 7 summarizes our results and shows further research possibilities.

2 Cartesian Genetic Programming

This section reintroduces the core principles of *Cartesian Genetic Programming* (CGP).

2.1 Representation

In CGP, a program is represented as a *feed-forward, directed, and acyclic* graph. Nowadays, it contains *nodes* arranged in a one dimensional grid with $c \in \mathbb{N}^+$ columnsd [28]. CGP takes an arbitrary amount of program inputs and feeds them forward through the graph to get the desired amount of program outputs.

There are three types of nodes present in a CGP graph: input, computational and output nodes. The *input nodes* are the first nodes of the program. They directly relay the program input to the other node types. The *computational nodes* are represented by multiple genes: One *function gene*, that specifies which function the node will apply on the given inputs, and $a \in \mathbb{N}^+$ *connection genes* that define the node's inputs. The value a is set to the highest arity of the defined function set. If a function needs less than a inputs, all unused connection genes will be ignored. The *output nodes* are typically the last nodes of the graph. They only receive the output previous node and redirect it as the output of the program. This node category consist of one connection gene, which refers to the computational or input node they take their output from.

Input and computational nodes can also be divided into active and inactive nodes. *Inactive nodes* are nodes that are not part of a path to any output nodes

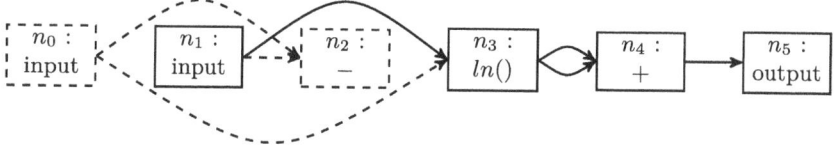

Fig. 1. A graph defined by a CGP genotype. The dashed nodes and connections are inactive.

and therefore do not contribute to the program output. Still, they are beneficial to the optimization process as they lead to genetic drift [33]. *Active nodes* are part of a path to any output nodes by one or more paths—therefore they contribute to the output of the program.

Figure 1 shows an example graph defined by a CGP genotype. It has $c = 6$ columns, takes two inputs and returns one output. The first two nodes n_0 and n_1 are input nodes and only relay the two program inputs. The nodes n_2, n_3 and n_4 are computational nodes. Node n_3 uses the function $ln()$ on its inputs, which has an arity of one. This leads to n_3 omitting the second input and only calculating with its first input. Therefore, only n_1, n_3 and n_4 are active nodes. At last, n_5 provides the output of the program by relaying the output of the computational node n_4. As a result, this CGP graph describes the following function:

$$f : \mathbb{R} \times \mathbb{R} \to \mathbb{R}$$
$$(n_0, n_1) \mapsto \ln(n_1) + \ln(n_1)$$

2.2 Common Evolutionary Operators of CGP

Most CGP variants use an elitist $(1+4)$-ES. It is commonly used in combination with neutral search to improve performance and convergence time [27]. In this context, *neutral search* describes the concept that if an offspring of the current parent has the same fitness value as the parent, it will always be chosen as the new parent. This allows for neutral drift to occur and improves the exploration of the search space [28].

As the mutation strategy, either a *probabilistic mutation* [16] or *Single* [14] are used. With the first operator, it simply iterates over all genes and mutates them with a predefined probability. This leads to children possibly not having mutated any active nodes and therefore not altering the program's output. Single, on the other hand, randomly selects genes and mutates them until an active node is mutated. This enforces a change in the phenotype and allows inactive nodes to be mutated. To improve readability, the aforementioned description of CGP will be called STANDARD in the following sections. There are also many more adaptations to the Evolutionary Algorithm of CGP like [10,13,18], which will not be used in this paper.

3 Ant-Based Metaheuristics

This section will describe the non-ES metaheuristics used in this paper to optimize CGP graphs. The integration and benchmarking of previously unexplored (or unpublished) options of ant-based metaheuristics is the key contribution of this work.

3.1 Ant System

Ant System (AS) [12] was the first proposed ant-based metaheuristic. AS is the simplest ant-based metaheuristic and every other ant-based metaheuristic can be seen as an extension of AS.

Initially, a given number of $m \in \mathbb{N}^+$ ants are randomly distributed among all nodes and each edge gets an initial pheromone level of $\tau_0 \in \mathbb{R}^+$, therefore $\tau_{ij}(0) := \tau_0$ for $i, j = 1, \ldots, \#\text{nodes}$. At each timestep, each ant selects the next node it will move to[1], based on the amount of pheromones present on the given edge and the length of the path to the next node. For the k-th ant, the transition probability to go from node i to node j at time t is defined as:

$$p_{ij}^k(t) := \begin{cases} \frac{[\tau_{ij}(t)]^{\alpha_{AS}} \cdot [\eta_{ij}]^\beta}{\sum_{c \in C_k} [\tau_{ic}(t)]^{\alpha_{AS}} \cdot [\eta_{ic}]^\beta} & if \ j \in C_k \\ 0 & otherwise \end{cases} \quad (1)$$

where C_k is the set of available nodes that ant k has not yet visited, $\eta_{ij} : -\frac{1}{d_{ij}}$ for d_{ij} being the distance between node i and j, $i, j = 1, \ldots, \#\text{nodes}$, and $\alpha_{AS}, \beta \in \mathbb{R}^+$ are hyperparameters that scale the importance of the pheromone versus the distance. The intensity of the pheromone trail left on edge (i, j) at timestep $t \in \mathbb{N}$ is represented by $\tau_{ij}(t)$. After all nodes have been visited, each ant lays pheromones on the connections it has used to construct its path to the inputs. The pheromone update is specified by the following two rules:

$$\tau_{ij}(t) := \rho \cdot \tau_{ij}(t - n) + \sum_{k=1}^m \Delta \tau_{ij}^k(t) \quad (2)$$

$$\Delta \tau_{ij}^k(t) := \begin{cases} \frac{1}{L_k} & if \ ant \ k \ used \ edge \ (i,j) \ in \ its \ latest \ tour \\ 0 & otherwise \end{cases}$$

where L_k is the fitness (defined by a task-appropriate metric) of ant k and $\rho \in [0, 1[$ is a hyperparameter that simulates the evaporation of pheromones on each edge.

---

[1] Note that the ants can only select nodes that are "further left" than the current location to avoid the creation of cyclical graphs.

3.2 Ant Colony System

The *Ant Colony System* (ACS) [11] is one of the many improvements of the Ant System. There are two main differences: At first, the authors changed the pheromone update rule and the state transitioning rule of ants as follows:

$$\Delta \tau_{ij}^k(t) := \begin{cases} \tau_0 & \text{if ant } k \text{ used edge } (i,j) \text{ in its latest tour} \\ 0 & \text{otherwise} \end{cases}$$

In addition to that, ACS introduced a *global update rule* where only the best ant is allowed to deposit pheromones. This best ant can either be the best ant that has been found so far, or the best ant of the current iteration of the algorithm. The pheromone update rule, with $\tilde{k}$ as the best ant, is described as following for all $i, j = 1, ..., n$:

$$\tau_{ij}(t) := (1 - \alpha_{ACS}) \cdot \tau_{ij}(t-n) + \Delta \tau_{ij}^{\tilde{k}}(t) \tag{3}$$

$$\Delta \tau_{ij}^{\tilde{k}}(t) := \begin{cases} \frac{1}{L_{\tilde{k}}} & \text{if } (i,j) \in \text{best tour} \\ 0 & \text{otherwise} \end{cases} \tag{4}$$

where $\alpha_{ACS} \in [0, 1[$ is another hyperparameter, that simulates pheromone evaporation on the edges. For the transition rule, they used the same function as the AS, but without the use of the α_{AS} hyperparameter given in AS. In addition to that, ACS introduces the hyperparameter $q_0 \in [0, 1]$, that determines a rate of *exploitation* vs *exploration*.

3.3 Max-Min Ant System

The *Max-Min Ant System* (MMAS) [31] is another enhancement of the Ant System. Its characteristic differences to the AS are that only the global best ant or the best ant of the iteration updates the pheromone trail. Furthermore, to avoid stagnation, the pheromone trails are limited to an interval $[\tau_{min}, \tau_{max}]$. Additionally, the pheromone trails are initialized to τ_{max} to get a higher exploration of the search space at the start of the algorithm. The state transition rule used in the MMAS is the same as in AS. MMAS updates its pheromones with the global update rule of ACS, as shown in Equation (3) and Equation (4) and omits the local update of all ants.

3.4 Ant System Local Best Tour

The last ant-based metaheuristic we tested is the *Ant System Local Best Tour* (ASLBT) [35]. The main idea behind ASLBT is to remove a global observer and let every single ant keep track of the best tour it has found so far. The pathfinding algorithm of ants is the same as in AS, which can be seen in Equation (1). The pheromone update rule is also the same rule as Equation (2) with the following addition:

$$\Delta \tau_{ij}^k(t) := \begin{cases} \frac{L_{best}^k(t)}{L_k(t)} & \text{if ant } k \text{ used edge } (i,j) \text{ in its latest tour} \\ 0 & \text{otherwise} \end{cases}$$

Table 1. Distance functions used by Ant-based metaheuristics.

Distance function	Mathematical definition
Manhattan Distance:	$\eta(i,j) := (i - j_1) + (i - j_2)$
Euclidean Distance:	$\eta(i,j) := \sqrt{(i - j_1)^2 + (i - j_2)^2}$
Logarithmic Distance:	$\eta(i,j) := \sqrt{ln(i - j_1) + ln(i - j_2)}$
Constant:	$\eta(i,j) := 1$

where $L_{best}^k(t)$ is the best fitness value ant k has found until timestep t and $L_k(t)$ is the current fitness value corresponding to ant k.

4 Including Ant-Based Metaheuristics Into CGP

After reintroducing ant-based metaheuristics, we now describe our method of integrating them into the CGP training algorithm. We use two different pheromone matrices for optimising the CGP Graphs. One for optimising the connections of the graph and one for optimising the functions of the individual nodes.[2]

All entries of the connection pheromone matrix are initialised with the respective default $\tau_0, \tau_{max} \in \mathbb{R}_0^+$ values as is typical for the different ant-based metaheuristics. Connections that are not allowed are initialised with 0 to stop ants from generating illegal solutions.

An ant creates a *genotype* by iterating over every node i and choosing the connection according to the probability

$$p_j := \frac{[\tau(i,j)]^{\alpha_{AS}} \cdot [\eta(i,j)]^{\beta}}{\sum_j [\tau(i,j)]^{\alpha_{AS}} \cdot [\eta(i,j)]^{\beta}}$$

where i, j are the two connections defined by the row index of the matrix. We used the parameter $\alpha_{AS} := 1$ for all transition probability calculations as it was proposed that way in [31] and [35] to limit the hyperparameter search space. The parameter α_{ACS} used in ACS still needs to be optimised. We implemented four different distance functions (see Table 1) to improve the optimizers capability. Given their mathematical definition, i is the current node's index, j_1 is the index of the first connection of node i, and j_2 is the respective second connection. To prevent the distance function from differentiating between different inputs, all input nodes got the same index for calculation of the distance (the highest index of all input nodes). Ideally, we would find one distance function that is good or even optimal for all cases. However, we did not find one function to be ideal as our hyperparametertuning selected different functions regularly.

---

[2] We also looked at having only one pheromone matrix that combines optimizing the graph and the individual nodes. This version showed worse performance on all tested datasets and was therefore discarded.

After the pheromones of all connections have been updated, each ant iterates through the function pheromone matrix and assigns functions according to the probability $p_j := \frac{[\tau(j)]}{\sum_j [\tau(j)]}$, where j is the column index of the matrix. Here, no distance function is used because the probability to use a specific function should not be dependent on a randomly assigned index.

The update of the pheromone level after each iteration is done according to the function defined by each algorithm. The only difference is, that for AS, ACS, and MMAS $\Delta \tau_{ij}^k := \frac{1}{1+f(a_k)}$, where $f(a_k)$ is equal to the fitness of ant k. For the ASLBT pheromone update $\Delta \tau_{ij}^k := \frac{f_{best}(a_k)}{f(a_k)}$ is used, where $f_{best}(a_k)$ is equal to the best fitness ant k has had so far. In all cases, $\Delta \tau_{ij}^k := 0$, if the connection or function was not used by ant k.

The available computation node functions for all problems are: $\sin()$, $\cos()$, $\tan()$, $\tanh()$, ReLu, Sigmoid, $\exp()$, $\ln()$, $\mathrm{abs}()$, $*(-1)$, $+$, $-$, $*$, $/$. Please note that the first ten functions are of arity one while the last four have an arity of two. Therefore, the function matrix consists of 14 different rows.

5 Ant-Based Cartesian Genetic Programming

Our work focuses on analysing the impact on the performance of CGP when the ES is substituted with different ant-based metaheuristics. There is already some existing research in this area, which we will present in the following.

Hara et al. [15] introduced the idea of using an adapted Max-Min Ant System [31] for mutating the connections of CGP nodes and named their system *Cartesian Ant Programming* (CAP). However, contrary to our work, they used fixed alternating function genes that were not mutated during training. Also, in comparison to standard Max-Min Ant Systems and our work, the distance between the nodes was not taken into consideration for the transition rule of ants. This allowed them to use a single ant that walks through the graph moving from one output node to one of the input nodes. After that, the ant backtracks until it reaches a node that has unconnected inputs and repeats the aforementioned step of choosing a single next connection. This was repeated until a complete CGP graph is built.

Kushida et al. [23] extended the work of Hara et al. [15] by introducing a function to use the inter-node distance for the transition rule of the ants. They also tried to enhance CAP by dynamically assigning functions to the computational nodes. For this, they used two different pheromone tables, one for the connections and another one for the functions. They updated them with the same rule as Hara et al. [15] and only used functions with arity two. Their approach made the ant choose the two connections of each node independent of each other, which is their main difference compared to our work. This independence might lead to ants not choosing the best order of input nodes for the given function because functions like subtraction or division are not associative.

At last, Luis et al. [24] introduced a rank-based ant algorithm for evolving a CGP graph. For this, they effectively used three different pheromone tables.[3] One is for the first connection of each node, another is for the second connection of each node, and the last is the function of each node. They also omitted the idea of Kushida et al. [23] to use a distance function to make the ants explore shorter connections first. Their method lead to a higher diversity in the population and therefore better adaptation to a dynamic environment. However, it did not lead to a better fitness of the trained model than the approach of Hara et al. [15].

6 Experimental Setup and Evaluation

We logged the mean fitness of the population, the standard deviation of the fitness, the mean absolute error on the regression problems, the best fitness found until the given iteration, the best fitness of the current population, the active nodes, and the number of function evaluations executed so far. To approximate the convergence time of each algorithm, we used the mean number of function evaluations it took to find the best solution during training *mean(F2B)*.

Additionally, in order to compare the different configurations based on solid statistical statements, we ranked the algorithms according to their final fitness values on the test data. Throughout the benchmarks, the fitness is always positive, therefore, a *t-distribution* can not model the data well [22]. Hence, we performed a Bayesian data analysis for the posterior distributions of our results. The model to compare the algorithms is based on the *Plackett-Luce model* described by Calvo et al. [6].[4]

6.1 CGP Variants and Configurations

To allow for a fair comparison a broad set of configurations have to be evaluated in addition to the ant-based metaheuristics. CGP is mostly used in combination with a (1+4)-ES with neutral search (we call it STANDARD; Sect. 2). However, the authors Kaufmann and Kalkreuth [20,21] found that a different parametrization of the $(\mu + \lambda)$-ES helps CGP achieve its full potential. Thus, we include $(\mu + \lambda)$-ES into our experiments to *ensure a fair comparison*. To go one step further, we also examine the impact of a (μ, λ)-ES as this approach is also very close to the $(\mu + \lambda)$-ES. Furthermore, the following three replacement strategies are examined in conjunction with both ES:

- Neutral Search [28]
- Random Selection
- Fitness Uniform Selection Scheme (FUSS) [17]

---

[3] By contrast, we use two pheromone tables for the functions and the connections, respectively.
[4] We used the Python library *cmpbayes* https://github.com/dpaetzel/cmpbayes.

For STANDARD, $(\mu + \lambda)$-ES, and (μ, λ)-ES, we tested both Single and probabilistic mutation.

To find the best hyperparameters for each metaheuristic, we used a *Tree-structured Parzen Estimator* implemented in the Python library *optuna* [1]. All configurations were tested four times with independent train–test splits and randomly chosen seeds. After finding the best hyperparameters for a benchmark, each CGP variant was run for 10 times with independent seeds and train–test splits for our evaluation. Each algorithm has a different set of hyperparameters that needs to be optimized. These hyperparameters are shown in Table 2.

Table 2. Hyperparameters tuned for each CGP version.

CGP variant	Hyperparameters
STANDARD	#nodes, mutation type $\in \{Single, probability\}$
$(\mu + \lambda)$, (μ, λ)	#nodes, μ, λ, mutation probability p, elitist selection scheme, mutation type $\in \{Single, probability\}$
AS, ASLBT	#nodes, β, m, τ_0, ρ, distance function
ACS	#nodes, β, m, τ_0, ρ, distance function, α, q_0, global/local best
MMAS	#nodes, β, m, τ_{min}, τ_{max}, ρ, distance function, global/local best

Table 3. An overview of the symbolic regression benchmarks used for testing. $U[a, b, c]$ means that c random samples are drawn from a uniform distribution in the range $[a, b]$. $E[a, b, c]$ defines a grid from a to b with spacing c.

Name	Variables	Equation	Data Set
Koza–3	1	$x^6 - 2 \times x^4 + x^2$	$U[-1, 1, 20]$
Pagie–1	2	$\frac{1}{1-x^{-4}} + \frac{1}{1-y^{-4}}$	$E[-5, 5, 0.4]$
Nguyen–7	1	$ln(x+1) + ln(x^2 + 1)$	$U[0, 2, 20]$

To decrease the search space of ACS we use $q_0 := 0.9$ as it was proposed in [11]. For the Max-Min Ant System, we use the dynamic recalculation of τ_{min} and τ_{max} proposed by Stützle et al. [31] to simplify the search space even further.

6.2 Benchmarks

To evaluate our metaheuristics, 21 different symbolic regression, regression, and classification benchmarks were tested. We used the *symbolic regression* benchmarks Koza–3, Pagie–1 and Nguyen–7 [34] (cf. Table 3). Furthermore, we present results on the *regression* problems Forest Fires [8] and Wine Quality (White) [7]

148   J. Trautwein et al.

and the *classification* problems Adult [4] and Chronic Kidney Disease [30].[5] The input data of every regression and classification dataset was normalized by implementing Min-Max Scaling on the range [0, 1]. The outputs of the regression problems were standardized by using the Z-score.

Each dataset was split randomly using a *Monte-Carlo Cross Validation* with an 80%–20% train–test split for ten runs each. For the regression and symbolic regression problems, the *mean squared error* is used as the fitness function. The classification problems are optimised with $1 - |MCC|$ as fitness function, where MCC is the *Matthews Correlation Coefficient*.

With these settings each CGP variant is classified as *solved* once the fitness value reaches less than 0.0001. Furthermore, each ant-based algorithm is seen as *converged* once there has not been an increase in fitness over the last 500 iterations. All algorithms are given a maximum of 100,000 iterations to finish their optimisation process which should ensure that convergence is achieved for even the slowest algorithm. Note that the number of iterations has to be multiplied by the population sizes to arrive at the number of function evaluations if a comparison would be made on budgets.

6.3 Results

We will now discuss our results on all different CGP variations. As our benchmark featured 21 different datasets from different types of learning tasks which generated a lot of data, we made a pre-selection and present the most interesting results in Table 4.[6] This article should be seen as a comprehensive benchmark that focusses on more than one application niche. As all datasets led to the generally same outcomes and findings with regards to the performance of the different optimizers, we are confident that out selection is made on sound assumptions and does not take away from a fair comparison.[7] However, we want to stress that the datasets not featured in the table are still important for our evaluation and discussion points.

---

[5] Additionally, we tested on the following additional regression and classification datasets (which are all publicly available as part of the UCI repository): Abalone, Air Quality, Appliances Energy Prediction, Bike Sharing Dataset - (Day & Hour), California Housing, Wine Quality (Red); Apnea-ECG, Bach Chorales Harmony, Car Evaluation, Diabetes. Their respective results as well as our *source code* can be found at https://github.com/trautwju/ACM_CGP.

[6] We made our selection for Table 4 based on a number of points: First, we considered datasets/learning tasks frequently used in the evolutionary computation community and featured in GP publications, e.g. Koza–3. Then, we limited the number of classification and regression datasets to give an approximately even representation. Last, we selected datasets that fulfil the criteria above and showed similar outcomes to others of the respective groups.

[7] Again, we refer to our GitHub https://github.com/trautwju/ACM_CGP and the supplementary material for the results on the remaining datasets.

Table 4. Results on a representative subset of our benchmark. We report the mean number of function evaluations until the best solution was found *mean(F2B)*, the mean fitness and its standard deviation $mean \pm std(fit)$, total number of nodes #*nodes*, and the probability of one configuration being the best p_{best}.

	CGP Variant	mean(F2B)	mean ± std(fit)	#nodes	p_{best}
Koza–3	STANDARD	49,003	0.00 ± 0.00	50	0.311
	$(\mu + \lambda)$-ES	67,528	0.00 ± 0.00	300	0.273
	(μ, λ)-ES	1,394,913	0.00 ± 0.00	650	0.270
	AS	212	0.01 ± 0.00	400	0.011
	ACS	126	0.02 ± 0.04	200	0.036
	MMAS	10,486	0.00 ± 0.00	250	0.054
	ASLBT	162	0.00 ± 0.00	500	0.046
Nguyen–7	STANDARD	152,306	0.00 ± 0.00	300	0.189
	$(\mu + \lambda)$-ES	261,586	0.00 ± 0.00	450	0.356
	(μ, λ)-ES	73,027	0.00 ± 0.00	50	0.396
	AS	771	0.18 ± 0.00	100	0.027
	ACS	1,201	0.59 ± 0.37	400	0.008
	MMAS	5,570	0.27 ± 0.22	400	0.018
	ASLBT	367	0.63 ± 0.47	800	0.005
Pagie–1	STANDARD	324,146	0.00 ± 0.00	950	0.301
	$(\mu + \lambda)$-ES	2,019,017	0.00 ± 0.00	450	0.417
	(μ, λ)-ES	2,979,659	0.01 ± 0.02	800	0.221
	AS	8,874	0.50 ± 0.17	350	0.006
	ACS	317	0.37 ± 0.21	100	0.016
	MMAS	10,795	0.27 ± 0.06	350	0.025
	ASLBT	410	0.42 ± 0.16	550	0.015
Forest Fires	STANDARD	253,605	0.41 ± 0.42	50	0.089
	$(\mu + \lambda)$-ES	1,731,673	0.61 ± 0.90	800	0.055
	(μ, λ)-ES	1,512,850	0.06 ± 0.00	550	0.647
	AS	384	0.21 ± 0.00	100	0.14
	ACS	192	2.98 ± 0.04	660	0.075
	MMAS	10,648	0.44 ± 0.86	700	0.114
	ASLBT	451	1.18 ± 1.39	300	0.041
Wine Quality (White)	STANDARD	389,552	0.74 ± 0.03	800	0.233
	$(\mu + \lambda)$-ES	2,850,189	0.75 ± 0.17	300	0.218
	(μ, λ)-ES	3,419,420	0.70 ± 0.07	550	0.372
	AS	155	0.98 ± 0.00	500	0.089
	ACS	669	0.93 ± 0.06	450	0.052
	MMAS	6,024	1.00 ± 0.03	200	0.015
	ASLBT	1,125	1.00 ± 0.06	200	0.021
Chronic Kidney Dis.	STANDARD	98,856	0.06 ± 0.06	400	0.255
	$(\mu + \lambda)$-ES	604,656	0.05 ± 0.02	600	0.261
	(μ, λ)-ES	1,050,944	0.02 ± 0.02	200	0.406
	AS	506	0.55 ± 0.31	250	0.008
	ACS	23	1.00 ± 0.00	350	0.003
	MMAS	9,467	0.27 ± 0.05	50	0.044
	ASLBT	590	0.38 ± 0.12	300	0.022
Adult	STANDARD	342,400	0.22 ± 0.00	350	0.265
	$(\mu + \lambda)$-ES	3,192,479	0.22 ± 0.00	600	0.350
	(μ, λ)-ES	2,358,825	0.22 ± 0.01	300	0.302
	AS	3,250	0.51 ± 0.26	500	0.005
	ACS	1,596	0.25 ± 0.06	550	0.055
	MMAS	12,090	0.31 ± 0.11	100	0.008
	ASLBT	1,520	0.40 ± 0.27	300	0.015

Symbolic Regression. On the symbolic regression benchmarks, the ant-based metaheuristics perform worse on every problem other than Koza–3. On Koza–3, all CGP variations apart from AS and ACS are able to solve the problem, but STANDARD, $(\mu + \lambda)$-ES, and (μ, λ)-ES still outperform all other metaheuristics.

On the other symbolic regression problems all ant-based algorithms clearly show a worse performance than STANDARD, $(\mu + \lambda)$-ES, and (μ, λ)-ES. ACS shows a relatively high standard deviation of the fitness value and a very low F2B as a locally optimal ant is found early on but never really improved upon. This is due to using the locally or globally best ant to update the pheromone table which apparently restricts the search space too much.

Still, the high standard deviation shows that it is possible for the ants to find better solutions. A possible improvement here is to implement some sort of restart or pheromone smoothing algorithm to reset the search and help ACS escape a local fitness minimum. MMAS shows a high standard deviation on Nguyen–7 which is unexpected because setting τ_{min} and τ_{max} should help with exploring the search space and our tuning process did not find values for these constants that achieved better results.

Regression. On the forestfire dataset, ASLBT did converge fast, but mostly ended up with a bad fitness score, although some runs did find good solutions quickly. Another interesting result is that AS outperforms STANDARD on Forest Fires. However, the wine quality white dataset shows that the ant-based metaheuristics often times cannot compete with the already established algorithms.

Classification. The classification problems show a similar result. Here, the ant-based metaheuristics can compete with STANDARD, $(\mu + \lambda)$ and (μ, λ)-ES in some runs but mostly reach a worse fitness score. However, the ant-based metaheuristics did converge considerably faster even when finding good results.

General Discussion. The additional implementations of the modified replacement schemes showed worse or at best only similar result to the established neutral search. Therefore, it seems to not have any positive impact to include further options which confirms the long-standing practice of using neutral search in CGP. The (μ, λ)-ES showed a significantly larger $mean(F2B)$ than the STANDARD or $(\mu + \lambda)$-ES on all problems apart from Nguyen–7, but showed no significant improvement in fitness other than on Forest Fires. The $(\mu + \lambda)$-ES also had a larger $mean(F2B)$ than STANDARD without an improvement in fitness. Probabilistic mutation outperformed Single most of the times with a well chosen mutation probability.

The optimal hyperparameters for each ant-based metaheuristic seem to be very dependent on the given problem, therefore making it impossible to decide on one generalist hyperparameter configuration. On most problems, the use of the globally best ant to update the pheromone tables for ACS and MMAS showed significantly better performance than using the iteration's best. Therefore, it seems feasible to remove this hyperparameter from the search space for future tuning. Looking at the different distance functions, most commonly the constant distance lead to the best result, closely followed by the logarithmic distance. When the number of nodes are examined, we can not see any trends between the different CGP configurations. Similarly to the other hyperparameters, the required number of nodes are completely dependent on the given problem statement in combination with its respective CGP variant.

Despite the high variations of achieved fitnesses during testing, the ant-based metaheuristics showed small standard deviations during the training process. This comes from the fact that they assume that the optimal solution of a problem is very close to the locally best found solution so far. Furthermore, the biased exploration, which all ant-based algorithms use, increases the possibility of generating the same solution in different iterations because there is no guarantee an active node has been changed. Therefore, it could be argued that the optimal graph in CGP requires a lot more exploration of the search space.

Early during training, MMAS does seem to suffer less from a lack of exploration which is in line with the original proposal of the system being better at exploring the search space. This is the case because τ_{min} makes the ants choose non-optimal connections and functions more often. The lack of exploration in later iterations could be caused by our use of the pheromone update function, which might put too much emphasis on good solutions early on and therefore leading to a loss of exploration of other solutions.

Another interesting discovery is that the *intensive* tuning of the pheromone evaporation rate ρ did not always positively influence the exploration and exploitation balance of the search space. Instead, exploitation seems to be too high as all ant-based metaheuristics except MMAS converge within the first few hundred iterations. This hypothesis is supported by the high standard deviation of their fitness values: Given better exploration, ant-based CGP *should* be able to converge towards a much better mean fitness value and do so more consistently. However, we can assume that—judging from the very high *mean(F2B)*'s of the ES variants—they might very well still be faster than their competition.

As stated at the beginning of this section, we also performed extensive statistical testing to confirm the results. Even though the raw numbers were quite expressive we added the tests to adhere to good scientific practices and encourage future researchers to do the same, especially when the results are not this clear. What can be seen from the tests (p_{best} in Table 4) is that the probability of an ant-based CGP being the best according to fitness is almost zero. In fact, we reach typically suggested thresholds for automated decision making (e.g. at least 80% according to Benavoli et al. [5]) to eliminate ant-based metaheuristics from consideration as the best optimizer of a CGP graph. The ES-based CGP approaches show similar probabilities to each other and overall, there is no clear picture, but we can discern a small tendency towards using the (μ, λ)-ES with elitism. However, we want to stress that this is also by far the slowest approach we tested.

7 Conclusion

In this work, we investigated the effect different ant-based metaheuristics have on CGP performance. We compared these algorithms with a standard CGP variant and the commonly used $(\mu + \lambda)$-ES and (μ, λ)-ES.

In our testing, we found that the ant-based metaheuristics show no significant benefit with regards to achieved fitness. While previous work [15,23,24] shows similar results of the Max-Min ant system on chosen symbolic regression problems, we came to the conclusion that using ant-based metaheuristics most of the time leads to an overall worse result. Even though the introduction of different distance functions lead to significantly more active nodes over the CGP graph, this increase of active nodes does not have any positive impact on the fitness and is barely visible in the number of nodes needed for the optimal solution.

One redeeming quality of ant-based metaheuristics is their fast convergence by a factor of 100 or more compared to ES-based CGP approaches (which could also lead to premature convergence). Thus, given the same training budget for all configurations, more CGP graphs optimized by ant-based metaheuristics can be generated. Paired with the high standard deviation of their fitness values, ant-based CGP has a high possibility to generate a good solution with only a fraction of the computational power needed. We would recommend that the ant-based CGPs will be run multiple times with far fewer iterations. Besides multiple independent runs, a possible strategy is to implement a random restart scheme where the pheromone matrices and the random seed are reset after some set budget was used. If the current best solution is now placed into an archive, it is still available after training. We assume that some of the solutions in the archive will be far off the optimal one, similarly to what we did experience in some of our runs. Nonetheless, there should be solutions that do fit the learning task well and that all of this can be achieved with a number of function evaluations equal to the ESs'.

Interestingly, a premature convergence contradicts the claims of some ant-based metaheuristics. For example, the *Max-Min Ant System* introduces hyperparameters that should improve exploration. This, however, seems to not be enough to improve graphs defined by CGP. Similar notions of ants often stuck in local optima were put forth by Prakasam and Savarimuthu [29]. The problem of locality of the search performed could also be investigated more in-depth. This could provide more insights into the behaviour of ant-based CGP, and why it performs badly.

Therefore, for additional future works, some improvements for ant-based metaheuristics that were proposed for optimization tasks could be tested. Scaling the value of β down over the course of iterations was proposed by Kushida et al. [23]. This leads to ants exploring shorter connections early and being able to choose longer connections later in the training, which might improve search space exploration. Furthermore, algorithms that help ACS escape local minima—like *pheromone trail smoothing* [31] or *2 Phase reinitialisation* [2]—should be tested to analyse if ant-based metaheuristics can improve the performance of CGP. At last, a different pheromone update function could be used to stop ants from landing in a local fitness minimum early on. Another possibility is to introduce ideas from *Artificial Bee Colony* (ABC) [19]—or even completely substitute the ant-based metaheuristics. Because the population of ABC features specialised individuals with different tasks (individuals for exploration, exploitation, etc.), ABC may be better suited for optimizing graphs defined by CGP [3].

Overall, we find that our rigorous benchmarking rejects the use of ant-based metaheuristics in the same way we would use ESs in CGP. Still, there are clear paths to exploit the orders of magnitude faster convergence speeds towards achieving better results than previously possible. This becomes especially pronounced when computation budgets are limited.

References

1. Akiba, T., Sano, S., Yanase, T., Ohta, T., Koyama, M.: Optuna: a next-generation hyperparameter optimization framework. In: Proceedings of the 25th ACM SIGKDD International Conference on Knowledge Discovery and Data Mining (2019)
2. Altiparmak, F., Karaoglan, I.: A genetic ant colony optimization approach for concave cost transportation problems. In: 2007 IEEE Congress on Evolutionary Computation, pp. 1685–1692 (2007). https://doi.org/10.1109/CEC.2007.4424676
3. Baykasoğlu, A., Özbakir, L., Tapkan, P.: Artificial bee colony algorithm and its application to generalized assignment problem. In: Chan, F.T., Tiwari, M.K. (eds.) Swarm Intelligence, chap. 8. IntechOpen, Rijeka (2007). https://doi.org/10.5772/5101
4. Becker, B., Kohavi, R.: Adult. UCI Machine Learning Repository (1996). https://doi.org/10.24432/C5XW20
5. Benavoli, A., Corani, G., Demšar, J., Zaffalon, M.: Time for a change: a tutorial for comparing multiple classifiers through Bayesian analysis. J. Mach. Learn. Res. **18**(1), 2653–2688 (2017)
6. Calvo, B., Ceberio, J., Lozano, J.A.: Bayesian inference for algorithm ranking analysis. In: Proceedings of the Genetic and Evolutionary Computation Conference Companion. GECCO '18, pp. 324–325. Association for Computing Machinery, New York (2018). https://doi.org/10.1145/3205651.3205658
7. Cortez, P., Cerdeira, A., Almeida, F., Matos, T., Reis, J.: Wine Quality. UCI Machine Learning Repository (2009). https://doi.org/10.24432/C56S3T
8. Cortez, P., Morais, A.: Forest Fires. UCI Machine Learning Repository (2007). https://doi.org/10.24432/C5D88D
9. Cui, H., Heider, M., Hähner, J.: Positional bias does not influence cartesian genetic programming with crossover. In: Affenzeller, M., et al. (eds.) Parallel Problem Solving from Nature – PPSN XVIII, pp. 151–167. Springer, Cham (2024). https://doi.org/10.1007/978-3-031-70055-2_10
10. Cui, H., Pätzel, D., Margraf, A., Hähner, J.: Weighted mutation of connections to mitigate search space limitations in cartesian genetic programming. In: Proceedings of the 17th ACM/SIGEVO Conference on Foundations of Genetic Algorithms. FOGA '23, pp. 50–60. Association for Computing Machinery, New York (2023). https://doi.org/10.1145/3594805.3607130
11. Dorigo, M., Gambardella, L.: Ant colony system: a cooperative learning approach to the traveling salesman problem. IEEE Trans. Evol. Comput. **1**(1), 53–66 (1997). https://doi.org/10.1109/4235.585892

12. Dorigo, M., Maniezzo, V., Colorni, A.: Ant system: optimization by a colony of cooperating agents. IEEE Trans. Syst. Man Cybern. Part B (Cybern.) **26**(1), 29–41 (1996). https://doi.org/10.1109/3477.484436
13. Fang, W., Gu, M.: FMCGP: frameshift mutation cartesian genetic programming. Complex Intell. Syst. **7**(3), 1195–1206 (2021). https://doi.org/10.1007/s40747-020-00241-5
14. Goldman, B.W., Punch, W.F.: Reducing wasted evaluations in cartesian genetic programming. In: Krawiec, K., Moraglio, A., Hu, T., Etaner-Uyar, A.Ş, Hu, B. (eds.) Genetic Programming, pp. 61–72. Springer, Heidelberg (2013)
15. Hara, A., Watanabe, M., Takahama, T.: Cartesian ant programming. In: 2011 IEEE International Conference on Systems, Man, and Cybernetics, pp. 3161–3166 (2011). https://doi.org/10.1109/ICSMC.2011.6084146
16. Harding, S., Graziano, V., Leitner, J., Schmidhuber, J.: MT-CGP: mixed type cartesian genetic programming. In: Proceedings of the 14th Annual Conference on Genetic and Evolutionary Computation. GECCO '12, pp. 751–758. Association for Computing Machinery, New York (2012). https://doi.org/10.1145/2330163.2330268
17. Hutter, M., Legg, S.: Fitness uniform optimization. IEEE Trans. Evol. Comput. **10**(5), 568–589 (2006). https://doi.org/10.1109/TEVC.2005.863127
18. Kalkreuth, R.: Two new mutation techniques for cartesian genetic programming. In: Proceedings of the 11th International Joint Conference on Computational Intelligence, IJCCI 2019, pp. 82–92. SCITEPRESS - Science and Technology Publications, Lda, Setubal, PRT (2019). https://doi.org/10.5220/0008070100820092
19. Karaboga, D., Basturk, B.: A powerful and efficient algorithm for numerical function optimization: artificial bee colony (ABC) algorithm. J. Global Optim. **39**(3), 459–471 (2007). https://doi.org/10.1007/s10898-007-9149-x
20. Kaufmann, P., Kalkreuth, R.: An empirical study on the parametrization of cartesian genetic programming. In: Proceedings of the Genetic and Evolutionary Computation Conference Companion. GECCO '17, pp. 231–232. Association for Computing Machinery, New York (2017). https://doi.org/10.1145/3067695.3075980
21. Kaufmann, P., Kalkreuth, R.: On the parameterization of cartesian genetic programming. In: 2020 IEEE Congress on Evolutionary Computation (CEC), pp. 1–8 (2020). https://doi.org/10.1109/CEC48606.2020.9185492
22. Kruschke, J.K.: Bayesian estimation supersedes the t test. J. Exp. Psychol. Gen. **142**(2), 573–603 (2013). https://doi.org/10.1037/a0029146
23. Kushida, J.I., Hara, A., Takahama, T., Nagura, S.: Cartesian ant programming with transition rule considering internode distance. In: 2016 IEEE 9th International Workshop on Computational Intelligence and Applications (IWCIA), pp. 101–105 (2016). https://doi.org/10.1109/IWCIA.2016.7805756
24. Luis, S., dos Santos, M.V.: On the evolvability of a hybrid ant colony-cartesian genetic programming methodology. In: Krawiec, K., Moraglio, A., Hu, T., Etaner-Uyar, A.Ş, Hu, B. (eds.) EuroGP 2013. LNCS, vol. 7831, pp. 109–120. Springer, Heidelberg (2013). https://doi.org/10.1007/978-3-642-37207-0_10
25. Margraf, A., Stein, A., Engstler, L., Geinitz, S., Hahner, J.: An evolutionary learning approach to self-configuring image pipelines in the context of carbon fiber fault detection. In: 2017 16th IEEE International Conference on Machine Learning and Applications (ICMLA), pp. 147–154 (2017). https://doi.org/10.1109/ICMLA.2017.0-165
26. Miller, J.F.: An empirical study of the efficiency of learning Boolean functions using a cartesian genetic programming approach. In: Proceedings of the 1st Annual

Conference on Genetic and Evolutionary Computation. GECCO'99, vol. 2, pp. 1135–1142. Morgan Kaufmann Publishers Inc., San Francisco (1999)
27. Miller, J.F.: Cartesian genetic programming. In: Miller, J.F. (ed.) Cartesian Genetic Programming. Springer, Heidelberg (2011). https://doi.org/10.1007/978-3-642-17310-3_2
28. Miller, J.F.: Cartesian genetic programming: its status and future. Genet. Program Evol. Mach. **21**(1), 129–168 (2020)
29. Prakasam, A., Savarimuthu, N.: Metaheuristic algorithms and polynomial turing reductions: a case study based on ant colony optimization. In: Proceedings of the International Conference on Information and Communication Technologies, ICICT, vol. 46, pp. 388–395. Procedia Computer Science (2015). https://doi.org/10.1016/j.procs.2015.02.035
30. Rubini, L., Soundarapandian, P., Eswaran, P.: Chronic Kidney Disease. UCI Machine Learning Repository (2015). https://doi.org/10.24432/C5G020
31. Stützle, T., Hoos, H.H.: Max-min ant system. Future Gener. Comput. Syst. **16**(8), 889–914 (2000). https://doi.org/10.1016/S0167-739X(00)00043-1
32. Suganuma, M., Kobayashi, M., Shirakawa, S., Nagao, T.: Evolution of deep convolutional neural networks using cartesian genetic programming. Evol. Comput. **28**(1), 141–163 (2020). https://doi.org/10.1162/evco_a_00253
33. Turner, A.J., Miller, J.F.: Neutral genetic drift: an investigation using Cartesian Genetic Programming. Genet. Program Evol. Mach. **16**(4), 531–558 (2015). https://doi.org/10.1007/s10710-015-9244-6
34. White, D.R., et al.: Better GP benchmarks: community survey results and proposals. Genet. Program Evol. Mach. **14**(1), 3–29 (2013). https://doi.org/10.1007/s10710-012-9177-2
35. White, T., Kaegi, S., Oda, T.: Revisiting elitism in ant colony optimization. In: Cantú-Paz, E., et al. (eds.) GECCO 2003. LNCS, vol. 2723, pp. 122–133. Springer, Heidelberg (2003). https://doi.org/10.1007/3-540-45105-6_11

Designing Lookahead Relocation Rules for the Container Relocation Problem with Genetic Programming

Marko Đurasević[1]($\boxtimes$), Mateja Đumić[2], Francisco Javier Gil-Gala[3], and Domagoj Jakobović[1]

[1] Faculty of Electrical Engineering and Computing, University of Zagreb, 10000 Zagreb, Croatia
{marko.durasevic,domagoj.jakobovic}@fer.hr
[2] School of Applied Mathematics and Informatics, J. J. Strossmayer University of Osijek, 31000 Osijek, Croatia
mdjumic@mathos.hr
[3] Department of Computing, University of Oviedo, Gijón, Spain
giljavier@uniovi.es

Abstract. The container relocation problem is an important combinatorial optimisation problem commonly found in warehouses and container ports. The goal of this problem is to retrieve all of the containers from the yard with the fewest container relocations between the stacks. Since the problem is NP-hard, various heuristics have been proposed to solve it, among which are relocation rules (RRs), simple constructive heuristics that incrementally construct the solution. However, it is quite difficult to design such RRs manually, so genetic programming has been applied to design new RRs automatically. A significant problem with RRs, whether manually or automatically designed, is that they usually have a limited view of the problem. This means that they will often make decisions that can negatively influence the future, meaning that the current decision would cause additional relocations. Therefore, this study investigates different relocation schemes that can be used within RRs to obtain rules with lookahead ability. These rules will enable containers to be relocated based on future information and, consequently, arranged better in the yard. For that purpose, three novel relocation schemes for automatically designed RRs are defined and evaluated on an existing problem set. The results demonstrate that integrating additional elements to evolve lookahead RRs can significantly improve the results.

Keywords: Container relocation problem · Genetic programming · Relocation rules

1 Introduction

The container relocation problem (CRP) is a problem that, since its first appearance in 1988 [31], has attracted the interest of numerous researchers. Its importance has increased in recent years because it has one of the leading roles in

© The Author(s), under exclusive license to Springer Nature Switzerland AG 2025
B. Xue et al. (Eds.): EuroGP 2025, LNCS 15609, pp. 156–172, 2025.
https://doi.org/10.1007/978-3-031-89991-1_10

modern warehouse and yard management. The authors in [19] claim that most international trade is done by ships, and usually, the goods are stored and transported in containers. Containers with goods are placed in a stacking area while waiting for loading onto the ship. Because of the limited storage space, containers are placed side by side and on top of each other.

Loading containers onto the ship must be done in a predetermined order, and the container can be retrieved only if it is located on top of its stack. Otherwise, it is necessary to relocate all containers above it that are blocking it to other stacks. Although rearranging the containers placed in the yard in the correct order prior to their retrieval is a possibility, it is estimated that 30%–40% of the outbound containers and 85%–90% of the inbound containers at European terminals lack the correct information about ships and terminals [33]. Therefore, the order in which the containers need to be retrieved is usually not known in advance [19], which makes it impossible to arrange containers in a way that would allow to retrieve all containers in a predetermined order without the need for relocating any containers, as there would be no blocked containers in the yard.

Different CRP variants can be found in the literature, which can be distinguished by way of storing containers (single bay [3,6,17,22,34], or multi-bay CRP [7,25,37]), restriction on moves (restricted [5,6,8,10,13,27,36] or unrestricted [3,17,23,26]), container IDs (unique [20–23,28,34,36] or duplicate [3,11,17,18]), and whether all containers need to be retrieved or not [26]. In a single-bay CRP, all containers are stored in one bay, while in multi-bay, there is more than one bay. In restricted CRP, only the containers blocking the target container can be relocated, while in unrestricted CRP, it is possible to relocate the containers from the stack aside from the one in which the target container is located. The purpose of such moves is to prevent potential relocations that could happen in the future. Furthermore, it is possible that containers have distinct or identical IDs. In the former case, the order in which the containers need to be retrieved is unique, whereas in the latter variant, containers with the same ID can be retrieved in any order, thus providing additional flexibility of the problem. Aside from the previous standard problem variants, other variants like dynamic [21], online [41] or stochastic CRPs [2,24], have also been recently investigated. These variants deal with situations where not all container IDs are known up front, or where new containers are being released into the system during time. The abundance in the different problem variants clearly demonstrates the importance of the problem in the real world.

As a result, the main objective in CRP is to find a minimum sequence of relocations that allows for the retrieval of all containers from the yard. Since CRP and most of its extensions are NP-hard problems [1], there are no exact algorithms that can be used to solve all problem instances in a reasonable amount of time [22,37]. As a result, numerous heuristic methods have been developed for solving different CRP variants [7,19,28]. Unfortunately, all of these methods usually require a substantial amount of time to solve CRPs, especially larger instances.

As an alternative to exact methods and metaheuristics, relocation rules (RRs) represent a simple and efficient way to solve CRPs [5]. RRs are constructive heuristics that work well in environments where time is crucial or unforeseen changes can occur. They are made up of two parts, a relocation scheme and a priority function. The relocation scheme defines the outline of the RR, whose purpose is to determine which container needs to be relocated next and to ensure that no constraints are broken, like relocating a container to a stack with a maximum height. To determine the stack to which the blocking container should be relocated, the relocation scheme uses the priority function that ranks all the potential destination stacks, and selects the one with the best priority. Therefore, the RR uses the relocation scheme to determine which container needs to be relocated next, and the priority function determines the best destination for relocating the container.

Usually, RRs were designed manually by human experts, and utilised a certain strategy for choosing the destination stack for blocking containers [5,22,35]. For example, one simple strategy would be to relocate the blocking container to the lowest stack. However, such simple RRs usually achieve inferior performance, which is mainly due to the fact that manually designing good RRs is a difficult task. Therefore, recent studies examined the application of genetic programming (GP) for automated development of new RRs for various CRP variants [39,40,43]. With such an approach it is possible to design more sophisticated strategies to determine how to relocate the selected container. It is shown that automatically developed RRs can achieve better results than the human-made ones.

Regardless, even automatically designed RRs suffer from a serious issue. The issue is that these RRs, especially those developed for the restricted variant, only consider where the current blocking container should be placed. However, they do not consider the placement of other containers or the effects which future potential relocations could have. Thus, one could say that these RRs have a quite myopic view on the problem, which hinders them from obtaining better results. This provides motivation to examine the potential of evolving RRs with lookahead, by introducing different modifications in the relocation scheme, making it possible to use the knowledge about future relocations or situations that can happen in the near future.

In this study we consider the single bay CRP with unique container IDs and introduce three novel relocation schemes for automatically designed RRs. These relocation schemes introduce the lookahead ability into RRs in different ways. The RRs evolved by using these novel relocation schemes are compared with human-made RRs, and RRs automatically evolved with GP but using existing restricted and unrestricted relocation schemes. The experiments show that the proposed relocation schemes can improve results, meaning that introducing such a lookahead property into RRs represents a viable approach by which the effectiveness of RRs can be significantly improved.

The rest of the paper is organized as follows: Sect. 2 gives a literature overview, while Sect. 3 introduces the formulation of CRP used in this paper.

In Sect. 4, RRs are introduced, and the process of automatically designing RRs with GP is briefly explained, while the lookahead RRs are described in Sect. 5. In Sect. 6, an experimental setup is given, and the main results are outlined. Finally, Sect. 7 provides a short conclusion and introduces new research directions.

2 Literature Overview

The single bay CRP was introduced in [31], while in [1] it was shown that this problem is NP-complete. Many different versions of CRP were investigated throughout the years, and numerous problem-solving methods were proposed [26].

The most commonly investigated version of CRP is the one in which containers are stored in one bay (single bay CRP), each container has a unique ID, and only the containers blocking the target container can be relocated. This version of the problem is called the restricted CRP. Because of NP-completeness, metaheuristic and heuristic methods are mainly used for solving CRPs.

Exact methods used to solve CRP include branch and bound [17,22], A* [9,27,37], decision trees [10,36], methods based on dynamic programming model [24] and (mixed) integer programming models [26,34]. Most papers with exact approaches also propose some (meta)heuristic approaches, and experiments show that the results achieved with heuristic approaches are only slightly worse than the ones achieved with exact methods. At the same time, the computational time required by heuristics and metaheuristics is significantly smaller than the time required for exact methods. One example can be found in [22], where the heuristic method ENAR was proposed and compared with the branch-and-bound method. In [34], the reshuffle index RR was proposed and validated in extensive numerical experiments.

Multi-bay CRP was for the first time studied in [25], in which problems with containers with duplicate IDs and using a restricted relocation scheme were solved with a three-phase heuristic. In this three-phase heuristic, each phase is used for different things: the first is used to construct a feasible solution, the second is used to reduce the length of the constructed solution, and the third is used for optimising the crane operation time. The second study that investigates the multi-bay variant is [37]. In this paper, iterative deepening A* combined with new lower-bound measures and heuristics achieved results better than the ones achieved with branch and bound methods. Furthermore, the four RRs were proposed in this work for the restricted version of CRP and two for the unrestricted version. Another RR was proposed in [5], called the Min-Max relocation rule, and used for solving a single bay problem with unique IDs and a restricted version of the relocation scheme. The results show that this rule, despite its simplicity, was efficient for minimizing the number of relocations, and it could find high-quality solutions quickly.

The GRASP method was used to solve single bay [19,32] and multi-bay [7] CRP with distinct priorities and a restricted scheme. Some other metaheuristics used for solving single bay with distinct priorities and restricted version are tabu search [21], and genetic algorithms [28]. While till recently, studies in which the unrestricted version was used or containers can have the same priorities were rare, there are few of them published in recent years [3,17,23,40].

The first attempt of automatically evolving RRs was made in [13]. In this study, the authors defined the structure of the RR manually but used a genetic algorithm to learn certain parameters of the rule to optimise its behaviour. However, this approach was quite restrictive since it required that the priority function that was used to rank the potential destination stacks is defined manually. In 2022, the first paper [39] in which the RRs were automatically developed with genetic programming (GP) was published. In it RRs were evolved for a single bay problem with distinct priorities for both the restricted and unrestricted version of the relocation scheme. After this paper, GP was used for evolving relocation rules for different problem versions: multi-bay with unique and duplicate IDs [40], single bay with distinct priorities but with energy consumption as objective [43], or the online container relocation problem [41]. Aside from this, several studies also examined ways in which these automatically designed RRs can be improved, such as ensemble learning [38] or rollout algorithms [42].

3 Container Relocation Problem

This paper considers the single bay CRP, in which all containers have unique IDs and must be retrieved from the yard. This version of CRP can be defined as follows: there are N containers placed in S stacks and H tiers. The maximum height of each stack needs to be less or equal to H. The assumption is that all containers in the yard are the same size because containers with different sizes are usually stored separately. Each container has a unique retrieval priority denoted with a number, where a smaller number means the priority is larger. This number is also used to represent the ID of the container.

There are two allowed moves in the yard: retrieval and relocation. Retrieval consists of picking up the container and putting it on the truck by which it would be transferred to the ship. The container can be retrieved only if its ID is currently the smallest in the yard and it is located on the top of its stack. Otherwise, a relocation move needs to be made. Relocation moves consist of picking up the container from the top of some stack and relocating it to a stack where there is place (the height is less than the maximum height). The container with the smallest ID in the yard is the one that needs to be retrieved next and is called the target container. When relocating containers from one stack to another, the stack from which relocation is made is called the origin stack, and the one to which the container is relocating is called the destination stack. The goal is to retrieve all containers in the fewest number of relocations possible, i.e., while optimizing the number of relocation moves. The feasible solution to this problem is any sequence of retrieval and relocation operations by which all containers can be retrieved from the yard in a predetermined order. A mathematical model of this problem can be found in [39].

Generally, we distinguish two types of CRPs, restricted (RE) and unrestricted (UN). In the restricted version, only containers blocking the target container can be relocated, while in the unrestricted version, containers from all stacks can be relocated. The restricted one considers only the next move, while the unrestricted version tries to reduce the number of relocations by relocating containers from the destination stack in the case in which it will become blocked by relocation that will happen. The unrestricted version has a lookahead ability in some way so it will be considered together with three novel relocation schemes proposed in this paper. Both relocation schemes - RE and UN will be explained in the following sections.

Figure 1 shows an example of a container yard. The container yard consists of five stacks and four tiers. In this container yard, 16 containers are stored; the priorities of containers range from 1 to 16, meaning that the container with the number 1 is the target container and needs to be retrieved next. After that, we need to retrieve the container with the number 2, which is placed in stack 1, and blocked with containers with numbers 7, 11, and 13. To retrieve container 2, we need to relocate those three containers. In the restricted version, we can relocate only those three containers before retrieving container 2. One possible solution is to relocate all three containers to stack 3, which has three empty places. In the unrestricted version, relocating containers from the other stacks is also possible. For example, we can relocate the container with number 3 to stack 5 because it has a smaller ID than the container on top of stack 4. In this way, we can avoid additional relocation if we, for example, relocate container 11 from stack 1 on top of container 3 while it is on stack 2.

4 Designing Relocation Rules with Genetic Programming

4.1 Relocation Rules

Relocation rules (RRs) are simple and fast constructive heuristics. They consist of two parts - the relocation scheme (RS) and priority function (PF). RS is the general framework by which a solution is built starting from scratch and creating a sequence of retrieval and relocation operations by which all containers from the yard will be retrieved in a predetermined order. It starts by checking if the target container is on the top of its stack; if it is, it is retrieved immediately. Otherwise, the containers blocking it must be relocated on another stack. Where to relocate blocking containers is a crucial decision in this problem, and for this part the RS utilises the PF.

The PF assigns a priority to each possible destination stack, and the container is relocated to the one with the largest (or smallest) value based on the PF used. In the literature, different PFs can be found, which, in the beginning, were manually defined [5,25,37], while recently GP was used to automate this process [39,40]. An example of a manually defined PF is the lowest position (TLP) [25] priority function, which assigns to each stack the priority equal to the number of containers currently placed in that stack and chooses the stack with the smallest number of containers in it.

RS is usually manually defined and must ensure the sequence it constructs is feasible. The basic version of RS is where only containers from the stack where

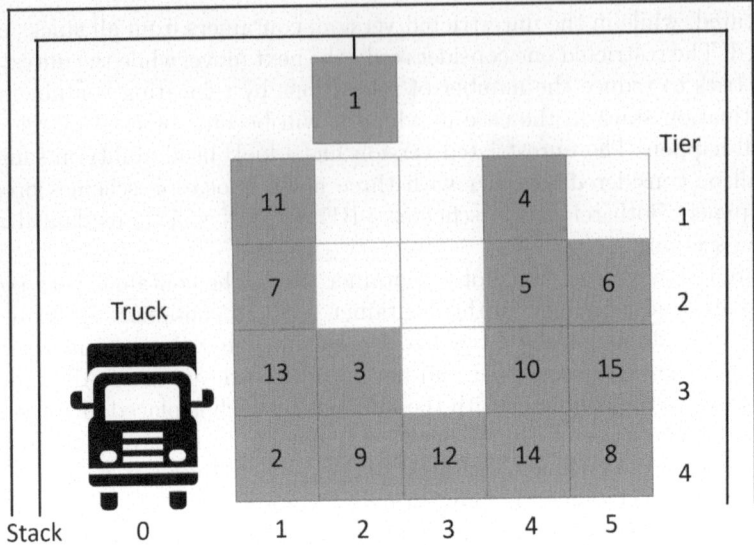

Fig. 1. Illustration of a simple container bay consisting of 5 stacks and 4 tiers

the target container is placed can be relocated. This RS is called a restricted relocation scheme, and its pseudocode is given by Algorithm 1. The CalculatePriority function from this algorithm uses a PF to determine the priorities of each stack on which container can be relocated.

Algorithm 1. Restricted relocation scheme
---
while *container yard not empty* **do**
    C = GetTargetContainer()
    S = GetStack(C)
    while *container C is not on top of stack S* **do**
        foreach *stack st, st != S and st not full* **do**
            | π_{st} = CalculatePriority(st);
        end
        Relocate top container from S to the stack with $min(\pi_{st})$
    end
    Retrieve container C
end

4.2 Designing Relocation Rules with Genetic Programming

The PF in RRs is usually represented as a mathematical function, and Genetic Programming (GP) can encode such an expression by using a tree-based representation of the solution. This is why GP is commonly used to evolve scheduling rules in different environments [4,12,14,15,29]. GP has been successfully used to evolve PFs for RRs for CRP in recent years [39,40,43].

When automatically evolving PFs, it is essential to define a proper set of terminals and functions that will be used as building blocks for PFs. Terminal nodes consist of the characteristics and properties of the yard and problem instances that can help decide which relocation move is the best. On the other hand, function nodes are mathematical functions that connect the terminals in some way. The terminals are used as leaf nodes in GP's individual and functions as non-leaf nodes. In a previous study [39], six terminals were detected as crucial for evolving good PFs: SH (height of the stack), EMP (number of available places in the stack), CUR (container ID of the container being relocated), RI (number of containers in the stack with a smaller ID than that of the selected container), AVG (average of container IDs in a stack) and DIFF (difference between smallest container ID in a stack with the ID of the container that is relocated). Furthermore, a previous study shows that addition, subtraction, multiplication, and protected division (return 1 if division by 0 occurs) are already enough for evolving good PFs [39].

5 Approaches for Evolving Lookahead Relocation Rules

A significant problem with RRs, regardless of whether manually or automatically designed PFs are used, is that they sometimes make decisions that cause additional relocations and negatively influence the objective function value. In this section, we will investigate four different modifications of the basic RE relocation scheme that will perform additional operations with the goal of reducing potential relocations in the future. Basically, the goal is to introduce a certain kind of lookahead into the RRs, which would allow them to perform moves that are beneficial for reducing the total number of relocations. For that purpose, one existing relocation scheme, called the unrestricted scheme, as well as three newly proposed schemes, the chain, well placed, and prerelocation schemes, will be used. All these modifications try to avoid additional relocations that can happen because of the relocation of containers blocking the target container on stacks that have one or more containers with smaller container IDs than the blocking container that would be moved.

The unrestricted (UN) relocation scheme, unlike the restricted RS, allows containers to be relocated from one or more stacks that do not contain the target container. The UN version of RS will check the IDs of containers from the stack to which the container should be relocated (destination stack). If any of the IDs are lower than the ID of the relocating container, it will try to relocate containers with a lower ID from the destination stack to another stack. However, the container from the destination stack will be relocated only if it does not block any other containers that are already on the stack on which it tries to relocate, i.e., it will relocate it only if there is a stack which contains containers with larger IDs than its ID.

The chain (CH) [20] relocation scheme does not consider only the next blocking container but also the one after this and tries to determine the best relocations for both. This is performed in a way that two scenarios are investigated. In the first scenario, the PF determines where the first blocking container will be moved. After the container is relocated, the PF is used to relocate the second blocking container. This is the same as how the standard RE scheme would perform the relocations. However, in the second scenario, the PF is first used to determine where the second blocking container would be moved and then where the first one would be moved. Thus, the second scenario makes the second blocking container the first one that will choose the destination stack. In the second scenario, it is important to take care of situations in which the second container would finish under the first one, which would be an illegal move. Because of this, in the second scenario, it is not possible to choose the same stack for both blocking containers. When both of these scenarios are considered, based on certain properties calculated from the resulting yard arrangement, it is determined which scenario should be executed. The second scenario will be used only if the following conditions are satisfied: the relocations in which the first and second containers are relocated such that the current yard's layout does not result in additional blocking of containers; relocation of the first container after the second one is relocated and relocation of the second one based on yard's layout in which first one is moved both do not cause additional blocking containers, or both cause additional blockages; and by relocating the second one first the difference between lowest container ID in the stack and the ID of relocated containers in stacks on which change happens will be smaller than the difference if the first one is relocated first.

When deciding where to relocate a container, the well-placed (WP) [9] relocation scheme first looks only at the stacks where the blocking container will be well placed. Being well placed means it will not block another container with a smaller ID. If there are such stacks, the priorities are calculated only for them and the one with the best priority is selected. Otherwise, the priorities are calculated for stacks that have empty places even though the container will not be well placed on them, and the one with the best priority values is selected, as in the original restricted RS.

The prerelocation (PR) [9] scheme, as the name says, will try to perform relocations before the blocking container relocates to the chosen destination stack. Therefore, before moving the blocking container, every other container on top of each stack will be considered to be relocated to the destination stack before relocating the blocking container. This candidate container will be moved to the destination stack only if it has a smaller ID than all the containers on the stack and a larger ID than the blocking container. Since there is a possibility that several such containers exist, the one with the smallest difference between its ID and the lowest container ID on the destination stack is relocated. This procedure is repeated as long as there are such containers, or until there is only one free place left on the destination stack. At that point the blocking container is relocated to the destination stack, placing it on top of all containers that were potentially relocated to that stack.

6 Experimental Study

6.1 Setup

To examine the effectiveness of the different RRs, the Caserta dataset was used [6]. This dataset consists of 840 instances containing between 3 and 10 stacks and 9 and 100 containers in the yard. The maximum allowed height that any stack can reach while the problem is solved equals $h + 2$, where h is the maximum height of the stacks at the start of the problem. The entire original Caserta dataset is used to evaluate GP's generalisation performance, meaning that this dataset is used to evaluate the solutions obtained by GP after the evolution process. However, an additional dataset has been generated and used by the GP to evaluate the solutions during the evolution. This dataset was generated the same way as the original Caserta dataset, with the same properties and number of instances.

To automatically generate RRs, we apply a standard GP algorithm [30]. In each iteration, this algorithm executes a 3-tournament selection operator. The better two individuals from the tournament are used in a crossover operator to produce a child individual, which is further mutated with a certain probability. This child individual is then placed into the population by replacing the worst individual in the tournament. This process is repeated until a predefined termination criterion is satisfied. The algorithm is executed with a population size of 1 000 individuals, a mutation probability of 0.3, a maximum tree depth of 5, and 50 000 function evaluations. Regarding the genetic operators, the algorithm uses the subtree, uniform, context preserving, size fair, one point crossover operators, and subtree, hoist, node complement, node replacement, permutation, and shrink mutation operators [30]. All parameters were fine tuned during a previous study [39].

During the evolution process of GP, each individual was evaluated on the training set, meaning the RR that the individual represents was used to solve all the instances in the training set. The individual's fitness is equal to the total number of container relocations that the RR performed on the training set. After GP finished with its execution, the best individual on the training set obtained in the final population was saved and evaluated on the test set to measure its generalisation performance. For each RR type under investigation, 30 independent executions of GP were performed to obtain an objective notion of the results. Thus, 30 RRs were obtained for each RS under consideration. Quantitive statistics (minimum, median, and maximum) are calculated for each method based on these 30 executions and are outlined in the next subsection. Furthermore, the Kruskal-Wallis test was performed to test whether the results obtained by the investigated methods are statistically significant, with a post hoc Dunn test and Bonferroni correction. The tests were performed with a significance value of 0.05. The tests were executed an a Windows 11 PC with an AMD Ryzen Threadripper

7980X 64-Cores and 256 GB of RAM. All the algorithms were implemented in C++ using the ECF framework [16].

To outline the effectiveness of automatically designed RRs, their performance will also be evaluated by comparing them with several manually designed RRs. These will include four restricted rules TLP [25], RI [25], Min-Max [5], PR3 [37], PR4 [37], two unrestricted rules PU1 and PU2 [37], and one chain rule CMM [20]. Since each of these RRs is defined in the literature by specifying both the RS and PF, they were used as they were defined originally, and were not investigated with different RS, although that would also be a possibility.

6.2 Results

Table 1 outlines the results obtained by RRs investigated in this study. The top 7 rows represent the results obtained by manually designed RRs. The table outlines each method's performance, together with the average size of RRs evolved by GP (denoted by the number of nodes in the expression) and the execution time of GP for the different variants (denoted in minutes).

First we can compare the manually and automatically designed RRs. From this we can clearly see that most of the evolved RRs achieve a better performance than any of the manually designed rules. The only exception is that the CH RR cannot always obtain results that are better than those of the manually designed RRs and on average it performs similarly as the PU2 rule. On the other hand, the PR RR can improve the performance over the best manually designed RR even by around 10%, whereas the othes provide an improvement of around 3%. This clearly demonstrates the limitations of the manually designed RRs that can be further improved by automatic development of PFs.

When comparing the automatically designed RRs between each other, we see that the ones using the CH scheme achieve the worst results, even compared to the basic RE scheme (by 2.6%). The other three schemes perform better than the RE scheme, with UN obtaining 1% better results, WP obtaining 0.6% better results, and PR obtaining 8% better results. Therefore, it is evident that the PR scheme achieves the overall best results among all the tested RSs. The difference between the results can even be better highlighted by Fig. 2, which outlines the results using a violin plot. This plot outlines that the results obtained using RE, UN, and WP are similarly distributed. However, the main difference can be observed between for the CH and PR RSs. For CH we observe that most of its solutions are worse than those obtained by the other RSs, whereas for the PR scheme we clearly see that the worst solutions obtained by it are still much better than the best ones obtained by any of the other RSs.

In order to determine whether the differences between the different RSs are significant, statistical tests have been performed. The results of these tests are outlined in Table 2. The results of these tests denote what was already evident from the previous results, i.e. that the CH RS achieve a significantly worse performance than any other RSs, that the PR RS achieves significantly better results than any other RS, and that the remaining RSs achieve the same performance.

Table 1. Results obtained by the different RR variants.

RR	Result				
TLP	35982				
RI	29524				
MM	28996				
CMM	28144				
PR4	25787				
PU1	25049				
PU2	24962				
	min	med	max	# nodes	Ex. time (min)
RE	23717	24297	24609	49	51
UN	23757	24053	24463	46	41
CH	24660	24953	25372	49	200
WP	23730	24145	24550	49	30
PR	**22172**	**22282**	**22382**	40	25

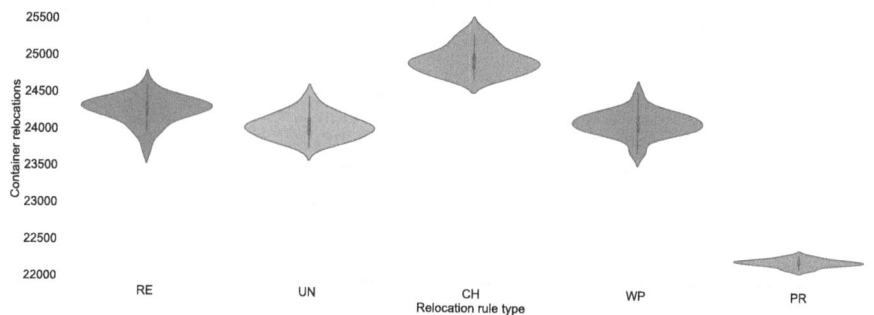

Fig. 2. Violin plots of the obtained results

Table 2. Results of the statistical tests for the tested RSs. The symbols $<$, $>$, and $\approx$ denote that the RS outlined in the row is significantly worse, better, or equal to the one denoted in the column, respectively.

	RE	UN	CH	WP	PR
RE	–	$\approx$	$>$	$\approx$	$<$
UN	$\approx$	–	$>$	$\approx$	$<$
CH	$<$	$<$	–	$<$	$<$
WP	$\approx$	$\approx$	$>$	–	$<$
PR	$>$	$>$	$>$	$>$	–

From the results it is surprising that the CH scheme did not improve the results over the basic RE scheme. However, this is likely due to the fact that for GP it is difficult to design a single PF that is used to determine the best relocation for both the first and the second decision. Furthermore, the same terminal nodes might not be suitable for performing both decisions. Thus, a potential resolution to this problem would be to use GP to evolve two PFs for each of the decisions. The UN and WP strategies achieve a similar performance, although UN achieves a slightly better result. The reason for this is that WP only restricts the moves to those that are potentially better (where the container would be well placed), but does not perform any additional relocations to improve the arrangement of the containers. Thus it restricts the RR to first perform better moves if possible. The UN scheme, on the other hand, allows additional relocations, which allow it to achieve slightly better solutions. Finally, the PR scheme likely achieves the best results since it allows other containers to be relocated to the destination stack to reduce potential future relocations. With such relocations it is possible that other containers that will soon have to be retrieved are unblocked.

Table 1 also outlines the average size of the generated PFs. We can see that by using the PR scheme the lowest expression size is obtained, consisting of an average of 40 nodes. The other schemes obtain solutions with a 15–20% more nodes in the expression. Aside from the complexity of expressions, it is also interesting to observe the duration of the training. In this case the CH scheme obtained the largest execution time by a wide range. This happens because it requires to simulate two situations, determine the better one, and executing it. This additional simulations introduce a significant overhead into the evaluation process of the solution. However, the other three schemes with unrestricted moves result in a shorter execution of GP compared to the basic RE scheme. Therefore, introducing unrestricted relocations does not increase the execution time of GP but rather improves it by around 20%–50%, depending on the method. This likely happens because the PF needs to be calculated for a lower number of relocations, which leads to a shorter execution time.

Figure 3 outlines the convergence of GP when evolving the different RR variants. The figure shows that the RE and CH RS start with the worst solutions, while WP and UN start with slightly better ones, and the PR starts with the best ones. This immediately suggests that the strategy incorporated into the RS is the best one, as random solutions generated initially will produce the best results. We also see that this RS has the fastest convergence among all the RSs since the convergence curve is quite steep at the start. On the other hand, CH, UN, and WP do not have such a steep learning curve. They converge quite quickly and improve only slightly after a certain point. This suggests that the PR RS also offers more potential for obtaining better solutions.

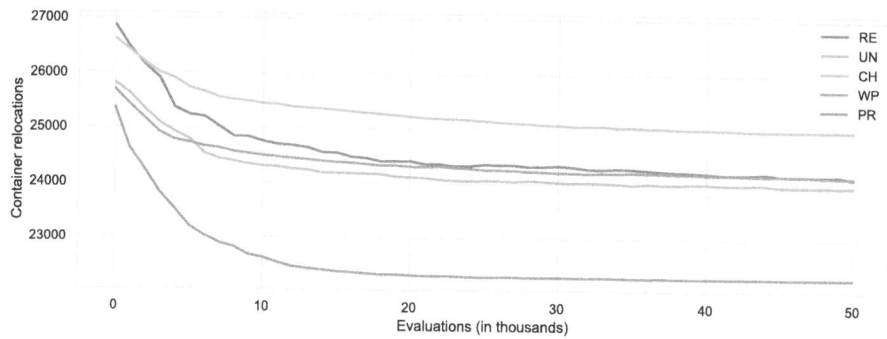

Fig. 3. Convergence plot of the methods on the test set.

7 Conclusion

This study investigated different RSs that can be used in the automated development of RRs to improve their performance. These RSs were designed by introducing lookahead into them to reduce potential future relocations when moving a container from one stack to another by considering which additional relocations could be performed. The results obtained in this study demonstrate that performing additional moves can be quite beneficial. Out of the four tested RS variants, the RS based on performing prerelocations achieved the best overall performance, significantly outperforming all other methods tested in the paper. This demonstrates that performing additional relocations, which consist of moving other containers to a better position in the yard before moving the selected container, represents the best strategy to decrease the overall number of relocations. Furthermore, this RS also resulted in the shortest training times and the least complex PFs, thus providing several benefits compared to other RSs.

The results obtained in this study open several potential future research directions. One is to continue investigating alternative relocation schemes since more sophisticated RSs seem to improve the results obtained by automatically generated RRs significantly. Another research direction would be to focus on alternative problem variants, such as dynamic, online, or stochastic container relocation problems, which introduce additional intricacies that make the problem even more challenging.

Acknowledgment. This work was supported by the Croatian Science Foundation under the project number IP-2022-10-5398, the European Union - NextGenerationEU under the grant NPOO.C3.2.R2-I1.06.0110., and the Spanish Government under projects MCINN-23-PID2022 and TED2021-131938B-I00.

References

1. Avriel, M., Penn, M., Shpirer, N.: Container ship stowage problem: complexity and connection to the coloring of circle graphs. Discret. Appl. Math. **103**(1–3), 271–279 (2000). https://doi.org/10.1016/s0166-218x(99)00245-0
2. Bacci, T., Mattia, S., Ventura, P.: The realization-independent reallocation heuristic for the stochastic container relocation problem. Soft. Comput. **27**, 4223–4233 (2023). https://doi.org/10.1007/s00500-022-07070-3
3. Boge, S., Knust, S.: The blocks relocation problem with item families minimizing the number of reshuffles. OR Spectrum **45**, 395–435 (2023). https://doi.org/10.1007/s00291-022-00703-x
4. Branke, J., Nguyen, S., Pickardt, C.W., Zhang, M.: Automated design of production scheduling heuristics: a review. IEEE Trans. Evol. Comput. **20**(1), 110–124 (2016). https://doi.org/10.1109/TEVC.2015.2429314
5. Caserta, M., Schwarze, S., Voß, S.: A mathematical formulation and complexity considerations for the blocks relocation problem. Eur. J. Oper. Res. **219**(1), 96–104 (2012). https://doi.org/10.1016/j.ejor.2011.12.039
6. Caserta, M., Voß, S., Sniedovich, M.: Applying the corridor method to a blocks relocation problem. OR Spectrum **33**(4), 915–929 (2011). https://doi.org/10.1007/s00291-009-0176-5
7. Cifuentes, C.D., Riff, M.C.: G-CREM: a GRASP approach to solve the container relocation problem for multibays. Appl. Soft Comput. **97**, 106721 (2020). https://doi.org/10.1016/j.asoc.2020.106721, https://www.sciencedirect.com/science/article/pii/S1568494620306591
8. Díaz, C., Riff, M.C.: New bounds for large container relocation instances using grasp. In: Proceedings - 2016 IEEE 28th International Conference on Tools with Artificial Intelligence, ICTAI 2016, pp. 343–349 (2017). https://doi.org/10.1109/ICTAI.2016.56
9. Expósito-Izquierdo, C., Melián-Batista, B., Marcos Moreno-Vega, J.: A domain-specific knowledge-based heuristic for the Blocks Relocation Problem. Adv. Eng. Inform. **28**(4), 327–343 (2014). https://doi.org/10.1016/j.aei.2014.03.003
10. Feng, Y., Song, D.P., Li, D., Zeng, Q.: The stochastic container relocation problem with flexible service policies. Transp. Res. Part B: Methodol. **141**, 116–163 (2020)
11. Forster, F., Bortfeldt, A.: A tree search procedure for the container relocation problem. Comput. Oper. Res. **39**(2), 299–309 (2012). https://doi.org/10.1016/j.cor.2011.04.004
12. Gil-Gala, F.J., Durasević, M., Sierra, M.R., Varela, R.: Evolving ensembles of heuristics for the travelling salesman problem. Nat. Comput. **22**(4), 671–684 (2023). https://doi.org/10.1007/s11047-023-09968-9
13. Hussein, M., Petering, M.E.: Genetic algorithm-based simulation optimization of stacking algorithms for yard cranes to reduce fuel consumption at seaport container transshipment terminals. In: 2012 IEEE Congress on Evolutionary Computation, CEC 2012, vol. 1, pp. 10–15 (2012). https://doi.org/10.1109/CEC.2012.6256471
14. Jacobsen-Grocott, J., Mei, Y., Chen, G., Zhang, M.: Evolving heuristics for dynamic vehicle routing with time windows using genetic programming. In: 2017 IEEE Congress on Evolutionary Computation (CEC), pp. 1948–1955 (2017). https://doi.org/10.1109/CEC.2017.7969539
15. Jakobović, D., Đurasević, M., Brkić, K., Fosin, J., Carić, T., Davidović, D.: Evolving dispatching rules for dynamic vehicle routing with genetic programming. Algorithms **16**(6), 285 (2023)

16. Jakobovic, D., Đurasević, M., Picek, S., Gašperov, B.: ECF: a C++ framework for evolutionary computation. SoftwareX **27**, 101640 (2024). https://doi.org/10.1016/j.softx.2024.101640
17. Jin, B., Tanaka, S.: An exact algorithm for the unrestricted container relocation problem with new lower bounds and dominance rules. Eur. J. Oper. Res. **304**(2), 494–514 (2023). https://doi.org/10.1016/j.ejor.2022.04.006, https://www.sciencedirect.com/science/article/pii/S0377221722002946
18. Jin, B., Zhu, W., Lim, A.: Solving the container relocation problem by an improved greedy look-ahead heuristic. Eur. J. Oper. Res. **240**(3), 837–847 (2014). https://doi.org/10.1016/j.ejor.2014.07.038
19. Jovanovic, R., Tanaka, S., Nishi, T., Voß, S.: A GRASP approach for solving the blocks relocation problem with Stowage Plan. Flex. Serv. Manuf. J. **31**(3), 702–729 (2018). https://doi.org/10.1007/s10696-018-9320-3
20. Jovanovic, R., Voß, S.: A chain heuristic for the blocks relocation problem. Comput. Ind. Eng. **75**(1), 79–86 (2014). https://doi.org/10.1016/j.cie.2014.06.010
21. Karpuzoğlu, O., Akyüz, M.H., Öncan, T.: A tabu search based heuristic approach for the dynamic container relocation problem. In: Doerner, K.F., Ljubic, I., Pflug, G., Tragler, G. (eds.) Operations Research Proceedings 2015. ORP, pp. 165–171. Springer, Cham (2017). https://doi.org/10.1007/978-3-319-42902-1_22
22. Kim, K.H., Hong, G.P.: A heuristic rule for relocating blocks. Comput. Oper. Res. **33**(4), 940–954 (2006). https://doi.org/10.1016/j.cor.2004.08.005
23. Kimms, A., Wilschewski, F.: A new modeling approach for the unrestricted block relocation problem (2023). https://doi.org/10.1007/s00291-023-00728-w
24. Ku, D., Arthanari, T.S.: Container relocation problem with time windows for container departure. Eur. J. Oper. Res. **252**(3), 1031–1039 (2016). https://doi.org/10.1016/j.ejor.2016.01.055
25. Lee, Y., Lee, Y.J.: A heuristic for retrieving containers from a yard. Comput. Oper. Res. **37**(6), 1139–1147 (2010). https://doi.org/10.1016/j.cor.2009.10.005
26. Lu, C., Zeng, B., Liu, S.: A study on the block relocation problem: lower bound derivations and strong formulations. IEEE Trans. Autom. Sci. Eng. 1–25 (2020). https://doi.org/10.1109/tase.2020.2979868
27. López-Plata, I., Expósito-Izquierdo, C., Moreno-Vega, J.M.: Minimizing the operating cost of block retrieval operations in stacking facilities. Comput. Ind. Eng. **136**, 436–452 (2019). https://doi.org/10.1016/j.cie.2019.07.045, https://www.sciencedirect.com/science/article/pii/S036083521930436X
28. Maglić, L., Gulić, M., Maglić, L.: Optimization of container relocation operations in port container terminals. Transport **35**(1), 37–47 (2019). https://doi.org/10.3846/transport.2019.11628
29. Nguyen, S., Mei, Y., Zhang, M.: Genetic programming for production scheduling: a survey with a unified framework. Complex Intell. Syst. **3**(1), 41–66 (2017). https://doi.org/10.1007/s40747-017-0036-x
30. Poli, R., Langdon, W.B., McPhee, N.F.: A field guide to genetic programming (2008). https://www.lulu.com/, http://www.gp-field-guide.org.uk. (With contributions by J. R. Koza)
31. Sculli, D., Hui, C.: Three dimensional stacking of containers. Omega **16**(6), 585–594 (1988). https://doi.org/10.1016/0305-0483(88)90032-1
32. da Silva Firmino, A., de Abreu Silva, R.M., Times, V.C.: A reactive GRASP metaheuristic for the container retrieval problem to reduce crane's working time. J. Heuristics **25**(2), 141–173 (2018). https://doi.org/10.1007/s10732-018-9390-0

33. Steenken, D., Voß, S., Stahlbock, R.: Container terminal operation and operations research - a classification and literature review. OR Spectrum **26**(1), 3–49 (2004). https://doi.org/10.1007/s00291-003-0157-z
34. Wan, Y.W., Liu, J., Tsai, P.C.: The assignment of storage locations to containers for a container stack. Nav. Res. Logist. **56**(8), 699–713 (2009). https://doi.org/10.1002/nav.20373
35. Wu, K.C., Ting, C.J.: A beam search algorithm for minimizing reshuffle operations at container yards. In: Proceedings of the International Conference on Logistics and Maritime Systems, pp. 703–710 (2010)
36. Zhang, R., Liu, S., Kopfer, H.: Tree search procedures for the blocks relocation problem with batch moves. Flex. Serv. Manuf. J. **28**(3), 397–424 (2015). https://doi.org/10.1007/s10696-015-9229-z
37. Zhu, W., Qin, H., Lim, A., Zhang, H.: Iterative deepening A* algorithms for the container relocation problem. IEEE Trans. Autom. Sci. Eng. **9**(4), 710–722 (2012). https://doi.org/10.1109/TASE.2012.2198642
38. Đurasević, M., Đumić, M., Gil-Gala, F.J.: Constructing ensembles of automatically designed relocation rules for the container relocation problem. In: 2024 IEEE Congress on Evolutionary Computation (CEC), pp. 1–8. IEEE (2024). https://doi.org/10.1109/cec60901.2024.10612112
39. Đurasević, M., Đumić, M.: Automated design of heuristics for the container relocation problem using genetic programming. Appl. Soft Comput. (2022). https://doi.org/10.1016/j.asoc.2022.109696
40. Đurasević, M., Đumić, M.: Designing relocation rules with genetic programming for the container relocation problem with multiple bays and container groups. Appl. Soft Comput. **150**, 111104 (2024). https://doi.org/10.1016/j.asoc.2023.111104
41. Đurasević, M., Đumić, M., Gil-Gala, F.J.: Designing relocation rules with genetic programming for the online container relocation problem. In: 2024 IEEE Congress on Evolutionary Computation (CEC), pp. 1–8. IEEE (2024). https://doi.org/10.1109/cec60901.2024.10611835
42. Đurasević, M., Đumić, M., Gil-Gala, F.J., Frid, N., Jakobović, D.: Improving the performance of relocation rules for the container relocation problem with the roll-out algorithm. In: Affenzeller, M., et al. (eds.) PPSN 2024. LNCS, vol. 15148, pp. 184–200. Springer, Cham (2024). https://doi.org/10.1007/978-3-031-70055-2_12
43. Đurasević, M., Đumić, M., Čorić, R., Gil-Gala, F.J.: Automated design of relocation rules for minimising energy consumption in the container relocation problem. Expert Syst. Appl. **237**, 121624 (2024). https://doi.org/10.1016/j.eswa.2023.121624

Evolved and Transparent Pipelines for Biomedical Image Classification

Camilo De La Torre[1,2](✉)[iD], Giorgia Nadizar[3][iD], Yuri Lavinas[1,2][iD], Robin Schwob[4][iD], Camille Franchet[5][iD], Hervé Luga[2,6][iD], Dennis Wilson[7][iD], and Sylvain Cussat-Blanc[1,2,8][iD]

[1] University Toulouse Capitole, Toulouse, France
camilo.de-la-torre-villacis@ut-capitole.fr
[2] IRIT - CNRS UMR5505, Toulouse, France
[3] University of Trieste, Trieste, Italy
[4] Centre Hospitalo-Universitaire de Toulouse, Toulouse, France
[5] Oncopole Claudius Regaud, Toulouse, France
[6] Université Toulouse Jean Jaures, Toulouse, France
[7] ISAE-SUPAERO, University of Toulouse, Toulouse, France
[8] Institut Universitaire de France, Paris, France

Abstract. This article presents an interpretable approach to binary image classification using Genetic Programming (GP), applied to the Patch-Camelyon (PCAM) dataset, which contains small tissue biopsy patches labeled as malignant or benign. While Deep Neural Networks (DNNs) achieve high performance in image classification, their opaque decision-making processes, prone to overfitting behavior and dependency on large amounts of annotated data limit their utility in critical fields like digital pathology, where interpretability is essential. To address this, we employ GP, specifically using the Multi-Modal Adaptive Graph Evolution (MAGE) framework, to evolve end-to-end image classification pipelines. We trained MAGE a hundred times with the best optimized key hyperparameters for this task. Among all MAGE models trained, the best one achieved 78% accuracy on the validation set and 76% accuracy on the test set. Among Convolutional Neural Networks (CNNs), our baseline, the best model obtained 84.5% accuracy on the validation set and 77.1% accuracy on the test set. Unlike CNNs, our GP approach enables program-level transparency, facilitating interpretability through example-based reasoning. By analyzing evolved programs with medical experts, we highlight the transparency of decision-making in MAGE pipelines, offering an interpretable alternative for medical image classification tasks where model interpretability is paramount.

Keywords: Image Classification · Genetic Programming · Interpretable Machine Learning

1 Introduction

Image classification has become an integral tool across various fields like robotics, transportation, security, and healthcare. Today, Deep Neural Network (DNN)

architectures dominate the field, owing to their ability to learn hierarchical feature representations directly from raw pixel data and achieve translation invariance [16,39]. However, despite these advances, Deep Learning (DL) methods are not without limitations. DNNs are difficult to interpret, and their decision-making process is blurred and opaque [4]. These models are typically "black boxes", offering little insight into how decisions are made, which can undermine trust in their results. Additionally, DNNs often require large labeled datasets and extensive computational resources to achieve competitive performance [35], making them challenging to deploy in scenarios with limited data [24]. Trained networks are also susceptible to data drifts and are not robust against adversarial data [12].

Certain fields, such as digital pathology, are considered critical due to their direct impact on real-life outcomes. In these areas, there is a strong demand for models that deliver accurate predictions and are fully transparent and interpretable [4]. Consequently, alternative Machine Learning (ML) approaches that prioritize interpretability and robustness are gaining increasing attention [2,31,32].

One such alternative is Genetic Programming (GP), a form of evolutionary computation that evolves programs to solve complex tasks [15]. GP leverages principles of natural selection, where populations of programs are evolved over generations to optimize a given objective. GP has been applied to image classification in various ways, from classifying images based on manual feature extraction data [29] to serving as a feature extractor for non-GP classifiers [5,11]. In addition, GP has also been used for end-to-end pipelines where both feature extraction and classification are automated, eliminating the need for manual feature engineering [3,7]. Finally, there is some emerging literature that explores the intrinsic interpretability of GP algorithms [22].

Given the demand for both performance and interpretability in critical domains such as medical image classification, and GP's potential to create interpretable image transformation pipelines, we investigate Multi-Modal Adaptive Graph Evolution (MAGE) [9,10], an extension of Cartesian Genetic Programming (CGP) [23] developed specifically to handle multiple data types. MAGE can evolve end-to-end image classification pipelines to perform both feature extraction and classification within a single evolutionary process. In this work, we show that MAGE achieves performance levels comparable to standard Convolutional Neural Networks (CNNs) while remaining fully interpretable by medical experts. Leveraging the "white-box" nature of CGP, medical professionals are able to assess the strengths and limitations of the best solutions, allowing them to delineate their coherence, robustness, and potential applicability.

In the following section, we review the application of GP in image classification, particularly its potential for creating end-to-end image processing pipelines. We also discuss successful classification methods applied to the PatchCamelyon (PCAM) dataset. We then detail our use of the MAGE framework for image classification, focusing on our choices for output representation and fitness function. This is followed by a description of the hyperparameter tuning performed for

MAGE and the CNN benchmarks conducted for comparison. Next, we present our results for both MAGE and CNNs, demonstrating MAGE's competitive performance. We also provide a medically guided interpretation of a pipeline's prediction to highlight the advantages of a fully decomposable and transparent approach. Finally, we discuss our findings and offer concluding remarks.

2 Related Work

2.1 GP and Image Processing

In recent years, the field of image classification has seen significant advances, especially with the use of DNNs, and specialized architectures well suited for classification [30]. However, the complexity and "black-box" nature of DNNs, as well as their need for large training data, motivate the study of alternative approaches [4,21,31]. One such approach is the use of GP techniques, which have shown promise in various image-processing tasks [14] and interpretability [22,27].

GP is a subfield of evolutionary computation in which populations of computer programs are evolved to solve complex problems by mimicking the principles of natural selection and genetic evolution [15]. GP encodes potential program solutions, which are iteratively refined over generations through genetic operations such as mutation, crossover, and selection. GP has been successfully applied in various stages of image classification tasks, offering flexibility in both feature extraction and classification processes. Initially, GP was employed to classify images based on manually extracted features, where it functioned as a classifier utilizing predefined input features to distinguish between image classes [29]. GP has also been utilized as a feature extraction method, generating meaningful features that are subsequently used by non-GP classifiers, such as support vector machines or decision trees, to complete the classification task [1,5,11]. GP has also been used as an end-to-end image classification pipeline, where both feature extraction and classification are automated within a single evolutionary process. In [3,7] GP automatically discovers feature representations while simultaneously evolving the final stage classifier responsible for making predictions. However, in both of these works, programs follow a fixed and rigid grammatical structure (filtering, aggregation, classification) instead of a dynamic one as in MAGE.

Among the different implementations of GP, CGP is notable for representing programs as a directed acyclic graph of operations, using a grid of nodes rather than the traditional tree-based representation found in standard GP [23]. In CGP, the nodes represent computational functions, and the connections between them define the program flow. Harding et al. [13] showed the versatility and generalizability of CGP for several use cases such as image denoising, cell counting and object detection. Leitner et al. [17] used CGP with high level functions to segment and classify rocks from Mars terrain. More recently, CGP has been used for genetic improvement of pre-defined programs for image segmentation [6]. Lastly, CGP achieved close to State-of-the-art (SOTA) performance at cell instance segmentation, by using computer vision operations and requiring little training data [8]. MAGE [9,10], a typed extension of CGP (akin to Strongly-typed GP [25] and Push GP [33]), handles natively data and operations of different types, enabling the efficient search of "image-to-class" pipelines. This recent advancement enables the application of CGP to a broader range of standard ML tasks, including classification, where multiple data types are involved.

2.2 The PCAM Dataset

We test our approach on the PCAM dataset [38], a balanced set of images derived from lymph node whole-slide images (WSI), labeled as either malignant or benign. Classification of the images in this dataset is challenging due to high variability, staining differences, and the limited information contained within small image patches.

The PCAM dataset is derived from the Camelyon16 (CAM16) dataset [19], a benchmark in the field of computational pathology, aimed at detecting metastases in WSIs of lymph node sections. The CAM16 dataset includes over 400 WSIs labeled for the presence or absence of metastases. The challenge associated with CAM16 lies in the different magnification levels and complex nature of histopathological images, which exhibit high heterogeneity in tissue types, staining protocols, and scanning devices [34].

The PCAM dataset used here consists of a balanced subset of color images (96×96 pixels) which are labeled in a binary fashion into malignant and benign patches [38]. A total of 327,680 images are available, 80 % are reserved for training (262 144 images), 10 % for validation, and 10 % for testing (32 768 images each). There is no overlap in WSI between these different splits. To generate the patches, PCAM follows the same splits as in CAM16, and WSI are randomly selected so as to enforce label balance across patches. A patch is labeled as a positive instance (malignant) if the center 32×32 pixel region contains, at least, one positively labeled pixel in the CAM16 WSI. According to PCAM's documentation[1], the exterior of the center region is independent from the center's label. That is, the presence (or not) of malignancy in the exterior of the center region should not influence the patch's label. However, this padding is preserved for convenience, in order to allow for the use of some convolutional layers that do not use *zero-padding*.

On this dataset, particular CNN architectures have shown remarkable performance. [38] proposed Group Equivariant Dense Networks in order to exploit the fact that histopathological images are symmetric under rotation, reflection, and translation. The model they proposed (P4M-DenseNet), achieved an accuracy score of 89.8 % on the PCAM test set. Using a Distance Metric Learning-based model, [37] proposed Proxy Neighborhood Component Analysis (ProxyNCA) with a Resnet34 Backbone initially trained on a weakly labeled dataset in digital pathology and then fully trained on the labeled PCAM dataset and achieved 90.47 ($\pm$ 0.59) % accuracy on the PCAM test set. Another work by the same

---

[1] https://github.com/basveeling/pcam.

authors [36] combined a Cross Entropy loss, a Contrastive loss and a Self Perturbation loss as the target of a standard Resnet34 model and achieved 90.36 (± 0.41) % accuracy on the PCAM test set.

3 Methods

In this section, we detail our approach to end-to-end image classification using GP, focusing on four main aspects: the choice of our output representation, the architecture of our multi-type MAGE algorithm, the Genetic Algorithm (GA) we opted for to account for dataset size and heterogeneity, and, lastly, the definition of our fitness function.

Most studies that explore end-to-end image classification pipelines using GP use a single output per individual and apply a fixed threshold to differentiate between the two classes [14]. In the present work, we opt for a dual-output approach, where each individual has two programs and each program produces one output.

In an end-to-end image classification pipeline, we expect the system to transform raw images, extract meaningful features, and utilize these characteristics to make accurate classification decisions. To evolve such programs, the operations within them must integrate and effectively combine different data types. To facilitate this search, we employed a type-aware GP approach, ensuring that only type-safe mutations are generated (i.e., mutations where the inputs align with the function's signature) [10]. Specifically, we used an extension of CGP called MAGE to automatically evolve image classification pipelines [10]. For the present use-case of image processing, MAGE genomes are structured into two distinct chromosomes: one dedicated to producing image outputs and the other to generating floating-point outputs. Certain functions perform image-to-image transformations, while others handle scalar-to-scalar operations. Type integration is facilitated by functions that allow floating-point values to parameterize image transformations, as well as those that convert image data into floating-point representations. As stated before, each evolved program generates two outputs: the final scalar node provides support for the negative class (benign), while the second-to-last scalar node supports the positive class (malignant). Label assignment is made according to the maximum value between the two outputs.

Usually, CGP and its extensions are optimized using the $(1+\lambda)$ Evolutionary Algorithm (EA). However, the size of the dataset compelled us to use batches of samples to assess the quality of each individual at each iteration. For generalization across iterations, we opted for a more classical GA which retains a portion of its individuals as parents for the next iteration. As such, at each iteration, the best e individuals are kept as the elite, and for the next iteration, each offspring will be a mutated copy of its parent, which is chosen with tournament selection between the previously selected elite individuals. As a final note, at each iteration, the best individual in the elite population, according to training metrics, is evaluated against the whole validation dataset for metric tracking.

One advantage of EAs over gradient optimization methods is their ability to directly optimize for the user-desired metric and, optionally, also through a smoother approximate of it. We saw performance improvements when minimizing the error rate ($Err = 1 - Acc$) in addition to a smoother metric like the Negative Log-Likelihood (NLL). Accuracy is defined as $Acc = \frac{TP+TN}{TP+FP+TN+FN}$ where TP,FP,TN,FN denote the number of true positives, false positives, true negatives and false negatives respectively. Since the PCAM dataset is balanced, we argue using error-rate is reasonable. The NLL is a common metric in binary classification defined as $NLL = -\sum_{i=1}^{N} y_i \log(\hat{y}_i) + (1-y_i)\log(1-\hat{y}_i)$ where for each instance i, y_i refers to ground truth classes (i.e., $\{0,1\}$) and $\hat{y}_i$ denotes the *softmax* score for the positive class. The range of NLL for a single instance i is $[0; \inf[$, to avoid over penalization for very wrongly predicted instances, we capped the NLL loss at 1 for each y_i : $NLL_{ceil} = \min\{NLL, 1\}$. As such, the fitness (or loss $\mathcal{L}$ since we want to minimize it) for each MAGE individual is calculated as $\mathcal{L} = \rho Err + (1-\rho) NLL_{ceil}$ where ρ is the coefficient that regulates the accuracy importance in the affine combination.

4 Experimental Setup

In this work, we explore end-to-end binary image classification using GP, specifically the MAGE framework. Given the dataset's heterogeneity, size, and complexity, we ask: can MAGE be used to construct interpretable classification pipelines without sacrificing substantial performance? To investigate this question, we start by performing hyperparameter tuning with Irace [20], then leverage the optimized parameters to perform a comprehensive comparison against CNNs. Following this, we collaborate with medical experts to interpret one of the most performant solutions. Code for this work is available here[2].

4.1 Hyperparameter Tuning of MAGE

Due to the use of batches in each iteration and the employment of a GA rather than a $(1+\lambda)$ GA, we had numerous hyperparameters to fine-tune. To efficiently handle this tuning process, we used the Irace [20] library. Table 1 shows the hyperparameters that we tuned using Irace.

For MAGE, we tuned the μ parameter to target between 1 and 4 active nodes for mutation, as well as the parameter n, the genome length, which ranged from 10 to 60 in increments of 10. Notably, for MAGE, genome length refers to the length per chromosome. The key GA parameters included the size of the truncated elite population e, which could vary between 2 and 100 individuals. This elite population was carried over from one iteration to the next. From this population, new individuals were generated through tournament selection of size t, with $t \leq e$. The number of offspring per iteration, new, was also tuned, ranging from 2 to 1000. Additionally, the loss parameter ρ, that balances accuracy and new, was tuned between 0 and 1. The batch size, textitbs, ranged from 5 to 2000. To ensure robust fitness statistics for each individual, from one generation

---

[2] https://github.com/camilodlt/evolved_pipelines_imagecls_2024.git.

Table 1. Hyperparameters for CGP, GA, and Batch Processing. Irace samples values from these ranges, with the condition that t is less than or equal to e.

Category	Parameters	Type	Values
CGP	Mutation rate (μ)	Ordinal	(1,2,3,4)
	Number of nodes (n)	Ordinal	(10, 20, 30, 40, 50, 60)
GA	Elite size (e)	Integer	[2,100]
	New individuals (new)	Integer	[2, 1000]
	Tournament size (t)	Ordinal	(2, 3, 5, 10, 15, 20)
	ρ	Real	[0, 1]
Batch Processing	Samples batch (bs)	Integer	[5, 2000]
	Sampling repetitions ($reps$)	Integer	[2,100]

to the next, multiple batches were used per iteration ($reps$ parameter), allowing us to compute either the median or mean fitness per individual across batches.

Lastly, we used "elitist-Irace"[3] with a budget of 300 experiments, two runs between statistical tests, and otherwise default parameters. Each run had a budget of 30M evaluations. The bootstrap configuration used was: $\mu = 1$, $n = 10$, $e = 2$, $new = 10$, $t = 2$, $bs = 200$, $reps = 5$ and $\rho = 0.5$.

4.2 CNN Baselines

In the following section, we compare MAGE against two CNN architectures. First, a Vanilla CNN network with two convolutional layers (with 16 and 32 output channels respectively), max-pooling, and two fully connected layers (with size 128 and 2 respectively) was used. To avoid overfitting, a dropout rate of 0.5 was used in the second to last layer. Second, a standard Resnet18 with approximately 12M parameters. For both networks, data augmentation is performed on the train split, using random horizontal flip and random vertical flip. For all splits, each image is normalized with means and standard deviations per channel calculated on the train split. The Adam optimizer was used for both with a learning rate of $1e^{-5}$.

We used different image channels for this experiment: Red, Green, Blue (RGB), Grayscale (GRAY), Hue, Saturation, Value (HSV), and Haematoxylin, Eosin, DAB (HED). HED channels, in particular, were included based on recommendations from histopathology experts who frequently use them in practice, indicating they may provide valuable information for the model. For the remainder of this work, channels are grouped like this for convenience: $Standard^* := $ RGB + HSV, $Standard^\dagger := Standard^* \cup$ GRAY, $All^* := Standard^* \cup$ HED and $All^\dagger := Standard^\dagger \cup$ HED. For MAGE we used $Standard^\dagger$ for hyperparameter

---

[3] https://cran.r-project.org/web/packages/irace/vignettes/irace-package.pdf.

tuning and *Standard*[†] and *All*[†] for benchmarking. We also trained the CNN baseline models using various color space combinations: GRAY, RGB, *Standard*[*], and *All*[*], with results reported for each setup.

5 Results

5.1 Hyperparameter Tuning

The Irace algorithm tested a total of 68 different configurations. Figure 1 illustrates the parameter combinations explored. Throughout the iterative process, Irace retained only the configurations that could not be statistically outperformed, progressively refining the search by favoring parameter distributions centered around these successful combinations. A consistent pattern emerged, with mutation rates of 1 or 2 proving reliable. Small genome sizes were particularly effective, with the smallest genome size (10 nodes) being predominantly favored. The number of elite individuals and offspring showed variability, though lower values for both parameters were generally preferred. Tournament sizes of 2 or 3 were consistently selected across Irace's iterations.

Given the limited evaluation budget for each run, robust selection in the face of batch variability required opting for large sample sizes with fewer repetitions or smaller samples with more repetitions. Irace predominantly exploited the latter strategy, choosing moderate batch sizes and evaluating individuals over multiple batches within a single iteration. Lastly, ρ values greater than 0 were consistently favored, indicating a tendency towards configurations that balanced optimization towards both NLL and accuracy. However, since ρ values mostly exceeded 0.5, there is a trend toward prioritizing accuracy as the primary optimization target.

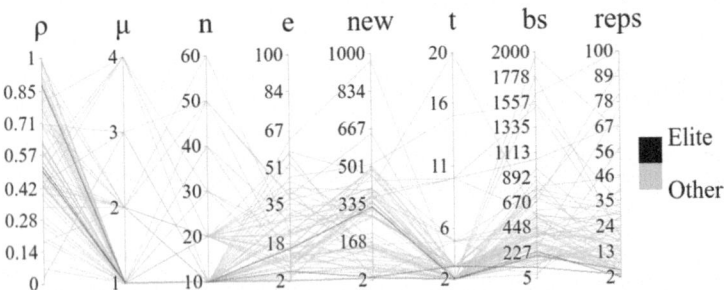

Fig. 1. All configurations tried by Irace. A line represents a combination of parameters. Black lines denote the three best combinations overall.

Figure 2 presents the final scores for all configurations, that is, the best validation accuracy achieved during each run. In this box plot, values are negative, as Irace minimizes the objective. The configurations are sorted by their average performance, with the best-performing configurations on the right and the worst on the left. Figure 2 shows that parameter selection has a significant impact on the final performance, with some configurations performing noticeably worse than others. Moreover, nearly all configurations exhibit substantial variability in performance. This suggests that multiple runs with the same parameter combination may be required to consistently achieve a highly competitive solution.

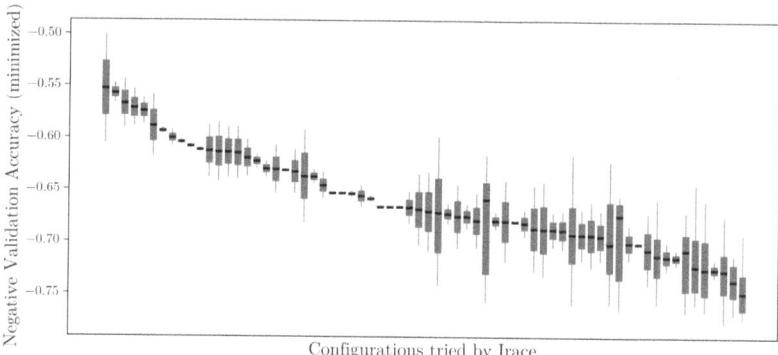

Fig. 2. Validation accuracy box plots per Irace configuration tested. Parameter configurations are sorted from worst to best on the horizontal axis.

The top three configurations, displayed in Table 2, are remarkably similar. These configurations share several common characteristics, including a mutation rate of a single active node, a small genome, a small tournament, and a low number of sampling repetitions (2, 3, or 5). The key differences among them lie in: the number of elite individuals (ranging from 2 to 16), the number of offspring (from 10 to 321), the size of each sampled batch (97 to 237 images), and the ρ value, which varied between 0.5 and 0.87.

These results show that multiple parameter combinations can achieve competitive performance. It is important to note, however, that the combinations found are influenced by the bootstrap parameter combination used and the evaluation budget per run, with more promising but evolutionary expensive configurations potentially requiring a higher budget. Further analysis of the tuning process is included in the supplementary materials.

Table 2. Best parameters combinations found and their mean accuracy score.

μ	n	e	new	t	bs	$reps$	ρ	Avg. Val. Acc. (Std.)
1	10	6	14	3	97	3	0.5188	**0.746** (0.02)
1	10	16	321	2	237	2	0.8759	0.73 (0.03)
1	10	2	10	2	200	5	0.5	0.736 (0.03)

5.2 Benchmarking Against CNNs

We conducted 100 runs with the best hyperparameter configuration found by Irace. Given the stochastic nature of each run, various image classification pipelines were generated. As illustrated in Fig. 2, there is notable variability in validation accuracy, even among the top-performing configurations. This variability emphasizes the necessity to conduct multiple independent runs to ensure the identification of high-quality solutions. Furthermore, the stochastic nature of the optimization process can be regarded as a beneficial characteristic. This is similar to how Quality-Diversity (QD) can be used to provide effective yet diverse solutions [26]. This diversity provides decision-makers with a range of high-performing pipelines to choose from which enables them to select classifiers based on specific criteria or preferences such as interpretability, simplicity or inference speed. Although interpretability can be considered a subjective criterion [28], diversity in interpretative classifiers is essential to build trust in algorithmic decisions.

To benchmark our method, we trained, from scratch, all CNN models under four different input sets, each for a maximum of 100 epochs: GRAY, RGB, $Standard^*$, and All^*. Table 3 reports the training and test accuracy of neural network weights which yield the highest validation accuracy for each input set. Table 3 also shows at which epoch (i.e., after how many processed images) did the model find these optimal weights. As for MAGE, for each run, the final solution is the pipeline with the highest validation accuracy across all iterations. In Table 3, for each input set, we show MAGE's best and average performance across the hundred runs. For the channel combination $All^\dagger$ we also show the performance of the most interpretable pipeline out of the top three.

On average, MAGE achieved a validation accuracy score of 74.2% with $Standard^\dagger$ and 74.8% with $All^\dagger$ inputs. Over all MAGE runs, the solution with the highest validation accuracy (MAGE^{val*}) reached 78% using $All^\dagger$ inputs. This model, MAGE^{val*}, achieved 76 % accuracy on the test set. Finally, the solution MAGE$^{interp.*}$ was chosen with medical experts as the most interpretable one out of the top three using $All^\dagger$, it achieved a validation accuracy of 76.8% and a test accuracy of 74%.

The results in Table 3 show that all CNN models overfitted the data, with training accuracy being consistently higher than validation accuracy. Among these, Vanilla CNNs demonstrated comparable validation accuracy performance across input sets, with the model trained on All^* achieving the highest validation score out of all models (84.5%) but only 77.1% accuracy on the test set. Resnet18 models trained on $Standard^*$ or All^* performed better on validation accuracy than those trained on RGB or GRAY (82.1%, 84.3% and 81.9, 80.2%, respectively). Globally, both the Resnet18 and Vanilla CNN performed similarly on the validation set, which is surprising, given the difference in model size and complexity. It was actually the smaller models that proved to be more generalizable, according to the test metrics, indicating a tendency for larger models to overfit the data.

Table 3. Best training, validation (Val.), and test accuracy scores for all models. For each method, the best model(s) found with respect to the validation set (i.e., weights for DNNs or program for MAGE) served to report all metrics. Standard deviation is reported for averages.

Model	Input	Train %	Val. %	Test %	# images (epoch)
Vanilla CNN	GRAY	88.0	82.7	80.9	22.8M (87)
	RGB	91.9	84.1	80.5	22.3M (85)
	$Standard^*$	92.3	83.6	78.5	24.9M (95)
	All^*	89.3	84.5	77.1	11.3M (43)
Resnet18	GRAY	89.9	80.2	77.1	2.6M (10)
	RGB	87.2	81.9	77.4	786K (3)
	$Standard^*$	99.	82.1	77.1	13.1M (50)
	All^*	95.6	84.3	79.	4.2M (16)
$MAGE^{val*}$	$Standard^\dagger$	74.2	76.7	73.3	1.1M
Avg. MAGE	$Standard^\dagger$	72.3 (1.9)	74.2 (2.0)	71. (2.5)	822K
$MAGE^{val*}$	$All^\dagger$	72.9	78.0	76.0	1.3M
$MAGE^{interp.*}$	$All^\dagger$	71.8	76.8	74.0	571K
Avg. MAGE	$All^\dagger$	72.3 (1.6)	74,8 (1.9)	72.0 (1.8)	947K

When comparing our method to CNNs, we observe that CNNs consistently achieve higher performance. Figure 3 shows this comparison with a common axis, namely, the number of images processed by each algorithm. However, despite being a few percentage points below CNNs in accuracy, our algorithm remains competitive. This is particularly noteworthy given the significant difference in complexity between the models. CNNs are highly sophisticated and resource-intensive, while our approach is built on a simpler, more transparent pipeline. This raises an important question: is the marginal gain in accuracy worth the added complexity? In certain applications, especially in fields requiring transparency, a slight trade-off in accuracy may be justified if it results in a more interpretable decision-making process.

In Fig. 3 we observe that across all MAGE runs, the best accuracy scores rapidly approach their peak within just a few generations, suggesting that our approach rapidly generalizes well to unseen data without even processing the dataset in its entirety. Such a characteristic is particularly desirable in scenarios where generating or querying abundant data is costly, or when data availability is limited. Table 3 further highlights that MAGE processes images significantly fewer times before converging than CNNs. While CNN typically require dozens of passes through the dataset except for the Resnet18 trained on RGB inputs, MAGE, on average, found its best solution after three passes over the dataset (947K images with $All^\dagger$ inputs). Additionally, the most interpretable model ($MAGE^{interp.*}$) required only slightly more than two passes. This data-efficiency makes our approach a strong candidate for an anytime algorithm.

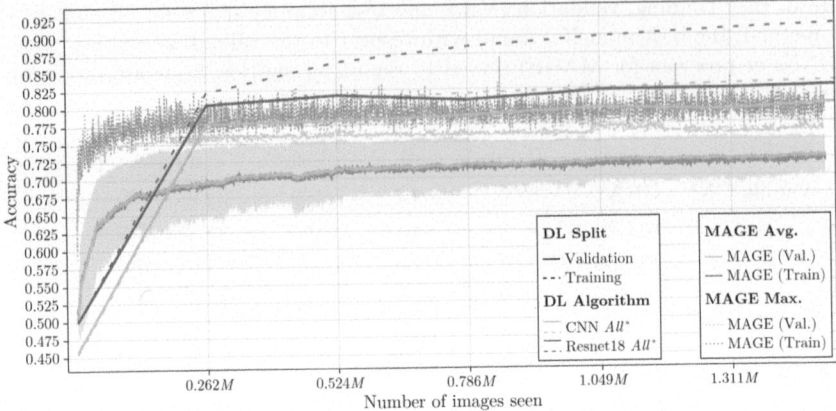

Fig. 3. Models' accuracy scores. For MAGE curves, colors indicate data splits, solid lines represent the average accuracy over 100 runs and ribbon denotes ±1 standard deviation. Dotted lines mark MAGE's maximum accuracy scores across all runs at each generation. The best-performing Vanilla CNN and ResNet18 models are displayed in distinct colors. Dashed lines correspond to training set accuracy, while solid lines denote validation accuracy. X-axis ticks are proportional to the dataset's size. (Color figure online)

6 Interpretability

Interpretability is essential, especially in domains such as digital pathology, where model predictions directly impact patient care and treatment decisions [4]. In these fields, the ability to understand a model's decision-making process increases trust and also ensures that the model bases its reasoning on medically relevant features rather than unintended biases.

The GP-evolved pipeline we present here is inherently interpretable. It achieves transparency through both simulatability and decomposability [18]. Simulatability allows users to follow the program's decision-making steps in a reasonable time. Decomposability enables users to examine individual components of the model, helping to break down its operations. Because GP-evolved programs are decomposable by design, they are good candidates for "explanation by example". A series of examples and counterexamples can be used to illustrate its rationale on a case-by-case basis, as well as its limitations.

From the five most performant models based on validation accuracy, medical experts chose the most interpretable one after analyzing their behavior on a series of cases. In Fig. 4 we examine a correct classification of a malignant case made by the model. By examining the sequence of operations, we can observe how specific patterns (e.g., tissue characteristics or absence of benign markers) guide the model toward a malignant classification. This example provides insight into the features the model considers for its classification.

The pipeline in Fig. 4 presents mainly two strategies for classifying patches. The first program, which detects benign characteristics, focuses uniquely on

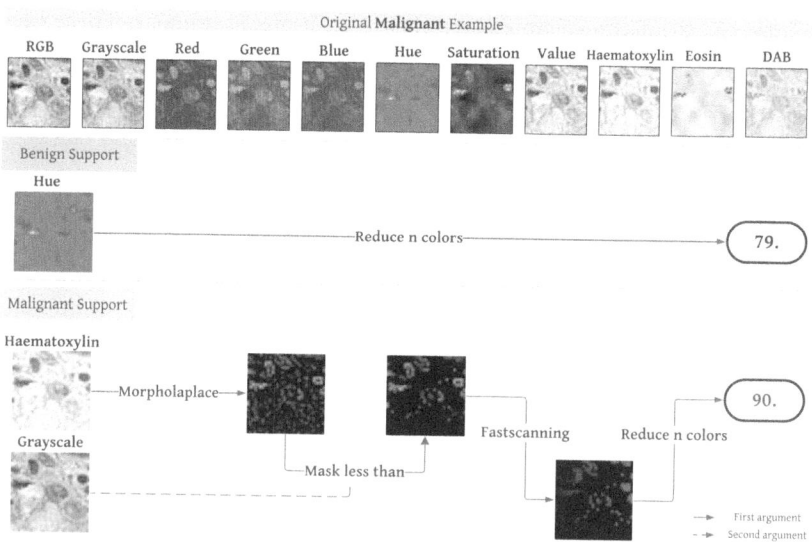

Fig. 4. Example of a correctly classified image by the chosen pipeline. Original channels are shown at the top. The RGB image is shown for understandability purposes, but it is not available as an input. The pipeline produces two outputs, the first (79.) is the support for the benign class and the second the support for the malignant class (90.). The highest number is chosen as the predicted class, in this case, a correct prediction of malignant tissue. Outputs are the results of a series of operations applied to inputs and intermediary outputs.

the heterogeneity of the Hue channel. This channel is very heterogeneous on images presenting a lot of very bright pixels and homogenous otherwise. As such, by counting the number of different unique pixel values, it can infer the predominance of non-nuclei information. The second program focuses on nuclei information revealed in the Haematoxylin channel. By applying an edge detection filter, the program can get a rough idea of the number and shape of cell nuclei. Background artifacts are removed after applying a mask that focuses on darker parts in the Grayscale channel. A segmentation algorithm, "fastscanning", detects close structures that do not differ too much in pixel intensity, assigning a different gray level to each one. Lastly, the program counts the number of such segments.

Although not shown here, a correctly classified benign instance is often characterized by a low or high value from program 1 (depending on bright pixels proportion) but, most importantly, a very low value from the second program. The latter output is explained by the fact that a patch populated by homogenous and numerous nuclei results in many detected edges. The "fastscanning" segmentation algorithm finds too many segments to fit them in Uint8 precision, hence, returning a value of 0 across all pixels. In other words, a homogeneous nuclei region produces an excessive number of segments, which overflows the image's precision and yields a single color, therefore, an output of 1, typically lower than the output from the first program. This behavior can be seen as a systematic conditional mechanism, similar to an if-else statement. Three benign and three malignant examples paired with the predictions from the three best models are available in the supplementary materials.

The full decomposition of a model also reveals its limitations. By analyzing the cases it classifies correctly and incorrectly, we can discern scenarios where the model may be less effective. For instance, the model reasons largely on the whole patch level. Because of this, if a clear malignant nucleus appears in an otherwise white patch, the model will be overconfident about its benign prediction.

Finally, interpretability can enable us to assess how closely the model aligns with the reasoning of medical professionals. For instance, experts recommended adding the HED channels, and these were frequently leveraged by the top-performing pipelines and improved their accuracy. Incorporating more expert feedback and domain-specific building blocks are important future steps.

7 Conclusion

In summary, this work demonstrates the development of an interpretable GP approach to binary image classification using the PCAM dataset. Specifically, we focused on the MAGE framework, an extension of CGP, for evolving multi-type and end-to-end image classification pipelines. The PCAM dataset presents a collection of image patches of lymph node biopsies in which malignancy must be detected. It is a challenging dataset because of its size and heterogeneity.

We performed hyperparameter tuning to identify the best strategy for enhancing program performance and improving generalizability to unseen images. We trained MAGE a hundred times with the best hyperparameters found. The MAGE solution with the highest validation accuracy (78%) achieved a test accuracy of 76%. In general, programs were remarkably robust to overfitting and their performance was consistent between them, with an average validation accuracy of 74.8% ($\pm 1.9\%$) and average test accuracy of 72% ($\pm 1.8\%$). We trained several CNN architectures from scratch in order to form a baseline of comparison. The best CNN model, on the validation set, achieved 84.5% validation accuracy and 77.1% test accuracy. For this dataset, the SOTA, which uses complex CNN architectures, sits at around 90% accuracy on the test set.

Although our approach sacrifices some accuracy points, it demonstrates that interpretable image classification is achievable with GP. The evolved programs are fully transparent, and through close collaboration with medical experts, we decomposed one of them to showcase the pipeline's decision-making process. This collaboration offered valuable insights into the program's logic, underscoring the importance of interpretable GP-based models as a viable alternative to 'black-box' approaches in medical image classification.

Acknowledgements. This project used compute resources from the CALMIP projects P21049 and P24042.

Disclosure of Interests. The authors have no competing interests to declare that are relevant to the content of this article.

References

1. Ain, Q.U., Al-Sahaf, H., Xue, B., Zhang, M.: Genetic programming for malignancy diagnosis from breast cancer histopathological images: a feature learning approach. IEEE Trans. Emerg. Top. Comput. Intell. (2025)
2. Amann, J., Blasimme, A., Vayena, E., Frey, D., Madai, V.I.: The Precise4Q consortium: explainability for artificial intelligence in healthcare: a multidisciplinary perspective. BMC Med. Inform. Decis. Mak. **20**(1), 310 (2020). https://doi.org/10.1186/s12911-020-01332-6
3. Atkins, D., Neshatian, K., Zhang, M.: A domain independent Genetic Programming approach to automatic feature extraction for image classification. In: 2011 IEEE Congress of Evolutionary Computation (CEC), pp. 238–245 (2011). https://doi.org/10.1109/CEC.2011.5949624, https://ieeexplore.ieee.org/document/5949624. iSSN 1941-0026
4. Barredo Arrieta, A., et al.: Explainable Artificial Intelligence (XAI): concepts, taxonomies, opportunities and challenges toward responsible AI. Inf. Fusion **58**, 82–115 (2020). https://doi.org/10.1016/j.inffus.2019.12.012, https://www.sciencedirect.com/science/article/pii/S1566253519308103
5. Bi, Y., Xue, B., Zhang, M.: Genetic programming with image-related operators and a flexible program structure for feature learning in image classification. IEEE Trans. Evol. Comput. **25**(1), 87–101 (2021). https://doi.org/10.1109/TEVC.2020.3002229, https://ieeexplore.ieee.org/document/9117044
6. Biau, J., Wilson, D., Cussat-Blanc, S., Luga, H.: Improving image filters with cartesian genetic programming. In: Proceedings of the 13th International Joint Conference on Computational Intelligence, pp. 17–27. SCITEPRESS - Science and Technology Publications, Valletta (2021). https://doi.org/10.5220/0010640000003063, https://www.scitepress.org/DigitalLibrary/Link.aspx?doi=10.5220/0010640000003063
7. Burks, A.R., Punch, W.F.: Genetic programming for tuberculosis screening from raw X-ray images. In: Proceedings of the Genetic and Evolutionary Computation Conference, GECCO 2018, pp. 1214–1221. Association for Computing Machinery, New York (2018). https://doi.org/10.1145/3205455.3205461
8. Cortacero, K., et al.: Evolutionary design of explainable algorithms for biomedical image segmentation. Nat. Commun. **14**(1), 7112 (2023). https://www.nature.com/articles/s41467-023-42664-x
9. De La Torre, C., Cortacero, K., Cussat-Blanc, S., Wilson, D.: Multimodal adaptive graph evolution. In: proceedings of the Genetic and Evolutionary Computation Conference Companion, GECCO 2024 Companion, pp. 499–502. Association for Computing Machinery, New York (2024). https://doi.org/10.1145/3638530.3654347
10. De La Torre, C., Lavinas, Y., Cortacero, K., Luga, H., Wilson, D.G., Cussat-Blanc, S.: Multimodal adaptive graph evolution for program synthesis. In: Affenzeller, M., et al. (eds.) PPSN XVIII. LNCS, vol. 15148, pp. 306–321. Springer, Cham (2024). https://doi.org/10.1007/978-3-031-70055-2_19
11. Fan, Q., Bi, Y., Xue, B., Zhang, M.: A multi-tree genetic programming-based ensemble approach to image classification with limited training data [research frontier]. IEEE Comput. Intell. Mag. **19**(4), 47–62 (2024). https://doi.org/10.1109/MCI.2024.3446148, https://ieeexplore.ieee.org/document/10709804
12. Goodfellow, I.J., Shlens, J., Szegedy, C.: Explaining and harnessing adversarial examples (2015). https://doi.org/10.48550/arXiv.1412.6572, arXiv:1412.6572

13. Harding, S., Leitner, J., Schmidhuber, J.: Cartesian genetic programming for image processing. In: Riolo, R., Vladislavleva, E., Ritchie, M.D., Moore, J.H. (eds.) Genetic Programming Theory and Practice X. Genetic and Evolutionary Computation. Springer, New York (2013)
14. Khan, A., Qureshi, A.S., Wahab, N., Hussain, M., Hamza, M.Y.: A recent survey on the applications of genetic programming in image processing. Comput. Intell. **37**(4), 1745–1778 (2021). https://doi.org/10.1111/coin.12459, https://onlinelibrary.wiley.com/doi/abs/10.1111/coin.12459
15. Koza, J.R.: Genetic programming as a means for programming computers by natural selection. Stat. Comput. **4**(2), 87–112 (1994). https://doi.org/10.1007/BF00175355
16. LeCun, Y., Bengio, Y., Hinton, G.: Deep learning. Nature **521**(7553), 436–444 (2015). https://doi.org/10.1038/nature14539, https://www.nature.com/articles/nature14539
17. Leitner, J., Harding, S., Forster, A., Schmidhuber, J.: Mars terrain image classification using cartesian genetic programming. In: Proceedings of the 11th International Symposium on Artificial Intelligence, Robotics and Automation in Space, i-SAIRAS 2012, pp. 1–8. European Space Agency (ESA) (2012)
18. Lipton, Z.C.: The mythos of model interpretability: in machine learning, the concept of interpretability is both important and slippery. Queue **16**(3), 31–57 (2018). https://doi.org/10.1145/3236386.3241340, https://dl.acm.org/doi/10.1145/3236386.3241340
19. Litjens, G., et al.: 1399 H&E-stained sentinel lymph node sections of breast cancer patients: the CAMELYON dataset. GigaScience **7**(6), giy065 (2018). https://doi.org/10.1093/gigascience/giy065, https://pmc.ncbi.nlm.nih.gov/articles/PMC6007545/
20. López-Ibáñez, M., Dubois-Lacoste, J., Pérez Cáceres, L., Birattari, M., Stützle, T.: The irace package: iterated racing for automatic algorithm configuration. Oper. Res. Perspect. **3**, 43–58 (2016). https://doi.org/10.1016/j.orp.2016.09.002, https://www.sciencedirect.com/science/article/pii/S2214716015300270
21. Marcus, G.: Deep Learning: a critical appraisal (2018). https://doi.org/10.48550/arXiv.1801.00631, arXiv:1801.00631
22. Mei, Y., Chen, Q., Lensen, A., Xue, B., Zhang, M.: Explainable artificial intelligence by genetic programming: a survey. IEEE Trans. Evol. Comput. **27**(3), 621–641 (2023). https://doi.org/10.1109/TEVC.2022.3225509, https://ieeexplore.ieee.org/abstract/document/9965435
23. Miller, J.F., Thomson, P.: Cartesian genetic programming. In: Poli, R., Banzhaf, W., Langdon, W.B., Miller, J., Nordin, P., Fogarty, T.C. (eds.) Genetic Programming. LNCS. Springer, Heidelberg (2000)
24. Miotto, R., Wang, F., Wang, S., Jiang, X., Dudley, J.T.: Deep learning for healthcare: review, opportunities and challenges. Brief. Bioinform. **19**(6), 1236–1246 (2018). https://doi.org/10.1093/bib/bbx044
25. Montana, D.J.: Strongly typed genetic programming. Evol. Comput. **3**(2), 199–230 (1995). https://doi.org/10.1162/evco.1995.3.2.199, https://dl.acm.org/doi/10.1162/evco.1995.3.2.199
26. Nadizar, G., Medvet, E., Wilson, D.: Searching for a diversity of interpretable graph control policies. In: Proceedings of the Genetic and Evolutionary Computation Conference, GECCO 2024, pp. 933–941. Association for Computing Machinery, New York (2024). https://doi.org/10.1145/3638529.3653987

27. Nadizar, G., Medvet, E., Wilson, D.G.: Naturally interpretable control policies via graph-based genetic programming. In: European Conference on Genetic Programming (Part of EvoStar), pp. 73–89. Springer (2024)
28. Nadizar, G., Rovito, L., De Lorenzo, A., Medvet, E., Virgolin, M.: An analysis of the ingredients for learning interpretable symbolic regression models with human-in-the-loop and genetic programming. ACM Trans. Evol. Learn. Optim. **4**(1), 1–30 (2024)
29. Nandi, R.J., Nandi, A.K., Rangayyan, R.M., Scutt, D.: Classification of breast masses in mammograms using genetic programming and feature selection. Med. Biol. Eng. Comput. **44**(8), 683–694 (2006). https://doi.org/10.1007/s11517-006-0077-6
30. Rawat, W., Wang, Z.: Deep convolutional neural networks for image classification: a comprehensive review. Neural Comput. **29**(9), 2352–2449 (2017). https://doi.org/10.1162/NECO_a_00990
31. Rudin, C.: Stop explaining black box machine learning models for high stakes decisions and use interpretable models instead. Nat. Mach. Intell. **1**(5), 206–215 (2019). https://doi.org/10.1038/s42256-019-0048-x, https://www.nature.com/articles/s42256-019-0048-x
32. Sheu, R.K., Pardeshi, M.S.: A survey on medical explainable AI (XAI): recent progress, explainability approach, human interaction and scoring system. Sens. (Basel Switz.) **22**(20), 8068 (2022). https://doi.org/10.3390/s22208068
33. Spector, L., Klein, J., Keijzer, M.: The Push3 execution stack and the evolution of control. In: Proceedings of the 7th Annual Conference on Genetic and Evolutionary Computation, GECCO 2005, Washington DC, USA, pp. 1689–1696. Association for Computing Machinery, New York (2005)
34. Srinidhi, C.L., Ciga, O., Martel, A.L.: Deep neural network models for computational histopathology: a survey. Med. Image Anal. **67**, 101813 (2021). https://doi.org/10.1016/j.media.2020.101813, https://www.sciencedirect.com/science/article/pii/S1361841520301778
35. Sun, C., Shrivastava, A., Singh, S., Gupta, A.: Revisiting unreasonable effectiveness of data in deep learning era. In: 2017 IEEE International Conference on Computer Vision (ICCV), pp. 843–852 (2017). https://doi.org/10.1109/ICCV.2017.97, https://ieeexplore.ieee.org/document/8237359. iSSN 2380-7504
36. Teh, E.W., Taylor, G.W.: Metric learning for patch classification in digital pathology (2019). https://openreview.net/forum?id=BJgtl1V6FN
37. Teh, E.W., Taylor, G.W.: Learning with less data via weakly labeled patch classification in digital pathology. In: 2020 IEEE 17th International Symposium on Biomedical Imaging (ISBI), pp. 471–475 (2020). https://doi.org/10.1109/ISBI45749.2020.9098533, https://ieeexplore.ieee.org/document/9098533. iSSN 1945-8452
38. Veeling, B.S., Linmans, J., Winkens, J., Cohen, T., Welling, M.: Rotation equivariant CNNs for digital pathology. In: Frangi, A.F., Schnabel, J.A., Davatzikos, C., Alberola-López, C., Fichtinger, G. (eds.) MICCAI 2018. LNCS, vol. 11071, pp. 210–218. Springer, Cham (2018). https://doi.org/10.1007/978-3-030-00934-2_24
39. Zeiler, M.D., Fergus, R.: Visualizing and understanding convolutional networks. In: Fleet, D., Pajdla, T., Schiele, B., Tuytelaars, T. (eds.) ECCV 2014. LNCS, vol. 8689, pp. 818–833. Springer, Cham (2014). https://doi.org/10.1007/978-3-319-10590-1_53

Unified Piecewise Symbolic Regression

Guillaume Doquet(✉)

Safran, Magny-les-Hameaux, France
guillaume.doquet@safrangroup.com

Abstract. Symbolic Regression (SR) searches for a closed-form mathematical expression describing the relationship between input and output features in data. The main theoretical draw of SR compared to traditional *black-box* regression techniques is that the learned models should be interpretable by design. However, typical SR methods struggle to discover sparse and accurate models when the shape of the output varies locally, depending on the values of some input features. Given that this is a common occurence in physics, SR should be able to learn *piecewise* models. We introduce a new piecewise SR framework called Unified Piecewise Symbolic Regression (UPSR). UPSR simultaneously partitions the input space into subregions and learns local regressors for each subregion, forming a global model unifying all subregions. We demonstrate its effectiveness on a large synthetic SR benchmark containing both piecewise and non-piecewise data structures. UPSR is shown to outperform state-of-the-art piecewise SR approaches, both qualitatively and quantitatively.

Keywords: Symbolic Regression · Genetic Programming · Interpretability

1 Introduction

Symbolic Regression (SR) consists in learning an explicit mathematical expression describing the relationship between input and output features in data. Formally, given input data X, output data Y and a loss function $\mathcal{L}$ measuring goodness-of-fit, the goal is to find $g^* = argmin_{g \in \mathcal{G}} \mathcal{L}(g(X), Y)$, where $\mathcal{G}$ denotes the vast search space of all closed-form functions. In practice, a set $\mathcal{M}$ of authorized mathematical *operators* (*i.e.* $\mathcal{M} = \{+, \times, sin, log, \sqrt{.}...\}$) is defined to help guide the search in $\mathcal{G}$, using *a priori* knowledge on which operators should appear in the output function. In order to encourage discovering an *interpretable* function structure, the search algorithm is typically augmented with a sparsity pressure mechanism, thus creating a tradeoff between regression accuracy and simplicity of the model.

By contrast, typical *black-box* machine learning techniques such as deep neural networks are usually concerned only with regression accuracy. Thus, although the precision of these regressors may be excellent, the learned parametric function $g(X; \theta)$ is hardly explainable by human experts, given that θ often consists of millions of neuronal weights and biases. Given the growing need for *explainable AI* [10, 16, 27] both in industrial and academic contexts, SR has in recent years gained in popularity for application across varied domains such as cosmology [4], medicine [14], material sciences

[31,35] or finance [28]. In physics, it is common practice to break down a complex phenomenon into smaller, easier-to-comprehend pieces. This typically happens by relying on one of the following two paradigms.

Stacking Submodels. The first approach is to build elementary submodels, each responsible for providing a partial explanation of the data. These submodels are added together to form a cohesive and readily understandable full model. Submodels can then be switched on/off depending on the local behavior of experimental data. For instance, in material sciences, the total strain of a metallic piece is often expressed as a sum of elastic, plastic and creep components [3] s.t. $\epsilon_{total} = \epsilon_{elastic} + \epsilon_{plastic} + \epsilon_{creep}$, and in the elastic regime $\epsilon_{plastic} = 0$.

Exclusive Submodels. The second approach is to first split the input space in interpretable subregions, then build a local model for each region. Here, submodels are not added together, rather selected via an *if/else* logic. For example, in fluid dynamics, the velocity u of a turbulent airflow near a wall [2] is typically considered *either* linearly proportional to the distance with the wall y in the so-called viscous layer s.t. $u \sim y$ if $y < K_{viscous}$, *or* logarithmically proportional in the inner layer s.t. $u \sim log(y)$ if $y > K_{inner}$.

Although the logic used differs slightly between these two approaches, the end result is the same : the final model is a *piecewise* closed-form equation. Accordingly, one would expect SR algorithms to also be able to produce piecewise regressors, mimicking the human expert for the sake of interpretability. We introduce an intuitive technique to enhance any suitable SR algorithm with the ability to produce piecewise models, following either *stacking submodels* or *exclusive submodels* paradigm (or both). The resulting algorithm is called *Unified Piecewise Symbolic Regression (UPSR)*. Section 2 discusses the relevant literature on piecewise SR. Section 3 describes our method, including important practical implementation details. Section 4 then contains the proposed experimental setup to assess the validity of the approach and the obtained results. Section 5 concludes the paper with final remarks and future perspectives.

2 Related Work

Many algorithms have been proposed for data-driven discovery of symbolic models, which are typically represented as *expression trees* [19]. The traditional heuristic to build and manipulate those expression trees relies on evolutionary algorithms (EAs) [21]. In EA-based SR, symbolic computations known as *mutations* are repeatedly performed to optimize the accuracy/simplicity tradeoff of expression trees, as implemented in open-source libraries such as gplearn [30].

Following the surge of research interest in SR, alternative *deep learning* techniques have recently been introduced. AI-Feynman [32,33] uses neural networks to discover simplifying properties in data such as symmetry or separability, attempting to restrict the search to small enough subspaces of $\mathcal{G}$ that low-order polynomial fits or even brute force exhaustive search perform well enough. Taking inspiration from natural language processing, DSR [22] views symbolic expressions as strings of words or tokens, the sequence of which can be learned by training a recurrent neural network.

Both traditional EA-based and deep learning-based SR have shown good performance on generic regression problems. However, none of the aforementioned algorithms are equipped to discover *piecewise* symbolic models. Consequently, when a simple piecewise model would be most adequate, these methods will instead produce either inaccurate or needlessly complex solutions. This behavior will be illustrated in Sect. 4 by using gplearn as a non-piecewise benchmark.

A comparatively small portion of SR literature is dedicated to learning piecewise models. Clustered Symbolic Regression (CSR) [18] separately learns each of the n modes $\{g_i(X_i, \theta_i)\}_{i=1}^n$ of a hybrid multi-modal dynamical system. This method requires prior expert knowledge of n, which one does not have access to in the general case. Similarly, Multi-level Symbolic Regression (MSR) [7] and Multi-View Symbolic Regression (MvSR) [25] were recently introduced to tackle situations where experimental data was collected from different sources $(X_i, Y_i)_{i=1}^n$ and the behavior of output data Y_i varies between sources. MSR and MvSR both look for an underlying function structure g common to all sources, assuming the only source of variability comes from the numerical values θ_i of its parameters, s.t. $\forall i \in 1, .., n, Y_i = g(X_i, \theta_i)$. This assumption of shared structure, although appropriate in the context of these studies, cannot be expected to hold in the general case.

The most intuitive approach for generic piece-wise SR is to *first* learn a Decision Tree (DT) to partition the input space into subregions, *then* learn separate submodels for each leaf in the DT. An early example of this paradigm is the well-known CART [17], where constant values are assigned to the leaves. Consequently, CART is restricted to piecewise constant models. Piecewise Linear regression Tree (PL-Tree) [15] thereafter proposes to replace the constant values with linear models. Finally, Piecewise Symbolic regression Tree (PS-Tree) [36] have recently proposed to further enhance the algorithm by generalizing to non-linear models for the leaves. Consequently, PS-Tree is the (*de facto*) state-of-the-art approach for generic piece-wise SR, and will be compared to our method in Sect. 4. A summary of the different DT structures is provided in Fig. 1 below.

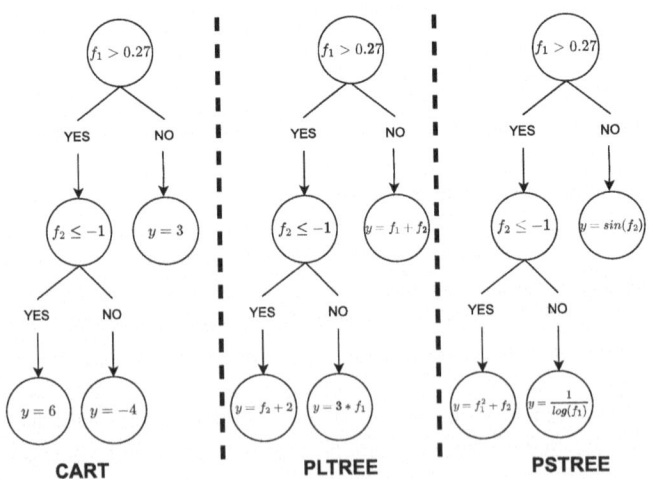

Fig. 1. Illustration of the differences between CART, PL-Tree and PS-Tree

3 Unified Piece-Wise Symbolic Regression

UPSR consists in enhancing *any suitable* EA-based SR algorithm to handle piecewise models by adding a new special operator, called *Piece*, to the existing set $\mathcal{M}$ of authorized mathematical operators. Section 3.1 first describes the components needed for an EA-based SR algorithm to be compatible with *Piece*. Section 3.2 thereafter introduces the *Piece* operator itself.

3.1 Evolutionary Algorithm Setup

Our preliminary studies have shown that EA-based SR should involve the following five essential components in order for the integration of *Piece* in $\mathcal{M}$ to perform adequately.

Interactions. In their most generic definition, EAs do not necessarily involve interactions between population members. *Genetic Programming (GP)* is generally understood as a subclass of EAs where *crossovers* allow exchanging information between individuals. In the context of piecewise SR, crossovers are vitally important, as they allow combining members of the population specializing on different parts of the data.

Elitism. Elitism in EA can be loosely defined as the idea of preserving the best individuals in a given generation for the next generation. This ability to memorize the most relevant local regressors is critical to the success of our approach.

Accuracy/Simplicity Trade-Off. In order to discourage partitioning the input feature space into too many subregions, it is important to involve a harsh and dynamic sparsity pressure, rather than a fixed maximum expression length of depth. Such a mechanism is present in *e.g.* the well-known NSGAII [5]: candidate regressors are sorted in Pareto fronts based on both regression accuracy and syntactic complexity of the formula. There are many existing variants of complexity [11,34]. We opted for a simple yet effective definition $C(g(X, \theta)) = |g| + |\theta|^2$, where $|g|$ denotes the minimal number of nodes needed to represent g with an expression tree. The second term is squared to emphasize that model parameters should be significantly more expensive than regular nodes, and adding too many parameters quickly becomes unaffordable.

Non-linear Optimization of Model Parameters. Tuning the numerical values of constants involved in SR models learned with GP is notoriously problematic [26]. The traditional way of determining their optimal values is via so-called *Ephemeral Random Constants (ERC)* [23], which values are initially sampled from a uniform distribution. ERCs can thereafter only change by combination through crossovers. This approach is typically inefficient for fine-tuning floating-point parameters, which are quite common in physics. Therefore, more recent algorithms tend to instead combine GP with local non-linear optimization [12,13] such as Levenberg-Marquardt [20]. UPSR also relies on this kind of hybridization of GP with an optimization algorithm, although we recommend opting for a *global* optimizer (see Sect. 3.2).

Special Composition Rules. It is common in the SR literature (both EA-based and non EA-based) to define special rules on some operators to avoid particularly undesirable outcomes [22]. For instance, one might want to forbid composition of trigonometric operators, with the reasoning that *e.g.* $cos(sin(cos(tan(f_i))))$ is practically never encountered in physics. In a similar fashion, we defined custom composition rules (see Sect. 3.4) to exploit the full potential of the *Piece* operator.

There are many open-source implementations (most of them in Python) of EAs to perform SR, such as the previously mentioned gplearn [30], as well as DEAP [8], karoogp [29] or PyGad [9]. However, customizing any of those libraries to accommodate the five components above is hardly practical. Therefore, we have in this work opted to build our own proprietary implementation from scratch in Python, with the aforementioned NSGAII as a starting reference point.

3.2 The *Piece* operator

The core idea behind *Piece* is to mirror the logic used by human experts in either *stacking submodels* or *exclusive submodels* paradigm. Accordingly, *Piece* has two modes, both of which rely on the well-known Heaviside function $H(x, \tau) := \mathbb{1}_{x \geq \tau}$. In *stacking* mode, *Piece* conditionally adds partial explanations to the full model, depending on the value of some input feature. Formally, given pre-mutation expression $g_{initial}(X, \theta)$, a particular input feature of interest f_i and a threshold parameter τ_i, the following symbolic modification is performed: $g_{initial} \leftarrow g_{initial} \times \mathbb{1}_{f_i \geq \tau_i}$. *Exclusive* mode instead combines two Heaviside functions with respective thresholds $\tau_{i,low}$ and $\tau_{i,high}$ s.t. : $g_{initial} \leftarrow g_{initial} \times \mathbb{1}_{f_i \geq \tau_{i,low}} \times \mathbb{1}_{f_i \leq \tau_{i,high}} = g_{initial} \times \mathbb{1}_{f_i \geq \tau_{i,low}, f_i \leq \tau_{i,high}}$. The mode selection is made randomly on a per-feature basis. The pseudocode describing the full behavior of *Piece* is provided in Algorithm 1. Additionally, an illustrative example is provided in Fig. 2.

Algorithm 1. Pointwise mutation through *Piece*

Input Symbolic expression $g_{old}(X, \theta_{old})$, set of all input features $\mathcal{F}$, probability $0 \leq p_e \leq 1$
Output Mutated symbolic expression g_{new}

1: **Initialize:**
   $\theta_{new} \leftarrow \theta_{old}$, $g_{new}(X, \theta_{new}) \leftarrow g_{old}(X, \theta_{old})$, and compute power set of $\mathcal{F}$, denoted $2^{\mathcal{F}}$
2: Select uniformly at random an element $F \in 2^{\mathcal{F}}$ (subset of input features)
3: **for** each feature $f_i \in F$ **do**
4:   Select *exclusive* mode with probability p_e or *stacking* mode with probability $1 - p_e$
5:   **if** *exclusive* mode **then**
6:     Create two new symbolic parameters $\tau_{i,low}, \tau_{i,high}$
7:     $g_{new}(X, \theta_{new} \cup \{\tau_{i,low}, \tau_{i,high}\}) \leftarrow g_{new}(X, \theta_{new}) \times \mathbb{1}_{f_i \geq \tau_{i,low}, f_i \leq \tau_{i,high}}$
8:     $\theta_{new} \leftarrow \theta_{new} \cup \{\tau_{i,low}, \tau_{i,high}\}$
9:   **else if** *stacking* mode **then**
10:    Create one new symbolic parameter τ_i
11:    $g_{new}(X, \theta_{new} \cup \{\tau_i\}) \leftarrow g_{new}(X, \theta_{new}) \times \mathbb{1}_{f_i \geq \tau_i}$
12:    $\theta_{new} \leftarrow \theta_{new} \cup \{\tau_i\}$
13: **end if**
14: **end for**
15: **return** g_{new}

By construction, the exclusive operator is a generalization of the stacking operator, given that $\mathbb{1}_{f_i \geq \tau_i} = \mathbb{1}_{f_i \geq \tau_{i,low}, f_i < +\infty}$. This is an intentional design choice: although the stacking mode is less powerful, it is more compliant with the sparsity pressure (adding only one new parameter instead of two), and is therefore more likely to survive elitist selection for future generations, thus helping keeping the complexity of the overall model as low as possible.

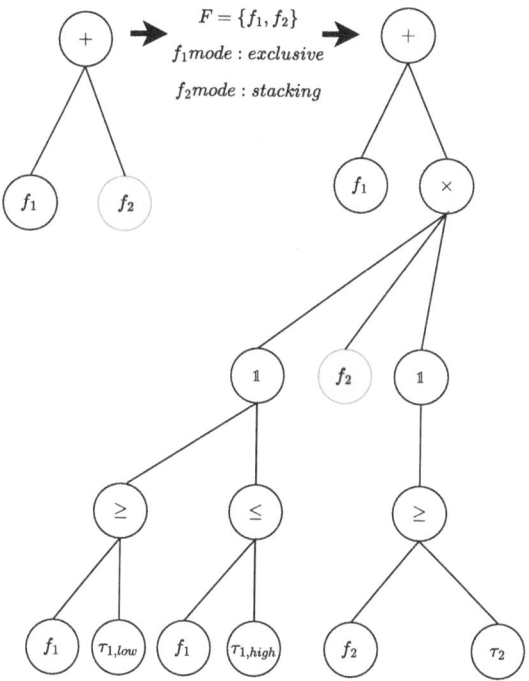

Fig. 2. Illustration of the effect of pointwise mutation through the proposed *Piece* operator. The node selected to operate on is in green.

As described in Sect. 3.1, once the symbolic structure of $g(X,\theta)$ has been fixed, a non-linear optimizer is tasked with finding $\theta^* = argmin_{\theta \in \mathbb{R}^m} \mathcal{L}(g(X,\theta), Y)$. Typically, this is performed with a local non-linear optimizer such as Levenberg-Marquardt [20], Nelder-Mead or Newton-Conjuguate Gradient. However, in our setup, not only is $\mathcal{L}(g(X,\theta), Y)$ clearly non-differentiable (given that $g(X,\theta)$ involves indicator functions), the landscape of $\mathcal{L}$ is non-smooth, with tall cliffs, shallow local basins, and local minima to avoid. As a result, all aforementioned local optimizers are either non-applicable (due to requiring gradient computation) or inefficient, as they consistently get trapped in local minima. It is therefore mandatory to rely on a global optimization scheme instead. We have opted for the Differential Evolution (DE) [24] algorithm, which is empirically efficient for optimizing all numerical values and does not require hyperparameter tuning. Moreover, DE can be trivially ran in parallel to asynchronously optimize all individuals in the population.

3.3 Improvements Offered by *Piece* over Decision Trees

By design, *Piece* defines an *implicit* decision tree-like structure, by contrast with typical piecewise SR where this structure is learned *explicitely*. This leads to the following crucial differences.

Flexibility. Boundaries in decision tree-based piece-wise SR are typically frozen in place at the end of the partition step (CART, PL-Tree). Admittedly, PS-Tree involves an adaptive partition mechanism to iteratively fuse overlapping submodels. However, this scheme will be shown in Sect. 4 to be unreliable and sensitive w.r.t. an hyperparameter controlling the maximum number of submodels. By contrast, in UPSR, numerical boundaries for each submodel are adaptive to the content of the submodels. If a submodel mutates to become better specialized in a smaller subregion, the associated boundaries will seamlessly narrow. Conversely, if a submodel gains in generalization power, the boundaries will accordingly widen. If the boundaries become too wide, then the sparsity pressure will encourage deleting these nodes from the model.

Sparsity. In order to encourage partitioning the input space in as few sub-regions as possible, DTs are typically constrained by setting a maximal tree depth, a maximal number of leaves, and a minimal number of datapoints to create a new leaf. This partition complexity control scheme is uncorrelated from the sparsity pressure applied to the submodels learned in the leaves. This is suboptimal in terms of interpretability, as one would expect a compromise between the complexity of the decision tree and the complexity of its leaves. The "Decision" operator introduced in [6], and later used in works interested in learning ensemble models with GP [1], suffers from the same limitation. By contrast, the submodels and the structure defined by the *Piece* operators in UPSR jointly contribute to a shared complexity budget. This naturally creates the desired complexity trade-off. Section 4 empirically shows that UPSR creates more frugal partitions than PS-Tree, without sacrificing regression accuracy.

Cohesiveness. In DTs, leaves are learned independently and kept separate from each other. This means that leaves can never be added together. Therefore, DT-based piecewise SR naturally follows only the *exclusive models* paradigm, and is ill-suited for the *stacking submodels* case (more in Sect. 4).

Cooperation. In DTs, leaves can not *cross- pollinate*, whereas crossovers and mutations constantly mix submodels together in UPSR, helping to capture salient patterns and keeping them for future iterations. In turn, this means that UPSR needs fewer generations to converge than DT-based SR (more in Sect. 4), compensating the relatively higher computational cost *per generation* (due to the global non-linear optimization step).

Partition interpretability. By design, *Piece* can only produce so-called *atomic splits* (enforced by special rules that will be introduced in Sect. 3.4), that is easily human-readable thresholds on the numerical values of some input features, without potentially complex nested subregions of interest like with the "Decision" operator from [6]. On the other hand, DT-based SR tends to produce submodels boundaries that are hardly understandable by human experts, as will be illustrated in Sect. 4.

3.4 Special *Piece* composition rules

As mentioned in Sect. 3.1, we define the following composition rules for the *Piece* operator :

Rule 1. No mutation or crossover can ever occur *inside Piece*. This means that once an instance of *Piece* has been introduced somewhere, it can never get its symbolic content modified. For instance, $g(X, \theta) \times \mathbb{1}_{sin(f_i) \geq log(\tau \times f_i)}$ breaks Rule 1.

Rule 2. The path from the root of the expression tree to an instance of *Piece* can only include the + and/or × operators. For instance, $\log(f_i) \times \mathbb{1}_{f_i \geq \tau_i} + f_j$ is a valid expression, but $\log(f_i \times \mathbb{1}_{f_i \geq \tau_i}) + f_j$ breaks Rule 2. Note that nested *Piece* instances such as $\mathbb{1}_{(f_i \times \mathbb{1}_{f_j \geq \tau_j}) \geq \tau_i}$ also break Rule 2 by definition.

Rule 3. Multiple instances of *Piece* cannot be applied to the same submodel unless there is no overlap between thresholded features, in which case they are combined together. Formally, if instances $Piece_1$ and $Piece_2$ would respectively involve feature subsets F_1 and F_2 and the same pre-mutation expression $g_{old}(X, \theta)$, then only one of them will apply to $g_{old}(X, \theta)$, unless $F_1 \cap F_2 = \emptyset$, in which case $Piece_3$ is created and applied to $g_{old}(X, \theta)$, with $F_3 = F_1 \cup F_2$. As an example, $g(X, \theta) \times \mathbb{1}_{f_i \geq \tau_i} \times \mathbb{1}_{f_j \geq \tau_j}$ with $j \neq i$ is valid and both indicator functions will be fused together. On the other hand, $g(X, \theta) \times \mathbb{1}_{f_i \geq \tau_1, i} \times \mathbb{1}_{f_i \geq \tau_2, i}$ breaks Rule 3.

The combination of Rule 1 and Rule 2 guarantee that the resulting partition is, as desired, delimited only by atomic splits. Rule 3 is theoretically redundant, as the sparsity pressure should handle undesirable expressions by dropping excess operators. However, we empirically found that hardcoding this rule helps speed up the evolution process quite significantly: rather than making mistakes and correct them afterwards, it is better not to do such mistakes in the first place. In practice, these rules are enforced by taking inspiration from deep learning-based SR [22]. The idea is to embed the rules inside the crossover/mutation operators themselves, by making it so that the probability of performing an illegal operation is zero. In UPSR, before performing any crossover or mutation on an individual, *legal* subtrees are tallied and operators *compatible* with each of these subtrees are listed. The actual crossover/mutation performed is then sampled uniformly only from these curated operations.

4 Experiments

We have performed a *qualitative* study (Sect. 4.2) and a *quantitative* one (Sect. 4.3). Section 4.1 first describes the dataset generation procedure for the quantitative study, as well as important hyperparameter setup common to both studies.

4.1 Experimental Setup

We have built a procedural piece-wise synthetic data generator. First, the input space is randomly partitioned in subregions. Then, symbolic submodels are randomly generated for each region. In *exclusive* data generation mode, submodels are independent from each other, whereas submodels are locally added together in *stacking* data generation mode (see Fig. 3). Then, we used this data generator to create a synthetic benchmark containing 100 datasets:

- 10 monovariate non-piecewise datasets.
- 10 multivariate (2 variables) non-piecewise datasets.
- 30 monovariate piecewise datasets with exclusive submodels (10 with 2 subregions, 10 with 3 subregions, 10 with 4 subregions).
- 30 monovariate piecewise datasets with stacking submodels (same distribution as for the exclusive case).
- 10 multivariate piecewise datasets with exclusive submodels and 4 regions.
- 10 multivariate piecewise datasets with stacking submodels and 4 subregions.

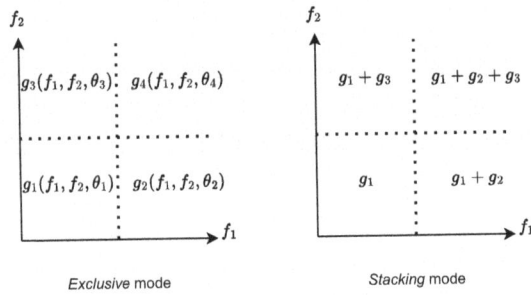

Fig. 3. Schematic illustration of the data generator in 2 dimensions. Left panel: *exclusive* submodels, right panel: *stacking* submodels.

For all datasets, lower and upper bounds for the input features and submodel parameters are set to $[-2, 2]$. The total number of datapoints in each dataset is set to 200, with the typical 75%/25% train/test split. In this benchmark, all submodels $g_i(\mathcal{F}, \theta)$ are linear, testing the ability of SR algorithms (all of which have access to non-linear operators) to avoid overfitting by keeping submodels as simple as possible. A larger benchmark also containing non-linear submodels is in preparation for future work. Note that even when the ground truth submodels are linear, it is clearly challenging to uncover the overall data structure, as illustrated in Fig. 4.

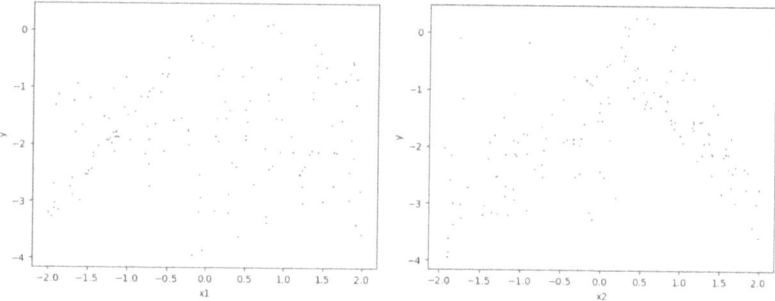

Fig. 4. Synthetic benchmark dataset with 4 *exclusive* multivariate linear submodels. Left panel: Output y (on the training set) as function of one variable. Right panel: y as function of the other variable.

After the data generation phase, 5 runs (with different random seeds) of PS-Tree, UPSR and GP-Learn were performed on each synthetic dataset, for a total of 500 generated symbolic models for each method. GP-Learn acts as a non piece-wise baseline and its hyperparameters are therefore all left at their default values. In order to provide a fair comparison between PS-Tree and UPSR, the population size for both methods is identical ($n_{pop} = 100$), and their respective maximum number of generations is set s.t. both methods converge and their average runtime is similar ($\approx$ 1 minute on our machine[1]): $n_{gen} = 10$ for UPSR, $n_{gen} = 100$ for PS-Tree. The maximum number of subregions in PS-Tree is set to $max\_leaf\_nodes = 15$. The important hyperparameters for UPSR and their fixed values are summarized in Table 1.

Table 1. List of important UPSR hyperparameters, which values are fixed for all experiments

Hyperparameter	Value
Population Size	100
Crossover Rate	0.8
Mutation Rate	0.7
Maximum generations	10
Accuracy metric	MSE
Complexity metric	$C(g(X, \theta)) = n_{nodes} + |\theta|^2$
Mutation actions	{$point\ (p = 0.75),\ hoist\ (p = 0.25)$}
Operator set $\mathcal{M}$	$\{+, \times, sin, cos, log, exp, Piece\}$
Exclusive mode probability p_e	0.5

---

[1] An Ice Lake Intel Xeon Gold 6342 CPU with 2×24 cores and 2.80 GHz. The crossover, mutation and optimization steps in each generation of UPSR are parallelized across the 48 cores.

4.2 Qualitative Results

Let us consider a simple monovariate dataset defined by symbolic relationship $y = |x|$ (absolute value), with $x \in [-2, 2]$. Given that the $|.|$ operator is absent from $\mathcal{M}$ in PS-Tree as well as UPSR, the simplest solution is for both methods to learn a piece-wise model s.t.:

$$y = \begin{cases} -x & \text{if } x < 0 \\ x & \text{if } x > 0 \end{cases}$$

Expectedly, this is what happens for UPSR, finding an exact solution (up to numerical variation of amplitude $< 1e^{-9}$ for the parameters) in 10 out of 10 runs, although with varying syntax, such as:

$$\begin{cases} y = x + (\theta_1 x * \mathbb{1}_{x \geq \tau_1, x \leq \tau_2}), \theta_1 \approx -2, \tau_1 \approx -2, \tau_2 \approx 0 \\ y = \theta_1 x + (\theta_2 x * \mathbb{1}_{x \geq \tau_1}), \theta_1 \approx -1, \theta_2 \approx 2, \tau_1 \approx 0 \\ y = (\theta_1 x * \mathbb{1}_{x \geq \tau_1, x < \tau_2}) + (x * \mathbb{1}_{x \geq \tau_3, x < \tau_4}), \theta_1 \approx -1, \tau_1 \approx -2, \tau_2 \approx \tau_3 \approx 0, \tau_4 \approx 2 \end{cases}$$

Note that all of the above expressions are *semantically* equivalent. *Syntaxic* stability is notoriously hard to enforce in SR algorithms, and is a direction of future work for UPSR (see Sect. 5).

We performed 5 runs of PS-Tree on this dataset for each value of $n_{max\_leaf\_nodes} \in [2, 5, 10, 20, 30, 40, 50, 60, 70, 80, 90, 100]$, and made the following observations.

Flexibility Issue. First of all, the number of submodels effectively discovered by PS-Tree is highly sensitive w.r.t. max_leaf_nodes, instead of staying stable around 2 (one for negative values, one for positive values), as showcased in Fig. 5. Moreover, the function $n_{leaves} = g(max\_leaf\_nodes)$ is not monotonic, further highlighting the volatility of the adaptive partitioning scheme in PS-Tree.

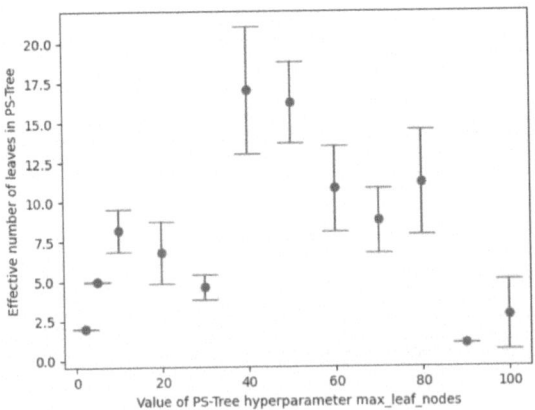

Fig. 5. Evolution of the number of submodels discovered by PS-Tree on $y = |x|$ w.r.t. its hyperparameter max_leaf_nodes

Sparsity Issue. The submodels discovered by PS-Tree are hardly sparse enough to be understandable by humans. As an example, here is a typical submodel :

$$-\frac{0.09 X_0^2 rand_{101}}{\sqrt{rand_{101}^2+1}} + 0.2 X_0^2 + \frac{0.4 X_0^2}{\sqrt{(-X_0+2rand_{101})^2+1}\sqrt{\tanh^2(rand_{101})+1}} + \frac{0.2 X_0}{\sqrt{X_0^2+1}\sqrt{(-X_0+2rand_{101})^2+1}}$$

$$-\frac{2.8\, rand_{101}}{\sqrt{\tanh^2\left(\sin\left(\frac{rand_{101}}{\sqrt{X_0^2+1}}\right)\right)+1}} - \frac{14.7\, rand_{101}}{\sqrt{\tanh^2\left(\frac{rand_{101}}{\sqrt{\tanh^2(\tanh(X_0))+1}}\right)+1}} - \frac{0.7\, rand_{101}}{\sqrt{\tanh^2\left(\frac{rand_{101}}{\sqrt{X_0^2+1}}\right)+1}}$$

$$-\frac{0.4\, rand_{101}\tanh(rand_{101})}{\sqrt{rand_{101}^2+1}\sqrt{\left(rand_{101}-\frac{rand_{101}}{\sqrt{X_0^2+1}}\right)^2+1}} - 27.6 - \frac{0.1\sin(rand_{101}+\sin(rand_{101}))}{\sqrt{\left(4X_0+\frac{rand_{101}}{\sqrt{X_0^2+1}}+\tanh(X_0)\right)^2+1}}$$

$$-\frac{0.3\sin(X_0-rand_{101})}{\sqrt{(3X_0+\tanh(X_0))^2+1}} - \frac{0.5\sin\left(X_0+2rand_{101}-\frac{rand_{101}}{\sqrt{X_0^2+1}}\right)}{\sqrt{(3X_0+\tanh(X_0))^2+1}} + \frac{0.1\sin\left(rand_{101}-\frac{rand_{101}}{\sqrt{X_0^2+1}}\right)}{\sqrt{\frac{rand_{101}^2}{X_0^2+1}+1}}$$

$$+\frac{0.2\sin\left(X_0-\frac{rand_{101}}{\sqrt{\tanh^2(\tanh(X_0))+1}}\right)}{\sqrt{rand_{101}^2+1}} - \frac{0.1\sin\left(2X_0+rand_{101}-\tanh\left(\frac{X_0}{\sqrt{(X_0+rand_{101})^2+1}}\right)\right)}{\sqrt{16 X_0^2+1}}$$

Each appearance of $rand_{101}$ corresponds to a separate ERC. The number of ERCs in PS-Tree is typically far larger than $|\theta|$ in UPSR. Submodels of such complexity can be found no matter the number of leaves, and for a given run, all leaves tend to have similar complexity.

Cohesiveness Issue. The ground truth $y = |x|$ has an obvious even symmetry (that is $y(x) = y(-x) \forall x \in \mathbb{R}$). Therefore, one would expect a submodel responsible for a negative x subregion to have similar structure as the corresponding submodel in the opposite positive x subregion (as is the case for the solutions typically discovered by UPSR). Visual inspection of the submodels reveal that this is not case for PS-Tree: we did not find any symmetry in the syntax of positive and negative submodels. Moreover, the submodel boundaries are themselves asymmetrical.

Partition Interpretability Issue. Similarly as for the submodels, the decision boundaries in PS-Tree are themselves hardly interpretable by humans. Below, we reproduce a typical path from the root of the DT to a leaf:

$$\left(\sin\left(\frac{2X_0^3}{\sqrt{rand_{101}^2+1}}+4X_0+\frac{X_0}{\sqrt{X_0^2+1}}+rand_{101}\right) \le -1\right) \wedge \left(\sin\left(X_0^2+3X_0\, rand_{101}+2X_0+3\, rand_{101}+\frac{rand_{101}}{\sqrt{X_0^2+1}}\right) \le -0.8\right)$$

Regression Accuracy. This qualitative analysis has shown that PS-Tree tends to produce models that are as close to black-box as SR can be, trading interpretability away for expressiveness. One would therefore expect close to perfect regression accuracy. Interestingly enough, this is not what we observe, even in the case $y = |x|$ (see Fig. 6 below).

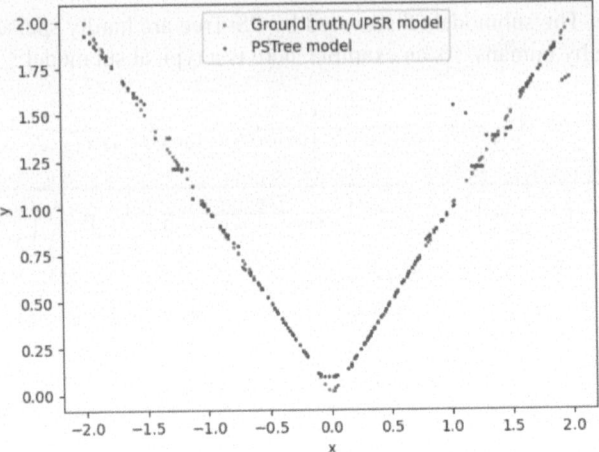

Fig. 6. Scatterplot comparing ground truth $y = |x|$ in red (found by UPSR) to a typical run of PS-Tree in blue (here with $max\_leaf\_node = 15$). Some submodels are quite accurate (e.g. for $x \approx 0.5$), while others underperform (e.g. for $x \approx 1.3$) (Color figure online)

4.3 Quantitative Results

Here, we evaluate each algorithm w.r.t. three criteria: their regression accuracy on the test set, the complexity of the learned models, and the complexity of their partition scheme relative to the complexity of the ground truth partition. For the regression accuracy, we report the typical test R^2 score. For the complexity of the models, we report the number of nodes in the resulting expression tree (for UPSR and gplearn). For PS-tree, this metric accordingly corresponds to the sum of the number of nodes in each leaf of the DT. As for the partition complexity, we define $\Delta_{regions} = |k_{regions} - \widehat{k_{regions}}|$, where $k_{regions}$ and $\widehat{k_{regions}}$ respectively denote the ground truth number of subregions and the number of subregions effectively discovered by the SR algorithm. For PS-Tree, $\widehat{k_{regions}}$ is by construction equivalent to the number of leaves in the DT. For UPSR, $\widehat{k_{regions}}$ is by design equivalent to the number of *Piece* instances in the model, plus one. For gplearn, $\widehat{k_{regions}} = 1$, given that this method only produces non-piecewise models. Given that all three algorithms in our benchmark are based on GP, which is notoriously sensitive to its initialization, we report aggregated results by taking the average, across all datasets, of the best run on each dataset.

Table 2 first summarizes the results for non-piecewise data. As expected, all three algorithms are able to discover a perfect regressor on all datasets (test $R^2 = 1$ with 0 variance), in both monovariate and multivariate cases. Moreover, UPSR does so with little reliance on *Piece* ($\Delta_{regions} < 1$), leading to simple and human-readable full models, similar to gplearn. By contrast, PS-Tree mistakenly partitions the input space into multiple subregions, and consequently learns complex submodels for each of these subregions. In particular, in the multivariate case, the resulting full model is more complex than UPSR/gplearn by two orders of magnitude (> 1000 nodes), and practically becomes a black-box. Moreover, the fact that, for PS-Tree only, the variance in complexity is higher than its mean highlights the volatility of the adaptive partition scheme. A tentative conclusion from these findings is therefore that, when tasked with performing SR on experimental data with no *a priori* knowledge on its structure, one should probably first rely on either traditional non-piecewise SR algorithms, or piece-wise SR with an implicit partition scheme (such as the one in UPSR) rather than an explicit one (like in DT-based SR).

Table 2. Aggregated results (standard deviation in parentheses, best performance in bold) on non-piecewise datasets.

Algorithm	Monovariate ($k_{regions} = 1$)			Multivariate ($k_{regions} = 1$)		
	Test R^2	$\Delta_{regions}$	Nodes	Test R^2	$\Delta_{regions}$	Nodes
PSTree	**1.00 (0.00)**	4.70 (1.16)	66.10 (12.63)	**1.00 (0.00)**	7.60 (6.20)	1129.60 (1569.29)
UPSR (ours)	**1.00 (0.00)**	**0.00 (0.00)**	5.33 (0.92)	**1.00 (0.00)**	0.60 (0.52)	12.50 (3.57)
gplearn	**1.00 (0.00)**	**0.00 (0.00)**	5.80 (2.53)	**1.00 (0.00)**	**0.00 (0.00)**	**7.80 (0.63)**

Table 3. Aggregated results on piecewise datasets with exclusive submodels.

Algorithm	Monovariate (*exclusive* submodels)			Multivariate (*exclusive* submodels)		
	Test R^2	$\Delta_{regions}$	Nodes	Test R^2	$\Delta_{regions}$	Nodes
PSTree	0.99 (0.02)	5.43 (2.64)	223.73 (237.60)	**0.91 (0.11)**	10.60 (1.27)	1574.20 (1536.57)
UPSR (ours)	**1.00 (0.00)**	**0.62 (0.71)**	27.59 (13.16)	0.87 (0.07)	**0.89 (0.71)**	59.78 (21.45)
gplearn	0.76 (0.44)	2.00 (0.00)	**26.90 (25.11)**	0.70 (0.28)	3.00 (0.00)	**37.80 (19.81)**

Table 4. Aggregated results on piecewise datasets with stacking submodels.

Algorithm	Monovariate (*stacking* submodels)			Multivariate (*stacking* submodels)		
	Test R^2	$\Delta_{regions}$	Nodes	Test R^2	$\Delta_{regions}$	Nodes
PSTree	0.97 (0.06)	5.71 (3.62)	196.00 (180.99)	**0.96 (0.01)**	10.20 (2.53)	4661.20 (3532.96)
UPSR (ours)	**1.00 (0.01)**	**0.65 (0.67)**	23.86 (8.92)	**0.96 (0.03)**	**1.10 (0.74)**	37.20 (11.21)
gplearn	0.82 (0.30)	2.00 (0.00)	**18.40 (12.57)**	0.76 (0.25)	3.00 (0.00)	**23.30 (15.67)**

Tables 3 and 4 thereafter report the results respectively for *exclusive* and *stacking* piece-wise datasets. Expectedly, gplearns falls significantly behind both PS-Tree and UPSR in terms of regression accuracy, as it is unable to capture the piece-wise nature of the data, thereby learning frugal but ineffective models. Furthermore, UPSR is shown to be efficient at retrieving the data partition structure ($\Delta_{regions} \approx 1$ in all situations). By contrast, PS-Tree significantly overestimates the data partition complexity ($\Delta_{regions} > 5$). In particular, it should be noted that for multivariate cases, PS-Tree routinely reached its maximum partition budget $max\_leaf\_nodes = 15$ (as evidenced by the high mean and low variance for multivariate $\Delta_{regions}$), indicating that the partition would certainly be even more complex if we pre-set a higher value for

max_leaf_nodes. Similarly as for non-piecewise data, the observed model complexity is at least one order of magnitude higher for PS-Tree compared to UPSR and gplearn, thus hardly interpretable by human experts.

Despite significantly lower partition and submodel complexities, UPSR is shown to achieve competitive or better performance compared to PS-Tree in terms of regression accuracy for all but the multivariate *exclusive* piece-wise case. The drop in UPSR regression performance from a *stacking* context to an *exclusive* one was expected: the implicit partition scheme in UPSR requires tuning double the amount of subregion boundary parameters for the *exclusive* case compared to the *stacking* one. Interestingly enough, the average PS-Tree model complexity approximately triples between multivariate *exclusive* and *stacking* cases. This is in line with our expectation that the explicit partition scheme in a DT is ill-suited for the *stacking* submodels paradigm.

5 Conclusion

In this paper, we have introduced a new framework for performing piece-wise Symbolic Regression (SR), in which input space partition and discovery of local regressors are simultaneous. The output is a single piece-wise model unifying all local regressors, hence the name *Unified Piecewise Symbolic Regression* (UPSR). In order to assess the potential of this approach, we have created an extensive and challenging benchmark of 100 synthetic datasets. UPSR has been shown to outperform a traditional non-piecewise SR method, as well as state-of-the-art Decision Tree-based piecewise SR, both qualitatively and quantitatively. In particular, UPSR produces models of significantly higher sparsity and interpretability than state-of-the-art piecewise SR. Both of these properties are especially valuable in the context of SR, allowing it to stand out from typical black-box methods.

We intend to further improve UPSR to be suitable for broader applications. The main priority for upcoming work is to assess the potential of UPSR on industrial noisy experimental data, where we expect the implicit partition scheme to be an advantage compared to traditional non-piecewise SR. A clear direction for future research is to link multiple instances of the *Piece* operator together, so as to impose continuity of the unified model when crossing the boundaries between subregions. Finally, it will be important to work on improving syntaxic stability, thereby increasing reliability of SR to provide new insights for human experts.

References

1. Bakurov, I., Castelli, M., Gau, O., Fontanella, F., Vanneschi, L.: Genetic programming for stacked generalization. Swarm Evol. Comput. **65**, 100913 (2021)
2. Bradshaw, P., Huang, G.P.: The law of the wall in turbulent flow. Proc. Roy. Soc. London Ser. A: Math. Phys. Sci. **451**(1941), 165–188 (1995)
3. Chaboche, J.L.: A review of some plasticity and viscoplasticity constitutive theories. Int. J. Plast **24**(10), 1642–1693 (2008)
4. Cranmer, M., et al.: Discovering symbolic models from deep learning with inductive biases. In: Advances in Neural Information Processing Systems, vol. 33, pp. 17429–17442 (2020)

5. Deb, K., Pratap, A., Agarwal, S., Meyarivan, T.: A fast and elitist multiobjective genetic algorithm: NSGA-II. IEEE Trans. Evol. Comput. **6**(2), 182–197 (2002)
6. Flasch, O., Friese, M., Zaefferer, M., Bartz-Beielstein, T., Branke, J.: Learning model-ensemble policies with genetic programming. Technical report (2015)
7. Fong, K.S., Motani, M.: Multi-level symbolic regression: function structure learning for multi-level data. In: International Conference on Artificial Intelligence and Statistics, pp. 2890–2898 (2024)
8. Fortin, F.A., De Rainville, F.M., Gardner, M., Parizeau, M., Gagné, C.: DEAP: evolutionary algorithms made easy. J. Mach. Learn. Res. **13**(1), 2171–2175 (2012)
9. Gad, A.F.: PyGAD: an intuitive genetic algorithm python library. Multimed. Tools Appl. **83**(20), 58029–58042 (2024)
10. Holzinger, A., Biemann, C., Pattichis, C.S., Kell, D.B.: What do we need to build explainable AI systems for the medical domain? (2017). https://arxiv.org/pdf/1712.09923
11. Kommenda, M., Beham, A., Affenzeller, M., Kronberger, G.: Complexity measures for multi-objective symbolic regression. In: Moreno-Díaz, R., Pichler, F., Quesada-Arencibia, A. (eds.) EUROCAST 2015. LNCS, vol. 9520, pp. 409–416. Springer, Cham (2015). https://doi.org/10.1007/978-3-319-27340-2_51
12. Kommenda, M., Burlacu, B., Kronberger, G., Affenzeller, M.: Parameter identification for symbolic regression using nonlinear least squares. Genet. Program Evolvable Mach. **21**(3), 471–501 (2020)
13. Kommenda, M., Kronberger, G., Winkler, S., Affenzeller, M., Wagner, S.: Effects of constant optimization by nonlinear least squares minimization in symbolic regression. In: Proceedings of the 15th Annual Conference Companion on Genetic and Evolutionary Computation, pp. 1121–1128 (2013)
14. La Cava, W.G., Lee, P.C., Ajmal, I., et al.: A flexible symbolic regression method for constructing interpretable clinical prediction models. NPJ Digit. Med. **6**(1), 107 (2023)
15. Lefakis, L., Zadorozhnyi, O., Blanchard, G.: Efficient regularized piecewise-linear regression trees. arXiv preprint arXiv:1907.00275 (2019)
16. Lepri, B., Oliver, N., Letouzé, E., Pentland, A., Vinck, P.: Fair, transparent, and accountable algorithmic decision-making processes: the premise, the proposed solutions, and the open challenges. Philos. Technol. **31**(4), 611–627 (2018)
17. Li, B., Friedman, J., Olshen, R., Stone, C.: Classification and regression trees (CART). Biometrics **40**(3), 358–361 (1984)
18. Ly, D.L., Lipson, H.: Learning symbolic representations of hybrid dynamical systems. J. Mach. Learn. Res. **13**(1), 3585–3618 (2012)
19. Meurer, A., et al.: SymPy: symbolic computing in python. PeerJ Comput. Sci. **3**, e103 (2017)
20. Moré, J.J.: The Levenberg-Marquardt algorithm: implementation and theory. In: Numerical Analysis: Proceedings of the Biennial Conference Held at Dundee, 28 June–1 July 1977, pp. 105–116. Springer, Heidelberg (2006)
21. Orzechowski, P., La Cava, W., Moore, J.H.: Where are we now? A large benchmark study of recent symbolic regression methods. In: Proceedings of the Genetic and Evolutionary Computation Conference, pp. 1183–1190 (2018)
22. Petersen, B.K., Landajuela, M., Mundhenk, T.N., Santiago, C.P., Kim, S.K., Kim, J.T.: Deep symbolic regression: recovering mathematical expressions from data via risk-seeking policy gradients (2019). https://arxiv.org/pdf/1912.04871
23. Poli, R., Langdon, W.B., McPhee, N.F.: A field guide to genetic programming (2008). http://www.gp-field-guide.org.uk
24. Price, K.V.: Differential evolution. In: Handbook of Optimization: From Classical to Modern Approach, pp. 187–214. Springer, Heidelberg (2013)
25. Russeil, E., De Franca, F.O., Malanchev, K., et al.: Multiview symbolic regression. In: Proceedings of the Genetic and Evolutionary Computation Conference, pp. 961–970 (2024)

26. Ryan, C., Keijzer, M.: An analysis of diversity of constants of genetic programming. In: European Conference on Genetic Programming, pp. 404–413. Springer, Heidelberg (2003)
27. Saeed, W., Omlin, C.: Explainable AI (XAI): a systematic meta-survey of current challenges and future opportunities. Knowl.-Based Syst. **263**, 110273 (2023)
28. Sheta, A.F., Ahmed, S., Faris, H.: Evolving stock market prediction models using multi-gene symbolic regression genetic programming. Artif. Intell. Mach. Learn. **15**(1), 11–20 (2015)
29. Staats, K., Pantridge, E., Cavaglia, M., Milovanov, I., Aniyan, A.: TensorFlow enabled genetic programming. In: Proceedings of the Genetic and Evolutionary Computation Conference Companion, pp. 1872–1879 (2017)
30. Stephens, T.: Genetic programming in python, with a scikit-learn inspired API: gplearn (2016). https://gplearn.readthedocs.io/en/stable/intro.html
31. Sun, S., Ouyang, R., Zhang, B., Zhang, T.-Y.: Data-driven discovery of formulas by symbolic regression. MRS Bull. **44**(7), 559–564 (2019). https://doi.org/10.1557/mrs.2019.156
32. Udrescu, S.M., Tan, A., Feng, J., Neto, O., Wu, T., Tegmark, M.: AI Feynman 2.0: pareto-optimal symbolic regression exploiting graph modularity. In: Advances in Neural Information Processing Systems, vol. 33, pp. 4860–4871 (2020)
33. Udrescu, S.M., Tegmark, M.: AI Feynman: a physics-inspired method for symbolic regression. Sci. Adv. **6**(16), eaay2631 (2020)
34. Vladislavleva, E.J., Smits, G.F., Den Hertog, D.: Order of nonlinearity as a complexity measure for models generated by symbolic regression via pareto genetic programming. IEEE Trans. Evol. Comput. **13**(2), 333–349 (2008)
35. Wang, Y., Wagner, N., Rondinelli, J.M.: Symbolic regression in materials science. MRS Commun. **9**(3), 793–805 (2019). https://doi.org/10.1557/mrc.2019.85
36. Zhang, H., Zhou, A., Qian, H., Zhang, H.: PS-tree: a piecewise symbolic regression tree. Swarm Evol. Comput. **71**, 101061 (2022)

Was Tournament Selection All We Ever Needed? A Critical Reflection on Lexicase Selection

Alina Geiger[(✉)], Martin Briesch, Dominik Sobania, and Franz Rothlauf

Johannes Gutenberg University, Mainz, Germany
{geiger,briesch,dsobania,rothlauf}@uni-mainz.de

Abstract. The success of lexicase selection has led to various extensions, including its combination with down-sampling, which further increased performance. However, recent work found that down-sampling also leads to significant improvements in the performance of tournament selection. This raises the question of whether tournament selection combined with down-sampling is the better choice, given its faster running times. To address this question, we run a set of experiments comparing ϵ-lexicase and tournament selection with different down-sampling techniques on synthetic problems of varying noise levels and problem sizes as well as real-world symbolic regression problems. Overall, we find that down-sampling improves generalization and performance even when compared over the same number of generations. This means that down-sampling is beneficial even with way fewer fitness evaluations. Additionally, down-sampling successfully reduces code growth. We observe that population diversity increases for tournament selection when combined with down-sampling. Further, we find that tournament selection and ϵ-lexicase selection with down-sampling perform similar, while tournament selection is significantly faster. We conclude that tournament selection should be further analyzed and improved in future work instead of only focusing on the improvement of lexicase variants.

Keywords: Tournament Selection · Lexicase Selection · Down-Sampling · Genetic Programming · Symbolic Regression

1 Introduction

Lexicase selection [24,39] was introduced just over a decade ago and has been quickly adapted in many domains in the field of genetic programming (GP). In comparison to traditional selection methods such as tournament selection, lexicase selection considers each training case individually and not in a compressed form (e.g. as fitness value). In program synthesis [18,22,38], the standard version of lexicase was already able to achieve significantly better results than, e.g., tournament selection or fitness-proportionate selection. With extensions such as ϵ-lexicase selection [27,28], this success could also be extended to symbolic regression [33], where standard lexicase cannot be used effectively.

A few years ago, a combination of lexicase and random down-sampling strategies [8,13,25] further improved performance. Instead of considering all available training cases in each generation of a GP run, down-sampling based selection methods use only a random subset, which significantly speeds up the execution time per generation [25]. Boldi et al. [4] improved this naïve random approach by introducing informed down-sampled lexicase selection which leverages population statistics to select subsets of relevant training cases.

However, recent studies [15,17] show that a combination of down-sampling with tournament selection reduces overfitting and bloat and also leads to high quality results. Considering this previous work, the question arises: was tournament selection all we ever needed?

This paper addresses this question in detail and compares ϵ-lexicase selection and tournament selection with both random and informed down-sampling. We analyze the performance, generalization behavior, diversity as well as size of the generated solutions on several synthetic and real-world symbolic regression problems. For the synthetic problems, we analyze different noise levels and different problem sizes to understand if performance is related to those problem characteristics. We compare all methods over the same number of generations instead of a given evaluation budget, because we assume that down-sampling is beneficial even with a much lower number of fitness evaluations.

We find that down-sampling improves the performance of both selection methods even when using only a small fraction of the fitness evaluations. The larger the problem is, the stronger the performance improvement. The increase in performance comes along with better generalization behavior when using down-sampling techniques. Moreover, we find that the improvements using down-sampling techniques are stronger for tournament selection than for ϵ-lexicase selection. Tournament selection combined with down-sampling performs similar to ϵ-lexicase selection, with tournament selection being considerably faster. With respect to population dynamics, down-sampling reduces code growth, especially when combined with tournament selection. Further, we find that down-sampling increases population diversity when combined with tournament selection. We conclude that down-sampling techniques close the gap between tournament selection and ϵ-lexicase selection for symbolic regression problems, making tournament selection the better choice.

We discuss related work in Sect. 2 and describe our experimental setup including benchmark problems and parameter settings in Sect. 3. Section 4 presents our results, followed by a discussion in Sect. 5. We conclude our work in Sect. 6.

2 Related Work

Tournament selection is a commonly used selection method, where n randomly chosen individuals compete in a tournament and the best individual is selected as a parent for the next generation [35]. One disadvantage of tournament selection is that the quality of individuals is measured in terms of an aggregated value, leading to a loss of information about the data structure [26]. In contrast, lexicase selection [24,39] evaluates individuals on each training case separately. This is beneficial during search, as individuals that solve certain training cases particularly well are preserved [20,21,34]. Additionally, it has been found that lexicase selection is able to maintain a high population diversity in comparison to tournament selection [19]. Lexicase selection has been successfully applied in many problem domains [1,7,22,24,30,31,37,38].

For symbolic regression problems with continuous-valued errors, ϵ-lexicase selection [27,28] significantly improved the solution quality compared to tournament selection and standard lexicase selection. ϵ-lexicase selection also considers individuals for selection that are near-elite on training cases by applying an ϵ-threshold to standard lexicase selection.

Recently, the combination of lexicase selection with down-sampling improved the solution quality even more in the domain of program synthesis [8,25] and symbolic regression [13–15]. A naïve down-sampling strategy is random down-sampling [16,17], where only a random subset of training cases is used in each generation to evaluate the quality of the individuals. It has been found that random down-sampling in combination with tournament selection improves performance, reduces overfitting and controls bloat in the domain of symbolic regression [17,29]. However, reduced overfitting was not (or only insignificantly) observed when combining random down-sampling with lexicase selection for program synthesis problems [23,36].

One drawback of random down-sampling is that the random selection of a subset of training cases may lead to the exclusion of important training cases for several generations. Therefore, Boldi et al. [4] proposed informed down-sampling, which creates more diverse subsets of training cases by using population statistics. Due to the success of informed down-sampling in program synthesis [4] and symbolic regression [15] its influence has been further analyzed for program synthesis problems [2,3,5].

Both random and informed down-sampling have been found to improve performance not only for lexicase selection but also (among others) for tournament selection [3,15]. Therefore, it is unclear if we need lexicase selection or if tournament selection combined with down-sampling is sufficient considering that tournament selection is significantly faster than lexicase selection.

3 Experimental Setup

In this section, we present the benchmark problems and GP parameter settings in detail.

3.1 Benchmark Problems

Our experiments are twofold: First, we study synthetic problems where we are able to modify the problem characteristics such as the level of noise and the number of features in order to better understand the effects of down-sampling. Second, we test our conclusions drawn on the synthetic problems on a representative set of real-world benchmark problems. The problem characteristics are described in Table 1.

Table 1. Benchmark problems with the number of features and the number of instances.

Problem	# Features	# Instances	Type
friedman1	10	100	synthetic
friedman2	4	100	synthetic
friedman3	4	100	synthetic
210_cloud	5	108	real-world
230_machine_cpu	6	209	real-world
207_autoPrice	15	159	real-world
505_tecator	124	240	real-world

For our experiments with synthetic problems, we generate Friedman datasets as described in [6,11]. For all Friedman datasets, we study noise levels of 0.0 (no noise), 0.05, 0.1, and 0.15 as in [10].[1] Further, we explore the influence of varying numbers of features. Therefore, we generate the FRIEDMAN1 problem with 5, 10, 25, and 50 features. It is important to note that only 5 features are actually correlated with the target output y.

In addition to the synthetic problems, we select 4 real-world regression datasets from PMLB [32] with less than 250 instances from diverse domains and with varying numbers of features. We focus on small datasets to finish our experiments in a reasonable amount of time, as we run GP over a large number of generations. Further, this allows us to better observe the generalization gap.

3.2 Parameter Settings

Our experiments are implemented using the DEAP Framework [9]. The parameter settings of the tree-based GP runs are shown in Table 2.

All experiments are performed using a population size of $N = 500$ and a generation limit of $G = 2,000$. We perform the evolutionary runs over a larger number of generations to better observe the generalization gap. Our setup differs from previous studies as evolutionary runs including down-sampling are usually given much more generations compared to runs without down-sampling [13, 25]. However, it has been found that down-sampling has positive effects beyond saving evaluation costs per generation such as overfitting and bloat control [17]. Therefore, we assume that down-sampling can be beneficial even when compared

---

[1] Noise is only applied to the instances used for training. The test set is noise-free.

Table 2. Parameter settings of our GP approach.

Parameter	Value
Population size	500
Generation limit	2,000
Primitive set	$\{\mathbf{x}, \mathrm{ERC}, +, -, *, \mathrm{AQ}, \sin, \cos, \mathrm{neg}\}$
ERC values	$\{-1, 0, 1\}$
Initialization method	Ramped half-and-half
Maximum tree depth	17
Crossover probability	95%
Mutation probability	5%
Runs	30

over a given number of generations. This means that the evolutionary runs with down-sampling use only a fraction of the fitness evaluations used by the ones without down-sampling.

For each problem, we split the number of instances into 70% training cases and 30% test cases. The down-sampling parameter is set to $d = 0.1$ [13], meaning that only 10% of the training cases are used to evaluate the quality of the individuals in each generation. For informed down-sampling, we set the parent sampling rate to $s = 0.01$ and the distance calculation scheduling parameter to $k = 10$ [4].

The fitness measure used to compare individuals in the selection process depends on the selection method. Tournament selection compares individuals based on an aggregated value. Therefore, we measure fitness for runs with tournament selection in terms of the mean squared error (MSE). We set the tournament size to $n = 7$. ϵ-lexicase selection selects individuals based on their performance on individual training cases. Hence, for runs with ϵ-lexicase selection, the squared errors of the individuals on the training cases are considered during selection.

In each generation, we save the individual with the lowest MSE on all training cases as the current best solution. This individual is then evaluated on the unseen test cases with respect to the MSE. Based on that, we calculate the generalization gap as the difference between the MSE on the test cases and the MSE on the training cases. Besides performance, we are interested in the influence of down-sampling on population dynamics. Therefore, we measure in each generation the diversity in the population in terms of error diversity, which is defined as the percentage of distinct error vectors in the population. Moreover, we track the median tree size in the population, where tree size is defined as the number of nodes per tree.

4 Results

In this section, we present our experimental results. First, we analyze performance, generalization, diversity, and size of the different selection approaches for synthetic regression problems. Second, we extend our findings to real-world problems. A statistical analysis of our results is provided in the supplementary material (see Appendix D) in our Zenodo repository [12].

4.1 Synthetic Problems with Noise

We first study the performance and population statistics of tournament and ϵ-lexicase selection with different down-sampling techniques on the synthetic Friedman problems with varying degrees of noise. We investigate no down-sampling (nds), random down-sampling (rds), and informed down-sampling (ids) variants of both tournament ($tourn$) and ϵ-lexicase (ϵ-lex) selection. All results are shown for 30 runs for each combination of down-sampling and selection method.

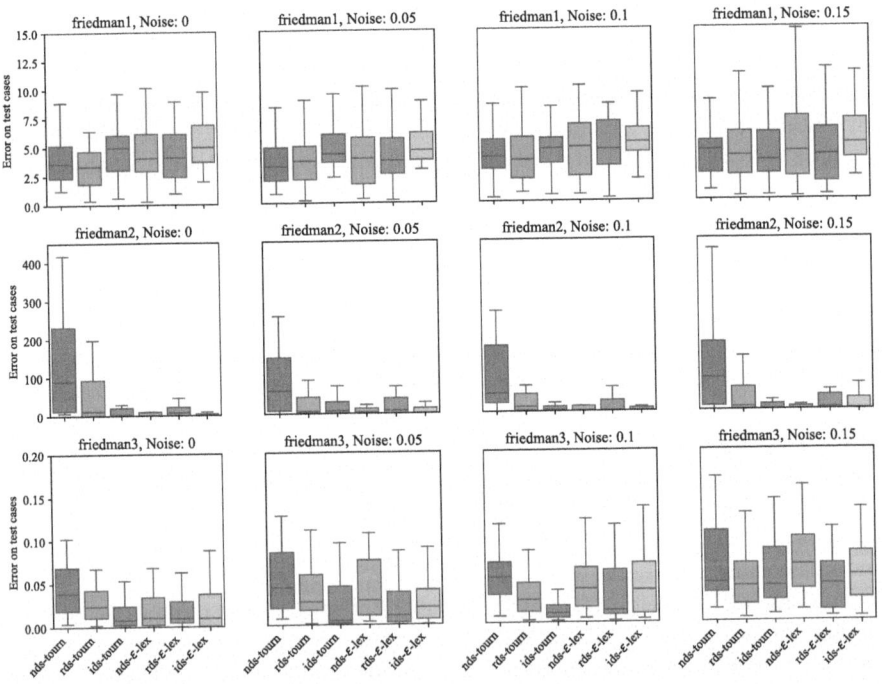

Fig. 1. Test fitness of the best individual for the synthetic problems across different noise levels. Outliers are not shown to improve readability.

Figure 1 displays the test error of the best individual found by the respective selection method in an evolutionary run.

We observe that the down-sampling methods often improve upon or perform equal to the selection method without down-sampling. Only for the FRIEDMAN1 problem with low noise the down-sampling variants are slightly worse. This is interesting considering that tournament and ϵ-lexicase selection without down-sampling have 10 times more fitness evaluations during training, clearly indicating that down-sampling is beneficial even beyond the saved computational effort. Another observation is that tournament selection without down-sampling performs significantly worse than ϵ-lexicase selection without down-sampling on the FRIEDMAN2 and FRIEDMAN3 problems. However, when combined with either random or informed down-sampling tournament selection performs equally well as ϵ-lexicase selection at all noise levels.

To gain a further understanding of the performance of the different selection methods and down-sampling methods we also investigate their generalization behavior. Figure 2 displays the generalization gap (difference between test and training error) of the solution per evolutionary run.

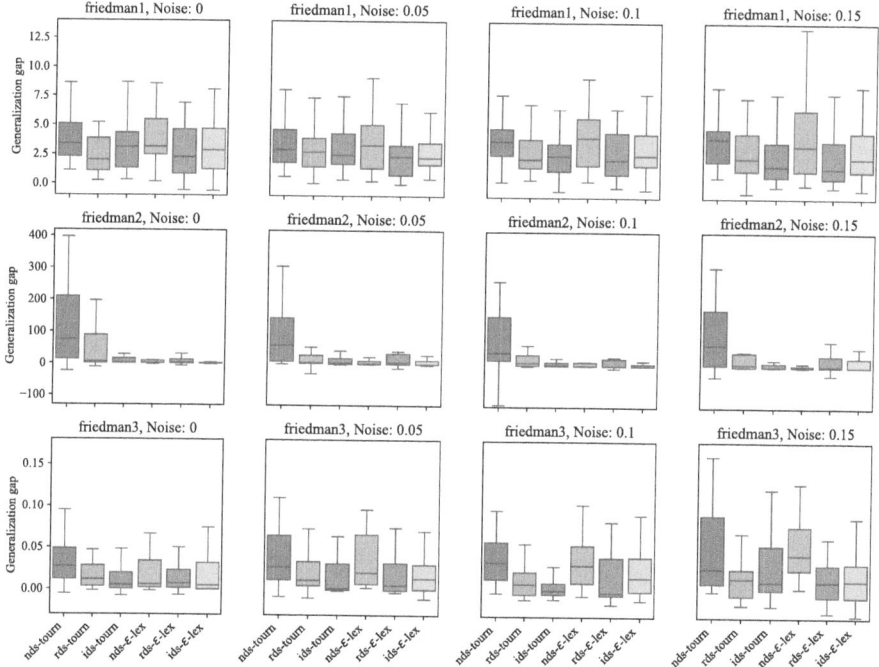

Fig. 2. Generalization gap between test and training fitness of the best individual for the synthetic problems across different noise levels. Outliers are not shown to improve readability.

In most settings, we observe that down-sampling decreases the generalization gap between training and test error for both selection methods, effectively working as a regularization technique. Especially for tournament selection this effect is apparent. This indicates that both random and informed down-sampling can reduce the generalization gap and negate the difference between both selection methods.

We also provide more details of the training and test error recorded over 2,000 generations in the supplementary material (see Appendix A) [12].

Another advantage of lexicase selection often mentioned in literature is high diversity preservation within a population [19]. Figure 3 displays the diversity (defined as the percentage of distinct error vectors in the population) over generations for the different selection and down-sampling approaches. The graphs are smoothed using a moving average with a window size of 20 generations. Line markers are added every 200 generations for better readability.

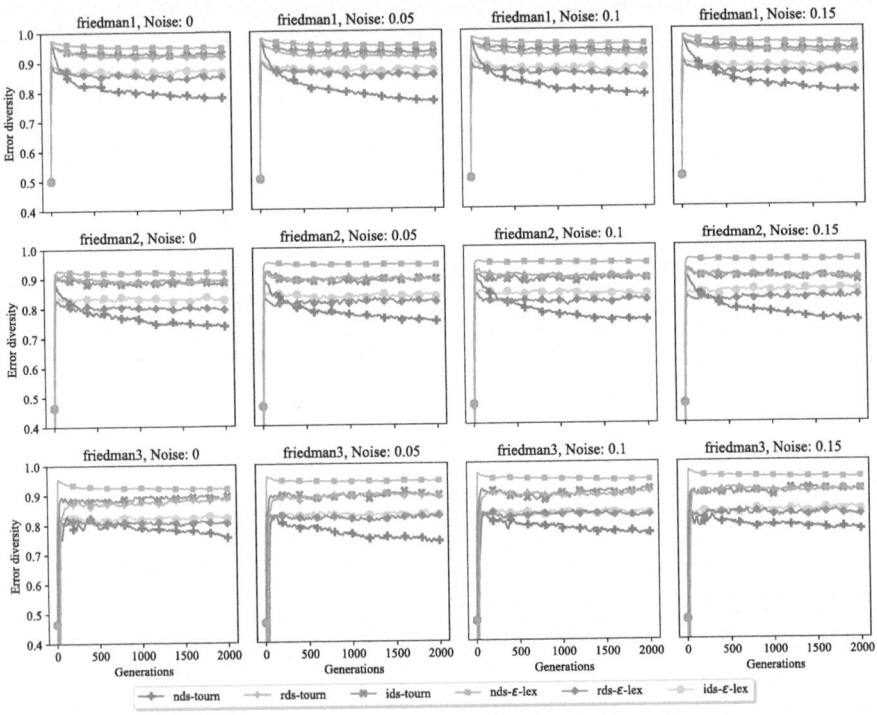

Fig. 3. Error diversity of individuals in a population over generations for the synthetic problems across different noise levels. Median over 30 runs is shown.

As expected, no down-sampling ε-lexicase selection exhibits the highest diversity and no down-sampling tournament exhibits the lowest diversity across all problems and noise levels. However, when paired with either random or informed down-sampling the diversity of tournament selection raises significantly, nearly reaching the level of ε-lexicase selection and even surpassing random and informed down-sampled ε-lexicase selection. This means that down-sampling enables tournament selection to also benefit from higher population diversity.

Lastly, Fig. 4 displays the median size of individuals (measured as the number of nodes) within a population. Once again, the graphs are smoothed using a moving average with a window size of 20 generations and line markers are added every 200 generations. We observe a very strong code growth for tournament selection without down-sampling across all problems and noise levels with a median size of up to 4 times as much tree nodes compared to no down-sampling ϵ-lexicase selection. However, when paired with either random or informed down-sampling this effect is negated similar to the difference in diversity. While the down-sampled ϵ-lexicase selection variants produce the smallest individuals, down-sampled tournament selection produces smaller individuals than no down-sampling ϵ-lexicase selection and comes very close to the down-sampling variants. We also observe that informed down-sampling controls code growth even better than random down-sampling for both tournament and ϵ-lexicase selection.

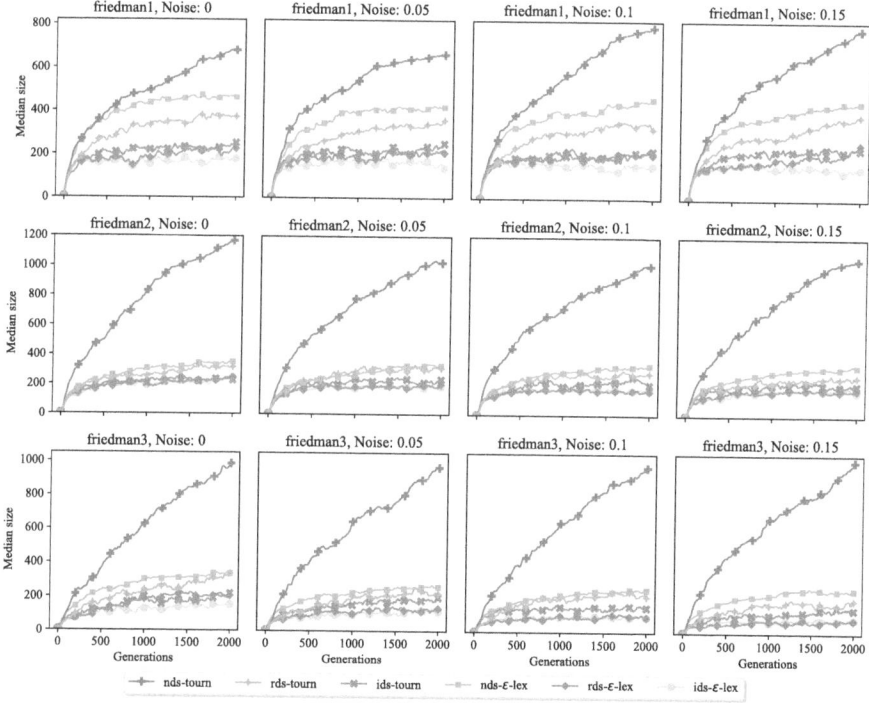

Fig. 4. Median size of individuals in a population (measured in tree nodes) over generations for the synthetic problems across different noise levels. Median over 30 runs is shown.

Overall we observe that for the synthetic Friedman problems both down-sampling techniques close the gap in performance between tournament and ϵ-lexicase selection. Additionally, down-sampling improves generalization and reduces code growth, especially for tournament selection. Lastly, down-sampling improves diversity for tournament selection and lowers diversity for ϵ-lexicase selection.

4.2 Synthetic Problems with Varying Numbers of Features

We are also interested in how the performance of the selection methods change when we adjust the problem size by increasing or decreasing the feature space. Therefore, we conduct experiments for the FRIEDMAN1 problem with varying numbers of features (5, 10, 25, 50) and no noise.

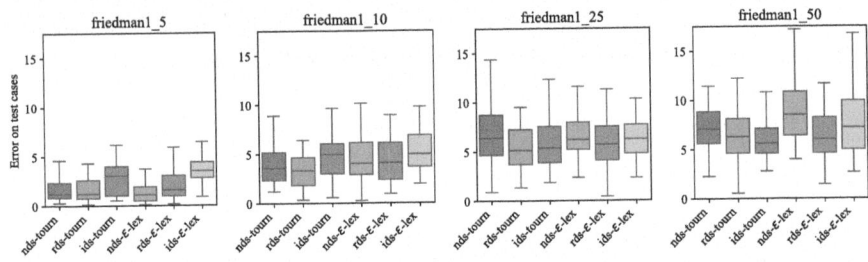

Fig. 5. Test fitness of the best individual for the synthetic FRIEDMAN1 problem with varying numbers of features (5, 10, 25, 50). Outliers are not shown to improve readability.

Figure 5 displays the test error of the best individual obtained by the different selection methods. We observe that for the problem instance with only 5 features (that are all correlated to the target variable) both tournament and ϵ-lexicase selection perform better without down-sampling and informed down-sampling performs worst. However, when increasing the problem size, down-sampling performs better than no down-sampling. For the problem instance with 25 features (and only 5 of those are correlated to the target variable), random and informed down-sampled tournament selection achieve the best median error on the test cases. In the largest problem instance with 50 features, informed down-sampled tournament selection achieves the best median test error.

We observe similar findings for the generalization gap displayed in Fig. 6. There are no major differences in the generalization gaps between tournament and ϵ-lexicase selection as well as no down-sampling and random or informed down-sampling for the small problem instance with 5 variables. However, when increasing the number of features both down-sampling techniques successfully reduce the generalization gap.

We still observe an increase in diversity and better code growth control for tournament when using down-sampling. However, there is no notable difference between problem sizes (see Appendix C in the supplementary material [12]).

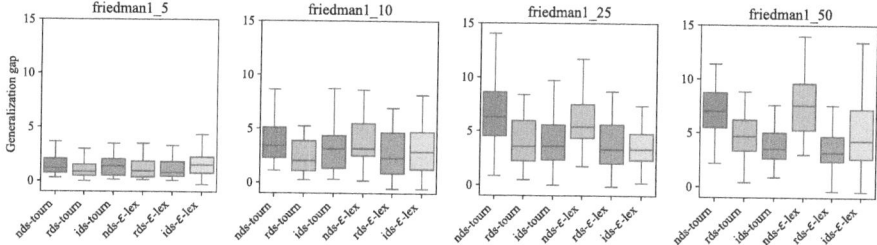

Fig. 6. Generalization gap between test and training fitness of the best individual for the synthetic FRIEDMAN1 problem with varying numbers of features (5, 10, 25, 50). Outliers are not shown to improve readability.

Overall, these findings indicate that the benefits of down-sampling are more apparent in larger problems and down-sampled tournament selection scales better than ϵ-lexicase selection with increasing problem size.

4.3 Real-World Problems of Varying Size

Finally, we extend our findings to four real-world problems of varying size. Figure 7 displays the test error of the best individual obtained by each selection method for the four different real-world problems. Both the CLOUD and MACHINE_CPU dataset are the smaller problems with 5 and 6 features respectively, followed by AUTOPRICE dataset with 15 features. The TECATOR dataset is the largest real-world dataset in our experiments with 124 features.

Similar to our results for the synthetic problems we observe that ϵ-lexicase selection performs slightly better than tournament selection for the two smaller problems (CLOUD and MACHINE_CPU) and there is no notable difference between no down-sampling and random or informed down-sampling for those datasets. For the larger AUTOPRICE problem, down-sampling slightly improves performance for both tournament and ϵ-lexicase selection with random down-sampling tournament selection achieving the best median test error. For the largest problem (TECATOR) especially informed down-sampling improves both tournament and ϵ-lexicase when compared to no or random down-sampling and informed down-sampled tournament selection achieves the best median test error.

A similar picture is observed for the generalization behavior on those problems as displayed in Fig. 8. There are no major differences in the generalization gap between no down-sampling and random or informed down-sampling for the CLOUD and MACHINE_CPU dataset and the differences between tournament and ϵ-lexicase selection are only marginal with ϵ-lexicase selection having a slightly lower generalization gap on the MACHINE_CPU dataset. However, on the AUTOPRICE dataset, down-sampling clearly reduces the generalization gap, especially for tournament selection. On the TECATOR dataset, informed down-sampling achieves the smallest generalization gap for both selection methods. Additionally, we observe a higher variance for ϵ-lexicase selection.

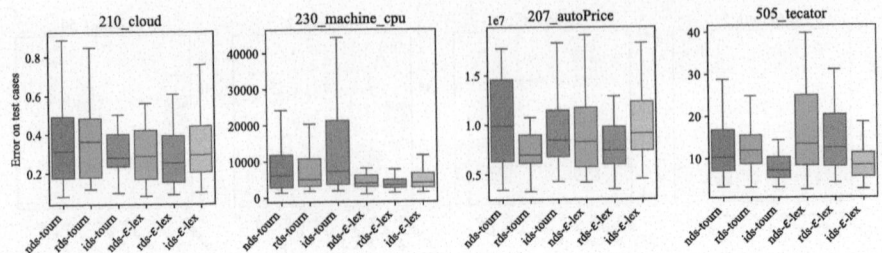

Fig. 7. Test fitness of the best individual for the real-world problems. Outliers are not shown to improve readability.

More details on the training and test error over generations are provided in the supplementary material (see Appendix B) [12], further confirming our results.

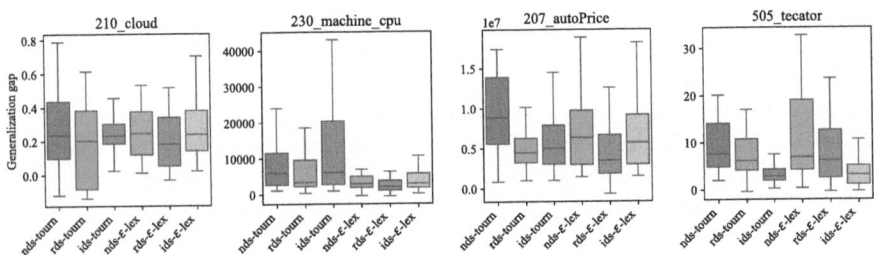

Fig. 8. Generalization gap between training and test fitness of the best individual for the real-world problems. Outliers are not shown to improve readability.

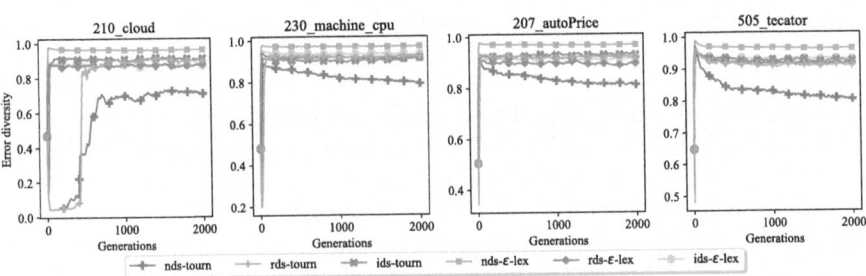

Fig. 9. Error diversity of individuals in a population over generations for the real-world problems. Median over 30 runs is shown.

Figure 9 displays the diversity over generations for the different selection methods. Similar to the synthetic problems we observe that ϵ-lexicase selection exhibits the highest diversity and tournament selection exhibits the lowest diversity. However, when paired with down-sampling the diversity of tournament selection is improved to the level of down-sampled ϵ-lexicase selection regardless of problem size, confirming our results observed for the synthetic problems.

We can also extend our findings regarding code growth control to the real-world datasets. As displayed in Fig. 10, down-sampling successfully reduces code growth for both tournament and ϵ-lexicase selection for all datasets. Especially for tournament selection this is crucial as tournament selection suffers from the most severe code growth in our experiments. With down-sampling (specifically informed down-sampling), however, the code growth of tournament selection mimics that of ϵ-lexicase.

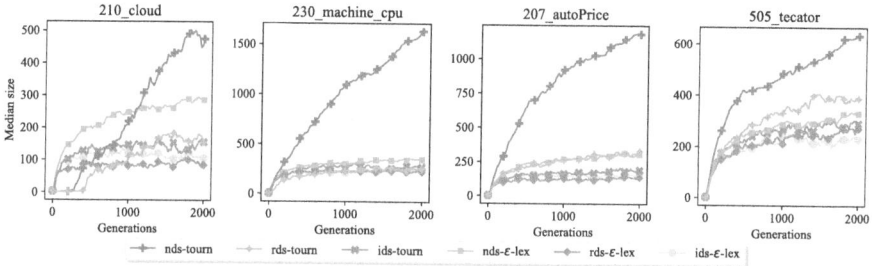

Fig. 10. Median size of individuals in a population (measured in tree nodes) over generations for the real-world problems. Median over 30 runs is shown.

Overall, we can confirm our findings obtained on the synthetic problems also for real-world problems: First, down-sampling is beneficial even when performed for the same number of generations; second, the benefit of down-sampling increases the larger a problem gets and informed down-sampling in particular performs best on larger problems; third, tournament selection with down-sampling scales better to larger problems than ϵ-lexicase; and fourth, down-sampling serves as code growth control and diversity preservation mechanism, providing tournament selection with the major benefits of ϵ-lexicase selection with no added cost.

5 Discussion

Our results demonstrate that tournament selection and ϵ-lexicase selection perform similar for symbolic regression problems when coupled with random or informed down-sampling techniques. Further, down-sampling also serves as a code growth control mechanism and preserves diversity for evolutionary runs with tournament selection to the degree ϵ-lexicase selection does. However, tournament selection has a way better worst case performance with $\mathcal{O}(TN)$ compared to lexicase selection with $\mathcal{O}(TN^2)$ for selecting N parents using T training cases [27]. This is also reflected in our empirical data: Tournament selection with down-sampling took around 40% less time than its ϵ-lexicase counterpart in our experiments.

However, our findings are limited to symbolic regression problems and might not generalize to other problem types. In the domain of program synthesis, other work suggests that lexicase still outperforms tournament selection even when combined with down-sampling [3].

Given the results we achieved with the combination of down-sampling and tournament selection as well as work in other domains, it becomes apparent that future research cannot only focus on a single selection method when providing new improvements but should evaluate their improvements holistically.

6 Conclusion

Prior work found that lexicase selection is superior to tournament selection in terms of performance and diversity maintenance [22,24,28]. However, down-sampling techniques [4,17] improved the performance of both selection methods and it has been found that tournament selection combined with informed down-sampling even surpasses the performance of ϵ-lexicase selection for symbolic regression problems in some cases [15]. With tournament selection being the faster selection method, we raised the question: Was tournament selection all we ever needed?

We approached this question by comparing ϵ-lexicase and tournament selection with random and informed down-sampling on synthetic as well as real-world symbolic regression problems in terms of performance, generalization behavior, diversity, and solution size. We found that down-sampling is even beneficial when comparing methods over the same number of generations. Further, we observed that down-sampling strongly reduces the generalization gap especially for larger problems. The performance improvements using down-sampling techniques are stronger for tournament selection. Additionally, we found that down-sampling with tournament selection and down-sampling with ϵ-lexicase selection perform similar, while down-sampled tournament selection runs significantly faster. Moreover, we found that down-sampling serves as a code growth control, which is especially relevant for tournament selection. Lastly, down-sampling strongly increases the population diversity for the evolutionary runs with tournament selection.

To answer our question, we argue that tournament selection might be all we ever needed, at least in the domain of symbolic regression, as it performed similar to ϵ-lexicase selection when combined with down-sampling, while being significantly faster than ϵ-lexicase selection. Therefore, we recommend future research to include tournament selection with down-sampling in their studies instead of only focusing on new lexicase variants. Further, tournament selection with down-sampling should be analyzed in other problem domains as well.

References

1. Aenugu, S., Spector, L.: Lexicase selection in learning classifier systems. In: Proceedings of the Genetic and Evolutionary Computation Conference, pp. 356–364. ACM (2019)
2. Boldi, R., et al.: The problem solving benefits of down-sampling vary by selection scheme. In: Proceedings of the Companion Conference on Genetic and Evolutionary Computation, pp. 527–530 (2023)
3. Boldi, R., et al.: Untangling the effects of down-sampling and selection in genetic programming. In: ALIFE 2024: Proceedings of the 2024 Artificial Life Conference. MIT Press (2024)
4. Boldi, R., et al.: Informed down-sampled lexicase selection: identifying productive training cases for efficient problem solving. Evol. Comput. 1–31 (2024)
5. Boldi, R., Lalejini, A., Helmuth, T., Spector, L.: A static analysis of informed down-samples. In: Proceedings of the Companion Conference on Genetic and Evolutionary Computation, GECCO 2023 Companion, pp. 531–534. Association for Computing Machinery, New York (2023)
6. Breiman, L.: Bagging predictors. Mach. Learn. **24**, 123–140 (1996)
7. Ding, L., Spector, L.: Optimizing neural networks with gradient lexicase selection. In: International Conference on Learning Representations (2021)
8. Ferguson, A.J., Hernandez, J.G., Junghans, D., Lalejini, A., Dolson, E., Ofria, C.: Characterizing the effects of random subsampling on lexicase selection. In: Genetic Programming Theory and Practice XVII, pp. 1–23. Springer (2020)
9. Fortin, F.A., de Rainville, F.M., Gardner, M.A., Parizeau, M., Gagné, C.: DEAP: evolutionary algorithms made easy. J. Mach. Learn. Res. **13**(1), 2171–2175 (2012)
10. de Franca, F., et al.: SRBench++: principled benchmarking of symbolic regression with domain-expert interpretation. IEEE Trans. Evol. Comput. (2024)
11. Friedman, J.H.: Multivariate adaptive regression splines. Ann. Stat. **19**(1), 1–67 (1991)
12. Geiger, A., Briesch, M., Sobania, D., Rothlauf, F.: Supplementary material of was tournament selection all we ever needed? A critical reflection on lexicase selection (2025). https://doi.org/10.5281/zenodo.14725848
13. Geiger, A., Sobania, D., Rothlauf, F.: Down-sampled epsilon-lexicase selection for real-world symbolic regression problems. In: Proceedings of the Genetic and Evolutionary Computation Conference, GECCO 2023, pp. 1109–1117. ACM (2023)
14. Geiger, A., Sobania, D., Rothlauf, F.: A comprehensive comparison of lexicase-based selection methods for symbolic regression problems. In: European Conference on Genetic Programming (Part of EvoStar), pp. 192–208. Springer (2024)
15. Geiger, A., Sobania, D., Rothlauf, F.: Lexicase-based selection methods with down-sampling for symbolic regression problems: overview and benchmark. arXiv preprint arXiv:2407.21632 (2024)
16. Gonçalves, I., Silva, S.: Experiments on controlling overfitting in genetic programming. In: 15th Portuguese Conference on Artificial Intelligence (EPIA 2011), pp. 10–13 (2011)
17. Gonçalves, I., Silva, S., Melo, J.B., Carreiras, J.M.B.: Random sampling technique for overfitting control in genetic programming. In: Genetic Programming, pp. 218–229. Springer, Heidelberg (2012)
18. Helmuth, T., Abdelhady, A.: Benchmarking parent selection for program synthesis by genetic programming. In: Proceedings of the 2020 Genetic and Evolutionary Computation Conference Companion, GECCO 2020, pp. 237–238. ACM (2020)

19. Helmuth, T., McPhee, N.F., Spector, L.: Effects of lexicase and tournament selection on diversity recovery and maintenance. In: Proceedings of the 2016 on Genetic and Evolutionary Computation Conference Companion, GECCO 2016 Companion, pp. 983–990. ACM (2016)
20. Helmuth, T., Pantridge, E., Spector, L.: Lexicase selection of specialists. In: Proceedings of the Genetic and Evolutionary Computation Conference, GECCO 2019, pp. 1030–1038. ACM (2019)
21. Helmuth, T., Pantridge, E., Spector, L.: On the importance of specialists for lexicase selection. Genet. Program Evolvable Mach. **21**(3), 349–373 (2020). https://doi.org/10.1007/s10710-020-09377-2
22. Helmuth, T., Spector, L.: General program synthesis benchmark suite. In: Proceedings of the 2015 Annual Conference on Genetic and Evolutionary Computation, GECCO 2015, pp. 1039–1046. ACM (2015)
23. Helmuth, T., Spector, L.: Problem-solving benefits of down-sampled lexicase selection. Artif. Life **27**(3–4), 183–203 (2021)
24. Helmuth, T., Spector, L., Matheson, J.: Solving uncompromising problems with lexicase selection. IEEE Trans. Evol. Comput. **19**(5), 630–643 (2014)
25. Hernandez, J.G., Lalejini, A., Dolson, E., Ofria, C.: Random subsampling improves performance in lexicase selection. In: Proceedings of the Genetic and Evolutionary Computation Conference Companion, GECCO 2019, pp. 2028–2031. ACM (2019)
26. Krawiec, K., O'Reilly, U.M.: Behavioral programming: a broader and more detailed take on semantic GP. In: Proceedings of the 2014 Annual Conference on Genetic and Evolutionary Computation, GECCO 2014, pp. 935–942. ACM (2014)
27. La Cava, W., Helmuth, T., Spector, L., Moore, J.H.: A probabilistic and multi-objective analysis of lexicase selection and epsilon-lexicase selection. Evol. Comput. **27**(3), 377–402 (2019)
28. La Cava, W., Spector, L., Danai, K.: Epsilon-lexicase selection for regression. In: Proceedings of the Genetic and Evolutionary Computation Conference 2016, GECCO 2016, pp. 741–748. ACM (2016)
29. Martínez, Y., Naredo, E., Trujillo, L., Legrand, P., Lopez, U.: A comparison of fitness-case sampling methods for genetic programming. J. Exp. Theor. Artif. Intell. **29**(6), 1203–1224 (2017)
30. Moore, J.M., Stanton, A.: Lexicase selection outperforms previous strategies for incremental evolution of virtual creature controllers. In: ECAL 2017, the Fourteenth European Conference on Artificial Life, pp. 290–297 (2017)
31. Moore, J.M., Stanton, A.: Tiebreaks and diversity: isolating effects in lexicase selection. In: The 2018 Conference on Artificial Life, pp. 590–597. MIT Press, Cambridge (2018)
32. Olson, R.S., La Cava, W., Orzechowski, P., Urbanowicz, R.J., Moore, J.H.: PMLB: a large benchmark suite for machine learning evaluation and comparison. BioData Min. **10**(1), 36 (2017)
33. Orzechowski, P., La Cava, W., Moore, J.H.: Where are we now? A large benchmark study of recent symbolic regression methods. In: Proceedings of the Genetic and Evolutionary Computation Conference, pp. 1183–1190. ACM (2018)
34. Pantridge, E., Helmuth, T., McPhee, N.F., Spector, L.: Specialization and elitism in lexicase and tournament selection. In: Proceedings of the Genetic and Evolutionary Computation Conference Companion, GECCO 2018, pp. 1914–1917. ACM (2018)
35. Poli, R., Langdon, W.B., McPhee, N.F.: A Field Guide to Genetic Programming. Lulu Press (2008)

36. Schweim, D., Sobania, D., Rothlauf, F.: Effects of the training set size: a comparison of standard and down-sampled lexicase selection in program synthesis. In: 2022 IEEE Congress on Evolutionary Computation (CEC), pp. 1–8. IEEE (2022)
37. Sobania, D., Rothlauf, F.: A generalizability measure for program synthesis with genetic programming. In: Proceedings of the Genetic and Evolutionary Computation Conference, pp. 822–829 (2021)
38. Sobania, D., Schweim, D., Rothlauf, F.: A comprehensive survey on program synthesis with evolutionary algorithms. IEEE Trans. Evol. Comput. **27**(1), 82–97 (2023)
39. Spector, L.: Assessment of problem modality by differential performance of lexicase selection in genetic programming: a preliminary report. In: Proceedings of the 14th Annual Conference Companion on Genetic and Evolutionary Computation, GECCO 2012, pp. 401–408. ACM (2012)

The Role of Stepping Stones in MAP-Elites: Insights from Search Trajectory Networks

Giorgia Nadizar[1](✉)[iD], Francesco Rusin[1][iD], Eric Medvet[2][iD], and Gabriela Ochoa[2][iD]

[1] Department of Engineering and Architecture, University of Trieste, Trieste, Italy
giorgia.nadizar@phd.units.it
[2] University of Stirling, Stirling, Scotland, UK

Abstract. MAP-Elites (ME) is a quality-diversity optimization algorithm designed to generate a diverse collection of high-performing solutions to complex problems by leveraging *"stepping stones"*. Stepping stones have been defined as intermediate solutions that, while not necessarily optimal themselves, contribute to the development of more effective final outcomes. A deeper understanding of the role of stepping stones in evolutionary optimization would be beneficial. To address this gap, we employ search trajectory networks (STNs), an analytical and visualization tool for studying the behavior of optimization algorithms. We refine the notion of stepping stones by incorporating the idea of betweenness centrality in networks. We consider a robotic navigation task with various controller representations (polynomials, artificial neural networks, and symbolic formulae encoded as trees), comparing the ME search process with that of a genetic algorithm, while also evaluating the differences across representations. Our findings show clearer evidence of stepping stones in ME, particularly when using more "direct" and "local" representations.

Keywords: MAP-Elites · Quality-Diversity · Search Trajectory Networks · Genetic Programming · Neuroevolution · Stepping Stones

1 Introduction

Stepping stones are physical or metaphorical elements that facilitate progress or advancement toward a goal. Historically, physical stepping stones were used as a practical solution to get safely across a body of water. People would use stones already present in the river or stream, but over time the use of purposefully placed stones became more common. They also had other purposes: in Japan, for example, stepping stones were placed in gardens and around temples to help practitioners slow down and focus their mind as they walked mindfully over them [9]. In the modern era, stepping stones have acquired more of a metaphorical meaning, across various domains.

G. Nadizar and F. Rusin—The first two authors contributed equally to this work.

© The Author(s), under exclusive license to Springer Nature Switzerland AG 2025
B. Xue et al. (Eds.): EuroGP 2025, LNCS 15609, pp. 224–239, 2025.
https://doi.org/10.1007/978-3-031-89991-1_14

In ecology, for instance, stepping stones refer to small patches of habitat that connect larger areas, allowing wildlife to move between isolated habitats [11,24]. Their role becomes fundamental especially in fragmented landscapes, where direct connections between larger habitats are absent, helping to maintain genetic diversity and population stability among species [30]. Yet, they might also act as "ecological traps" leading the individuals to dead-ends [21].

In the context of evolutionary algorithms (EAs), stepping stones serve analogous functions in promoting connectivity and facilitating progress toward complex goals. Namely, for EAs, they are intermediate solutions that lead to higher-performing outcomes. They represent various configurations or designs that may not directly correlate with the final desired solution, but are essential for navigating the search space effectively [6].

It has been postulated that quality-diversity (QD) algorithms, such as ME [15], can avoid premature convergence on sub-optimal solutions because they excel at generating stepping stones by promoting a diverse range of solutions that can be built upon later [3]. However, the currently accepted definition of stepping stones is too wide to be useful for ascertaining their actual contribution to the success of EAs, including QD algorithms. Namely, it makes it hard to verify the claim that "QD algorithms are better at promoting stepping stones" [15,18]. A narrower—and potentially more useful—definition should acknowledge the fact that stepping stones are not only solutions in the ancestry of a high-performing solution, but (i) they are key point of traversal towards the solution which (ii) should often be neglected by a purely fitness-directed search, due to their generally lower performance.

Building on this intuition, we leverage STNs [19] for detecting stepping stones. STNs are graphical representations that capture the paths taken by algorithms through the solution space during optimization processes that can help in understanding how different solutions contribute to finding optimal or near-optimal results. In this work, we consider STNs based on ancestry, i.e., we track the ancestry of the final best solution, and we partition solutions based on their behavior, i.e., two solutions belong to the same STN node if their behavior is the same (with some approximation). We thereby propose to identify stepping stones utilizing the betweenness centrality (BC) measure—a measure in graph theory that quantifies the importance of a node within a network based on how many shortest paths between other nodes traverse it.

We assess the approach on a navigation task, where the behavior of a solution depends on its trajectory. Our evaluation considers two kinds of agent, three controller representations, and two arenas with increasing levels of deceptiveness. We compare the STNs of ME with those of a standard genetic algorithm (GA), and confirm the emergence of stepping stones in ME in correspondence to behaviors that are temporarily less advantageous but ultimately solve the task. Moreover, we highlight the strong relation between representation and stepping stones: the more direct and with higher locality the representation, the more crucial fostering the emergence of stepping stones (as done in ME) becomes to solve the task.

2 Related Work

The idea of stepping stones first emerged in QD algorithms, which promote both fitness and solution diversity [3]. Novelty search (NS) [10] introduced searching for novelty over fitness, promoting exploration of diverse areas which may contain stepping stones leading to higher-performing solutions later in the process. Later, Mouret and Clune [15] introduced ME, which balances exploration (through behavioral diversity) and exploitation (through fitness).

A critical aspect in identifying stepping stones is understanding the relationship between behavior and fitness [20,28]: while descriptors that are not well-aligned with fitness measures can make the search inefficient [16], key intermediate solutions—the so-called stepping stones—often differ significantly from the final result [14]. This problem became apparent when researchers tried to replicate results from human-guided evolution in Picbreeder [25]. Pure objective-based search failed, as evolution stagnated in local optima [31]. In contrast, ME achieved notable results [6], underscoring the deceptiveness of the original fitness objective [5].

Solutions encoding and their locality also play a key role [26]. Especially for robotic tasks [2], generative encodings rarely benefit from the diversity boosts provided by QD algorithms, due to their low locality in the behavior space [27]. Veenstra et al. [29] confirmed this for 2D robot evolution, where direct encodings benefited more from stepping stones fostered by ME than indirect ones.

In an analogous scenario of robot body-brain evolution, Nordmoen et al. [18] also studied the ancestry of some elite solutions in terms of descriptors. Their study is closely linked to ours, yet our work differs from theirs in that (a) we consider multiple agents and controller representations, (b) we analyze the ancestry on aggregates of runs with a dedicated tool (namely, STNs), and (c) we also proposed a metric for identifying stepping stones.

3 Background: Search Trajectory Network (STNs)

STNs are a graph-based, data-driven modeling and visualization tool to analyze the behavior of evolutionary and other heuristic optimization algorithms [19]. In these graph models, nodes are *locations* of *representative solutions* in the search space and edges connect consecutive locations in the search process. The definition of location depends on the type and size of the search space. For continuous optimization, a location can be defined as a hypercube with a prefixed numerical precision [19]. In combinatorial optimization, an individual solution can be a location [19], but for large search spaces, some clustering of solutions into a single location is required to achieve models of manageable size [23]. When studying complex domains, which have a clear distinction between genotype and phenotype (or behavior) spaces, locations can be defined in the phenotype/behavior space rather than in the (substantially larger) genotype space. This has been investigated for both neuroevolution [22] and genetic programming (GP) [4,8]. The definition of representative solution is associated with the type of algorithm [19]. For population-based algorithms, the best individual in each generation is usually considered as representative solution. For single-point (local search) based meta-heuristics, the incumbent solution is the natural choice of representative.

For analyzing stepping stones in the present study, we use STNs with locations in the behavioral space as in previous work [4,8,22]. However, our analysis requires a different definition of a representative solution. Specifically, we consider genealogical ancestry, that is, the historical sequence of parents and offspring of the best solution at the end of runs. The ancestors of the best solution at each generation are, therefore, taken as the representative solutions.

3.1 Model Definition

To define a STN model, we need to specify the nodes and edges. The relevant definitions are given below.

Representative solution. A solution to the optimization problem at a given iteration that represents the status of the search algorithm. For studying stepping stones, we choose representative solutions as the ancestors of the final best solution. Since we here consider mutation-only EAs, there is exactly one ancestor of the final best solution at each iteration.
Location. A non-empty subset of solutions that results from a predefined partitioning of the search space. In our study, locations aggregate solutions (genotypes) with similar behaviors, where behaviors are characterized by some *descriptors*.
Nodes (N). Locations of the corresponding representative.
Trajectory. A sequence of locations following representative solutions across iterations of the EA.
Edges (E). Directed, connecting two consecutive locations in the search trajectory.
STN. Directed graph $STN = (N, E)$, with nodes N and edges E.

4 Finding Stepping Stones with STNs

We propose using STNs for identifying and visualizing stepping stones in the evolutionary process. This section describes our approach.

4.1 Data Collection and Model Construction

To construct the STN models, we conduct 10 runs for each considered configuration (a configuration being given by the combination of problem, representation, and EA). Once a run is completed, we take the best individual and track its ancestor-offspring relations back to one initial solution, hence obtaining a trajectory. We use the sequence of ancestors in the trajectory as representative solutions, with edges capturing ancestor-offspring relationships between nodes.

We aggregate data collected from 10 runs to construct a single STN model for each configuration. Notice that different runs can traverse the same nodes. However, in the aggregated STN model, each unique node appears only once, and shared nodes among runs can be identified by multiple incoming edges.

4.2 Stepping Stones Identification

The definition of stepping stone solutions according to the QD literature (see Sect. 2), does not require any additional property apart from being in the genealogical ancestry of the final best solution. However, in this article we propose refining this definition by highlighting stepping stones as those "central" or "important" solutions in the genealogical path to good solutions. To measure node centrality, we resort to the body of knowledge in complex networks [17], where centrality metrics are used to analyze the importance of nodes and understand how information flows and influence spreads. In particular, we take inspiration from *betweenness centrality*, which measures how often a node lies on the shortest path between two other nodes. Nodes with high betweenness centrality play a crucial role in information flow and act as bridges between different parts of the network.

We are interested in genealogical paths from start to final best solutions. Therefore, we identify stepping stones as those STN nodes that appear with high frequency in shortest-paths between initial solutions and the best solution. Specifically, we select as stepping stones those nodes above the Φ-th percentile in terms of their frequency of appearance in these paths. In this work we set $\Phi = 75$ to capture values sensibly above the median, although it might be worth investigating the impact of other choices in the identification of stepping stones.

4.3 STN Metrics and Visualizations

Once the network model is constructed, we can characterize it with several metrics. We select a set of metrics commonly used in STN analysis [19,22], and add a new metric capturing the number of stepping stones. Namely, we consider (i) the *number of nodes* (N. nodes), (ii) the *number of stepping stones* (N. step. st.), (iii) the *number of edges* (N. edges), (iv) the *number of paths from start nodes to the best node in successful runs* (N. paths), and (v) the *proportion of worsening edges*, i.e., edges leading to nodes with worse fitness (ρ_w). The definition of a successful run (appearing in (iv)) varies depending on the problem and can be based on behavioral criteria or a specified fitness threshold.

Besides the considered metrics, network visualization is also crucial for grasping their structure and underlying patterns. Node-edge diagrams, where nodes are points and edges lines, can help to inspect network structure, flow patterns, bottlenecks and critical nodes. Key characteristics of nodes and edges can be emphasized through visual attributes like color, size, and shape; and the graph-layout (nodes coordinates in the plane) can be selected to highlight the network structure. We provide more details about our visualization choices in Sect. 6.

5 Experiments

5.1 Problems

We work on a two-dimensional navigation task, where an agent is placed in an arena with some walls and has to reach a certain position, called the "target".

An arena is a square with coordinates ranging from 0 to 1, containing some walls, i.e., segments the agent is not allowed to pass through. Therefore, to reach the target the agent is required to find a path that does not intersect the walls.

We consider two arenas, *maze* and *blocky*. The former contains two horizontal walls that form an S-shaped passage (Fig. 2). The latter adds two vertical extensions which make the passage harder to pass through (Fig. 3). In both cases, the target is in position $(0.5, 0.15)$ and the agent starting position is $(0.5, 0.75)$: the agent hence needs to run through the passage to reach the target. We consider a run to be "successful" if the final position of the agent lies in a radius of 0.1 from the target. In both arenas, we perform simulations that last 30 s, with ticks of 0.1 s.

We consider two kinds of agents: a punctiform one ("point") and a circular robot with two wheels ("robot"). They differ in how they perceive the environment and in what actions they can perform.

Point Agent. The punctiform agent is assumed to have negligible size and no orientation and its state in the arena is therefore fully described by its position, which is a vector in $[0,1]^2$. At each time step, this agent perceives its current position and can take a step in any direction, resulting in its coordinates shifting of some value (Δ_x, Δ_y); both Δ_x and Δ_y cannot exceed in module a fixed value $\Delta_{\max} = 0.05$; also, if the movement would make the agent collide with a wall, it stays still instead. Based on existing works on similar tasks [16], we rescale the inputs to $[-1, 1]$ to ease the optimization.

The controller for this kind of agent is a function from $[-1,1]^2$ to $[-1,1]^2$, with the two inputs representing the current position and the two outputs representing the step components after being multiplied by $\Delta_{\max}$.

Robot Agent. The robot is a circular agent with two wheels, whose coordinated movement can make it move and steer; this agent does have an orientation, therefore its state in the arena is described by a combination of the position of its center and an additional value in $[-\pi, \pi]$ indicating its angle. It also has a measurable radius, meaning the environmental constraints are more restrictive with respect to the point; if its movement would collide with a wall it does not move like its punctiform counterpart, but it can still rotate. The robot does not perceive its own state (position and orientation). Instead, it perceives the distance from the target and the angle it forms with it; also, it is equipped with six proximity sensors, each of which can detect the presence of the closest wall in a given direction, placed at regular angles around half of the robot circumference (the frontal part). Here too we rescale all inputs to $[-1, 1]$.

The controller for this kind of agent is a function from $[-1,1]^8$ to $[-1,1]^2$, where the outputs are later multiplied by $v_{\max} = 0.05$ (expressed in distance per timestep) and represent the speed of the two wheels.

We observe that the controller for the robot agent is comparatively "less direct" than that of the point agent. This is because (i) the environment is perceived indirectly through sensors, and movement is achieved through actuators, i.e., the wheels, rather than directly. As a result, two similar controllers for the point agent are likely to produce very similar outcomes, whereas for the robot agent, two similar controllers might lead to more distinct outcomes.

5.2 Controller Representations

To model the controller, we consider three alternatives: (i) two third degree *polynomials* (one for each output) in the given variables; (ii) an *artificial neural network (ANN)*; (iii) two symbolic formulae represented as *trees*, like in tree-based genetic programming (TGP). In all cases, the functions process the sensory inputs to compute the actuation values; if the latter ones lie outside the target domain of $[-1, 1]^2$, we clip them.

As for the two agents, we remark that different controller representations have different degrees of "directness", which in this case broadly corresponds to the notion of locality. In the following we order representations from the most direct/local one to the least direct/local one.

Polynomials. This controller uses two third degree polynomials, one for each element of the output. The optimizable parameters of the polynomials are the coefficients; the polynomials require a total of 20 parameters for point navigation and 330 parameters for robot navigation.

ANN. The ANN controller is a multi-layer perceptron (MLP), i.e., a fully connected feed-forward neural network, with a fixed architecture consisting of a single hidden layer with 4 neurons in point navigation and 16 neurons in robot navigation, and 2 outputs, using tanh as activation function. In this case the optimizable parameters are the MLP weights; a total of 22 for point navigation and 178 for robot navigation.

Symbolic Formulae. To represent the formulae, we employ two regression trees, i.e., two trees where each leaf is either a variable or a constant in $\{0.1, 1, 10\}$ and each non-terminal node is an operation in $\{+, -, \times, \div^*, \log^*\}$, where $\div^*$ and $\log^*$ are the division and logarithm operation when the input is in the domain, and 1 or 0, respectively, otherwise. We ensure during construction that each node has a number of children equal to the arity of the corresponding operation. For this representation, the entire formulae are optimizable.

5.3 Optimization

We optimize our solutions using two mutation-only EAs with broad applicability in terms of representation of the solution, GA and ME.

A GA iteratively evolves a fixed size population of n_pop solutions. First, solutions are initialized according to the chosen representation. Then, at each iteration, the GA builds an offspring of n_pop solutions by selecting parent individuals with a tournament selection (with $n_\mathrm{tour} = 10$ individuals) and applying upon them a representation-dependent mutation operator. After that, the GA merges the parents and offspring and applies truncation selection to retain only the best n_{pop} solutions to form the population at the next iteration. The GA stops iterating when at least n_{evals} solutions have been evaluated.

ME is a QD algorithm that creates a map of high-performing solutions at each point in a space defined by user-specified dimensions of diversification [15]. ME works by storing solutions in an *archive*, also known as *repertoire* or *map*. At first n_{pop} solutions are randomly generated, whereas successively new solutions are generated by applying mutation to solutions sampled from the archive—n_{pop} solutions are sampled and mutated at each iteration. The decision if a solution is stored in the archive is based on two elements: the fitness of the solution and its descriptors. Namely, if there is no element in the archive corresponding to the same descriptors, the solution is stored. Conversely, if there is already an element, the decision is based on local competition, i.e., the best solution among the existing one and the new one is kept. As for the GA, ME stops iterating when at least n_{evals} solutions have been evaluated.

Both ME and GA require a *fitness function* f, i.e., a function to measure the performance of a given solution; as our goal is always for the agents to quickly reach the target and stay on it, but also considering that a relatively continuous function is beneficial to the evolution, we pick the average distance between the agent and the target during the simulation, which we aim at minimizing.

ME also requires two descriptors: we use the x and y coordinates of the final position of the agent, as these numbers provide a decent overview of the behavior of the agent itself. In order to make them useful for indexing the individuals in the archive, we discretize them in $h = 10$ bins in the interval $[0, 1]$ and make them return the bin index.

Concerning the representation-dependent parameters, we use the following values. For the ANN and polynomial controllers, which are defined by a vector of parameters in $\mathbb{R}^q$, we generate the initial population by randomly sampling each element of each vector in $[-1, 1]$ with uniform probability. As mutation, we use the element-wise Gaussian mutation applied with $\sigma = 0.2$. For the tree-based controller, we initialize the population through an adaptation of the ramped-half-and-half procedure to the case of pair of trees, as done in [12]. As mutation, we employ the standard sub-tree mutation. In both cases, we enforce the tree height to be between 4 and 10.

In all our experiments, we set $n_{\text{pop}} = 200$ and $n_{\text{evals}} = 100000$. For each configuration we perform 10 independent runs, to ensure consistency of results and to collect a reasonable amount of data for the STNs. We perform the experiments with JGEA [13].

6 Results and Discussion

6.1 Performance Analysis

Before analyzing the STNs, we show in Fig. 1 the progression and the final distribution of the fitness of the best individual of the population.

In the point navigation problem with the maze arena, polynomials and tree-based controllers optimized with ME outperform everything else, totaling 9 and

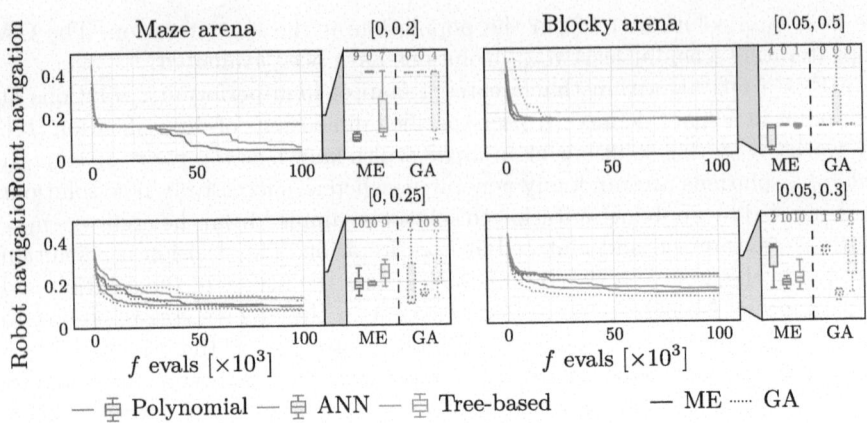

Fig. 1. Progression of the fitness of the best individual during the evolution (median across 10 runs) and distribution of the final best fitness values for each problem, controller representation, and EA. We annotate boxplots with the number of successful runs.

7 successful runs, respectively. Conversely, most other agents are completely unable to get over the top wall and just converge on the local optimum given by clashing on it right under the target. This happens even more frequently in the blocky arena, where only a small percentage of controllers optimized with ME actually manage to reach the target.

The results are very different for robot navigation, where most runs are successful across both arenas. Interestingly, for the maze arena GA even outperforms ME with ANNs in terms of fitness, leading to agents that reach the target faster. In addition, in this problem polynomials are the worst performing controllers (with very few successes in the blocky arena), while ANNs are the best (almost always solving the task), in stark contrast with point navigation. Although the larger search space of polynomials compared to ANNs (with 330 vs. 178 parameters) for the robot could account for this, we rule it out as the cause, given the plateau observed in the fitness progression plots. We thus conclude that it is the controller representation itself that is more or less suited to addressing this task.

Overall, these first results highlight how the effectiveness of a search algorithm is deeply intertwined with the directness of the representation when the fitness landscape is, to some extent, deceptive: while GA works well with a pretty indirect one, ME outperforms it heavily with a more direct one. We elaborate further on this examining the STNs in the following.

6.2 STN Analysis

Figure 2 and 3 show the STNs resulting from our experiments; the problems and the representations are ordered from more direct on the top to less direct at the bottom. Our visualizations give shapes and colors to nodes according to their type (Start, Medium, End, Regular, Stepping Stone, Target) and line-styles to edges for highlighting if they are worsening or not (i.e., if they lead to a node with worse fitness or not). The size of nodes and the opacity of edges is proportional to their sampling frequency, that is, the number of times they appear across the 10 runs. We use two alternative graph layouts, which reveal complementary views of the search process. First, we take advantage of the behavior descriptors, which give the final position of the agent in the arena, as node coordinates (left plots in Fig. 2, all plots in Fig. 3). Second, we use a force-directed graph layout (with stress majorization [7]) for the x coordinate, and set node fitness as the y coordinate (right plots in Fig. 2).

ME vs. GA. From the visualizations we can make several comparative observations between ME and GA. First, we clearly see the impact of the diversity promotion of ME, as the graphs are far more spread out with respect to GA. This is probably also a consequence of the fact that in GA solutions that perform poorly get quickly evolved out of existence: therefore they are unable to spawn other individuals, which in turn makes it impossible for them to be part of the ancestry of the final best individual. Instead, ME always keeps in its archive some individuals with non-optimal performance, which can later mutate into better individuals.

Second, from the plots we can note that GA rarely presents intermediate steps, as the solutions that manage to get over the wall directly reach the target, in contrast with ME, which can be seen roaming around the space before eventually converging. This makes it evident why a more indirect representation is beneficial to GA: the lower locality means that each step in the EA can perform some pretty big jumps phenotypically (and behaviorally), which may result in a lucky solution that manages to overcome the wall. In fact, we can note that the majority of stepping stones highlighted for the GA correspond to local optima where the algorithm temporarily converges until one lucky mutation occurs (e.g., below the top wall). This is not the case for ME, which overcomes the wall by slowly reaching new points and manages not to get trapped in local optima thanks to its intrinsic structure. This is corroborated by the location of stepping stones, which tend to appear in points of connection between different optima, either local or global (e.g., in the corridor between the two walls).

Last, we can confirm the differences between point and robot navigation already highlighted in Sect. 6.1. Looking at the GA STNs of point navigation we clearly note that the directness of the problem rarely allows to escape the local optima below the walls, which become either stepping stones or, more frequently, end nodes. For this task the enforced exploration of ME is thus required for success. Regarding robot navigation instead, we can see that each combination has managed to solve the problem at least once—both with GA and ME— even though convergence speed and consistency vary. The indirectness of the representations also results in heavier exploration, as can be seen by the fact that the maps for both ME and GA are more covered, with ME in particular.

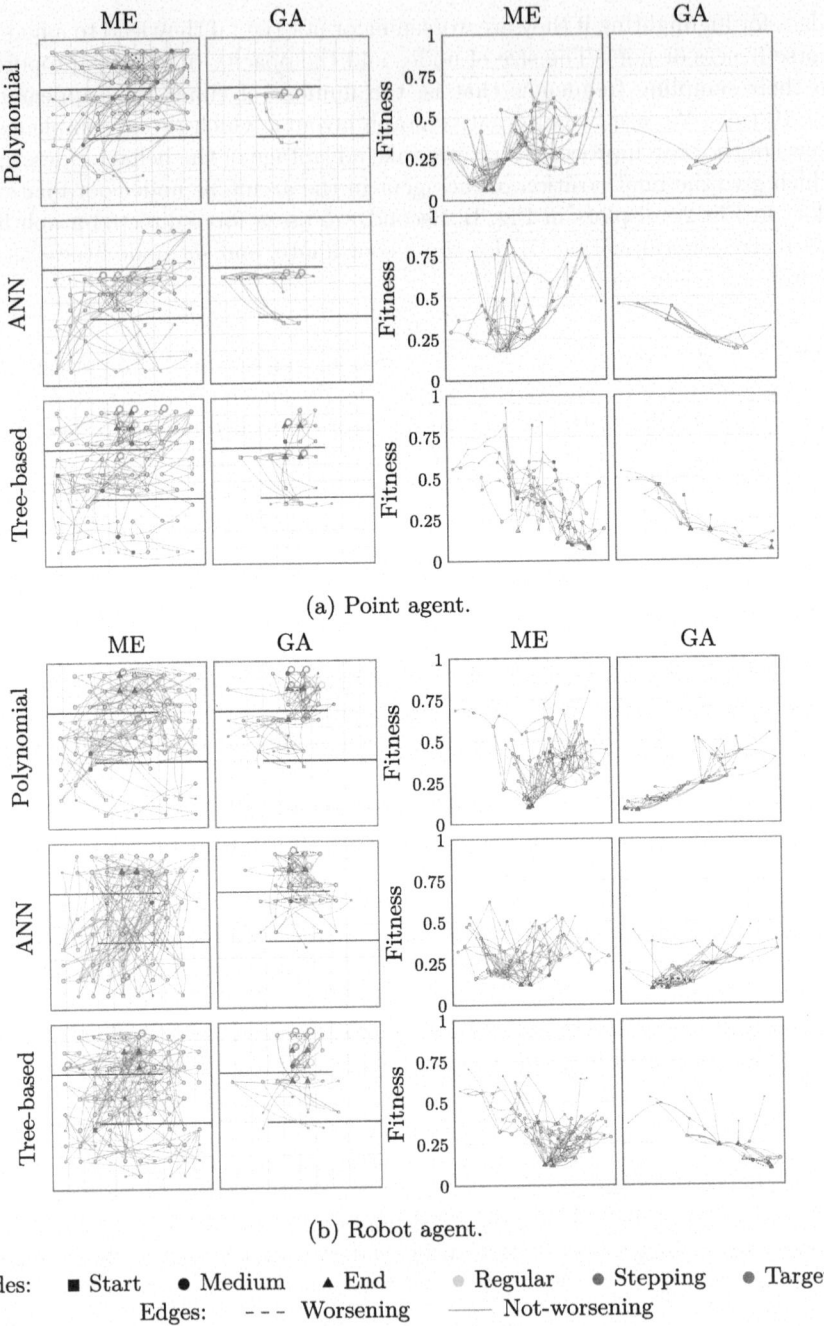

Fig. 2. STN analysis of the search process of 10 evolutionary runs with ME and GA with different agents and controller representation (one per row) for the maze arena (shown as an overlay in the left half of the plots). The first two columns show the STNs with the nodes located according to their descriptors; the last two columns show the same STNs with the y given by the fitness and the x given by the force-directed layout.

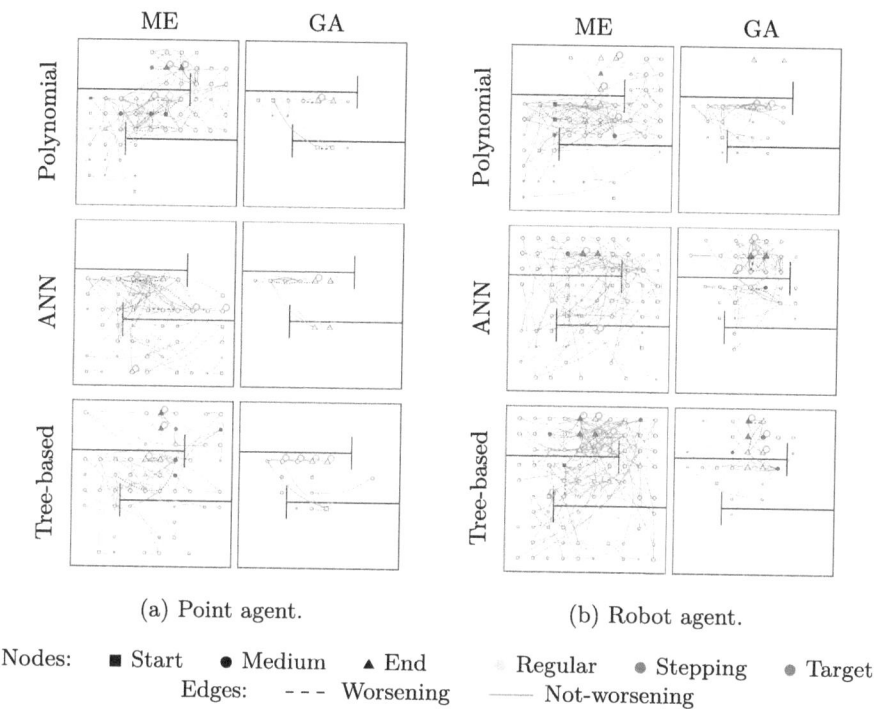

Fig. 3. STN analysis of 10 evolutionary runs with ME and GA with different agents and controller representations (one per row) for the blocky arena (shown as an overlay in all the plots). The STN nodes are located according to their descriptors. The force-directed STNs are omitted for space reasons.

ME and Stepping Stones. Focusing on ME only, it is worth reasoning further on stepping stones and on the locations where they appear. We start our analysis with point navigation, where we can see that both polynomial and tree-based representations present some stepping stones, marked in red in Figs. 2 and 3. In the maze arena, polynomials show the majority of them in points that intuitively are expected to be a step towards overcoming the wall, while tree-based is more unclear, making it look like a stepping stone in this case is more like a point that the solutions commonly pass through—possibly by chance. The role of stepping stones for the case of maze solved with polynomial is particularly evident in the force-driven layout: nodes linked to them have better (lower) fitness, showing that stepping stones lay along a path where the fitness of solutions needs to worsen before further improving. The trend for the blocky arena is similar to that of the maze for polynomials, where most of the stepping stones lie in the corridor between the two walls—a necessary traversal point towards the solution. For the tree-based representation, instead, there are fewer stepping stones, with only one located on the expected best path towards the solution. This highlights the more chaotic exploration within this representation.

ANNs instead show how ME explores most of the space it can reach, but it rarely manages to overcome the wall (never actually reaching the target), strengthening the claim, already made in the past, that ANNs are not well-suited to be evolved with standard ME [1]. Also, the lack of stepping stones in this representation seems to correlate with the inability to solve the problem, as the optimization seems to fail to find the steps towards a good solution.

Regarding robot navigation, in the maze arena, polynomial and tree-based both seem to operate like tree-based does in point navigation, meaning that the stepping stones are present, but look more like points the evolution regularly traverses rather than needed steps. Differently, for ANN the behavior is more similar to that of the polynomials for the point, with a stepping stone emerging in the correspondence a step needed to overcome the top wall.

Finally, more "chaotic" stepping stones emerge in the blocky arena, which shows the impact of the indirectness of the representation in this kind of problem. The greater difficulty given by the blocky arena combined with higher levels of indirectness creates a situation where the exploration does not follow a clear pattern, randomly roaming the space until a solution is found. We stress that this does not hold for point navigation, where successful combinations of algorithm and representation produce more coherent stepping stones, i.e., in more expected locations, in both arenas, as the representation is more direct.

STN Metrics Analysis. To conclude, Table 1 reports the STN metrics for our experiments. Apart from the already mentioned stepping stones, we can see that the runs that are at least somewhat successful have a higher number of nodes and edges, showing a bigger amount of exploration. Furthermore, the metrics show again that ME is far more explorative than GA, as even with robot navigation, where GA is still effective and sometimes outperforms ME, ME has far more edges and nodes. Last, ρ_w, the proportion of worsening edges, seems to be pretty constant at around 0.30 for ME with larger variability for GA. This is particularly relevant as the considered problems all share a deceptive fitness landscape, where worsening transitions are needed to reach the final goal, unless the new solution lies on another—better—local optimum (e.g., over the wall). The consistency of ρ_w for ME indicates how this algorithm systematically safeguards worsening transitions, often finding the arrival point, i.e., the worse solution, to be a stepping stone. Conversely, GA does not rely on this mechanism. In this case, the worsening transitions correspond to elements with worse performance which still make it into the elite—thus the larger variability in ρ_w.

Summary. In summary, our analysis highlights the importance of stepping stones in more direct representations, where they are essential to overcoming local optima that would otherwise trap evolution. Conversely, in less direct representations, stepping stones are more likely to emerge as incidental points of traversal, often in regions where they are not strictly necessary for escaping local optima.

Table 1. STN metrics for the search process of 10 evolutionary runs with ME and GA with different agents, arenas, and controller representations. The background color is darker in each row for larger values.

		Point						Robot					
		ME			GA			ME			GA		
		Poly	ANN	TGP	Poly	ANN	TGP	Poly	ANN	TGP	Poly	ANN	TGP
Maze	N. nodes	64	47	75	9	10	21	72	68	79	30	27	23
	N. step. st.	4	0	6	0	0	2	3	1	3	2	1	3
	N. edges	239	140	213	20	35	55	230	204	245	97	110	57
	N. paths	18	0	16	0	0	24	18	16	20	14	14	14
	ρ_w	0.34	0.26	0.31	0.10	0.26	0.15	0.30	0.29	0.32	0.29	0.35	0.18
Blocky	N. nodes	54	46	52	12	8	13	53	65	76	21	32	26
	N. step. st.	5	0	4	0	0	0	4	1	4	0	1	3
	N. edges	202	156	146	33	29	34	182	180	255	64	119	65
	N. paths	18	0	16	0	0	0	9	20	20	0	14	14
	ρ_w	0.34	0.34	0.29	0.18	0.21	0.09	0.33	0.27	0.32	0.25	0.29	0.23

7 Concluding Remarks

In this paper, we explored the emergence of stepping stones in ME, a quality-diversity optimization algorithm. We leveraged STNs—an analytical and visualization tool for examining the behavior of optimization algorithms—and we refined the concept of stepping stones relying on the notion of betweenness centrality in networks. We tested this approach on a robotic navigation task using various controller representations: polynomials, artificial neural networks, and symbolic formulae encoded as trees, comparing the performance of ME and of genetic algorithm. Our findings reveal that in more direct representations, i.e., representations with higher locality, stepping stones are not merely incidental points of traversal of evolution, but play a key role in overcoming local optima to drive optimization. In contrast, with less direct representations, stepping stones function more as common traversal points than essential aids for escaping local optima. These results suggest that the effectiveness and the utility of ME varies with the directness of the representation.

Acknowledgments. This study was carried out within the PNRR research activities of the consortium iNEST (Interconnected North-Est Innovation Ecosystem) funded by the European Union Next-GenerationEU (Piano Nazionale di Ripresa e Resilienza (PNRR) - Missione 4 Componente 2, Investimento 1.5 - D.D. 1058 23/06/2022, ECS_00000043).

References

1. Colas, C., Madhavan, V., Huizinga, J., Clune, J.: Scaling map-elites to deep neuroevolution. In: Proceedings of the 2020 Genetic and Evolutionary Computation Conference, pp. 67–75 (2020)
2. Cully, A., Clune, J., Tarapore, D., Mouret, J.B.: Robots that can adapt like animals. Nature **521**(7553), 503–507 (2015)
3. Cully, A., Demiris, Y.: Quality and diversity optimization: a unifying modular framework. IEEE Trans. Evol. Comput. **22**(2), 245–259 (2017)
4. De La Torre, C., Cussat-Blanc, S., Wilson, D., Lavinas, Y.: On search trajectory networks for graph genetic programming. In: Proceedings of the Genetic and Evolutionary Computation Conference Companion, pp. 1681–1685 (2024)
5. Dean, J., Cheney, N.: Towards stepping stones in goal-directed search: improving image perceptual similarity by combining evolution, backpropagation, and Fourier transforms. In: ALIFE 2024: Proceedings of the 2024 Artificial Life Conference. MIT Press (2024)
6. Gaier, A., Asteroth, A., Mouret, J.B.: Are quality diversity algorithms better at generating stepping stones than objective-based search? In: Proceedings of the Genetic and Evolutionary Computation Conference Companion, pp. 115–116 (2019)
7. Gansner, E.R., Koren, Y., North, S.: Graph drawing by stress majorization. In: Graph Drawing, pp. 239–250. Springer, Heidelberg (2005)
8. Hu, T., Ochoa, G., Banzhaf, W.: Phenotype search trajectory networks for linear genetic programming. In: European Conference on Genetic Programming (Part of EvoStar), pp. 52–67. Springer (2023)
9. Keane, M.P.: Japanese Garden Design. Tuttle Publishing (2012)
10. Lehman, J., Stanley, K.O.: Abandoning objectives: evolution through the search for novelty alone. Evol. Comput. **19**(2), 189–223 (2011)
11. Lynch, A.J.: Creating effective urban greenways and stepping-stones: four critical gaps in habitat connectivity planning research. J. Plan. Lit. **34**(2), 131–155 (2019)
12. Medvet, E., Nadizar, G.: GP for continuous control: teacher or learner? The case of simulated modular soft robots. In: Genetic Programming Theory and Practice XX, pp. 203–224. Springer (2024)
13. Medvet, E., Nadizar, G., Manzoni, L.: JGEA: a modular java framework for experimenting with evolutionary computation. In: Proceedings of the Genetic and Evolutionary Computation Conference Companion, pp. 2009–2018 (2022)
14. Mouret, J.B.: Evolving the behavior of machines: from micro to macroevolution. Iscience **23**(11) (2020)
15. Mouret, J.B., Clune, J.: Illuminating search spaces by mapping elites. arXiv preprint arXiv:1504.04909 (2015)
16. Nadizar, G., Medvet, E., Wilson, D.: Searching for a diversity of interpretable graph control policies. In: Proceedings of the Genetic and Evolutionary Computation Conference, pp. 933–941 (2024)
17. Newman, M.: Complex Networks, 2nd edn. Oxford University Press (2018). ISBN 9780198805090
18. Nordmoen, J., Veenstra, F., Ellefsen, K.O., Glette, K.: Map-elites enables powerful stepping stones and diversity for modular robotics. Front. Robot. AI **8**, 639173 (2021)
19. Ochoa, G., Malan, K.M., Blum, C.: Search trajectory networks: a tool for analysing and visualising the behaviour of metaheuristics. Appl. Soft Comput. **109**, 107492 (2021)

20. Pugh, J.K., Soros, L.B., Szerlip, P.A., Stanley, K.O.: Confronting the challenge of quality diversity. In: Proceedings of the 2015 Annual Conference on Genetic and Evolutionary Computation, pp. 967–974 (2015)
21. Rocha, É.G., Brigatti, E., Niebuhr, B.B., Ribeiro, M.C., Vieira, M.V.: Dispersal movement through fragmented landscapes: the role of stepping stones and perceptual range. Landscape Ecol. **36**(11), 3249–3267 (2021). https://doi.org/10.1007/s10980-021-01310-x
22. Sarti, S., Adair, J., Ochoa, G.: Neuroevolution trajectory networks of the behaviour space. In: International Conference on the Applications of Evolutionary Computation (Part of EvoStar), pp. 685–703. Springer (2022)
23. Sartori, C.C., Blum, C., Ochoa, G.: An extension of STNWeb functionality: on the use of hierarchical agglomerative clustering as an advanced search space partitioning strategy. In: Genetic and Evolutionary Computation Conference, GECCO 2024. ACM (2024). https://doi.org/10.1145/3638529.3654084
24. Saura, S., Bodin, Ö., Fortin, M.J.: Editor's choice: stepping stones are crucial for species' long-distance dispersal and range expansion through habitat networks. J. Appl. Ecol. **51**(1), 171–182 (2014)
25. Secretan, J., Beato, N., D Ambrosio, D.B., Rodriguez, A., Campbell, A., Stanley, K.O.: Picbreeder: evolving pictures collaboratively online. In: Proceedings of the SIGCHI conference on human factors in computing systems, pp. 1759–1768 (2008)
26. Tarapore, D., Clune, J., Cully, A., Mouret, J.B.: How do different encodings influence the performance of the map-elites algorithm? In: Proceedings of the Genetic and Evolutionary Computation Conference 2016, pp. 173–180 (2016)
27. Tarapore, D., Mouret, J.B.: Evolvability signatures of generative encodings: beyond standard performance benchmarks. Inf. Sci. **313**, 43–61 (2015)
28. Templier, P., Grillotti, L., Rachelson, E., Wilson, D., Cully, A.: Quality with just enough diversity in evolutionary policy search. In: Proceedings of the Genetic and Evolutionary Computation Conference, pp. 105–113 (2024)
29. Veenstra, F., Olsen, M.H., Glette, K.: Effects of encodings and qualitydiversity on evolving 2D virtual creatures. In: Proceedings of the Genetic and Evolutionary Computation Conference Companion, pp. 164–167 (2022)
30. Wang, H., et al.: Stepping stone strategy: a cost-effective way to address habitat fragmentation of endangered wildlife in montane forest. Ecosyst. Health Sustain. **9**, 0073 (2023)
31. Woolley, B.G., Stanley, K.O.: On the deleterious effects of a priori objectives on evolution and representation. In: Proceedings of the 13th Annual Conference on Genetic and Evolutionary Computation, pp. 957–964 (2011)

Micro-step Time-Series Regression: Insights from System Identification Using Symbolic Regression

Hengzhe Zhang[1], Alberto Tonda[2,3](✉), Qi Chen[1](✉), Bing Xue[1], Evelyne Lutton[2,3], and Mengjie Zhang[1]

[1] Centre for Data Science and Artificial Intelligence and School of Engineering and Computer Science, Victoria University of Wellington, PO Box 600, Wellington 6140, New Zealand

[2] UMR 518 MIA-PS, INRAE, Université Paris-Saclay, 91120 Palaiseau, France
{alberto.tonda,evelyne.lutton}@inrae.fr

[3] UAR 3611 Institut des Systèmes Complexes de Paris Île-de-France (ISC-PIF) CNRS, 75013 Paris, France

Abstract. Time-series forecasting is widely applied across various domains, yet most approaches rely on predefined time steps given by each problem. Based on observations from dynamic systems with known ground truth, we identify that large-step forecasts can lead to substantial errors due to insufficient modeling of continuous dynamics. To address this, we propose a micro-step time-series regression technique that decomposes predictions into smaller intervals, so that genetic programming-based feature construction can capture finer temporal patterns to improve the prediction performance. Specifically, we employ linear interpolation to allow the evolutionary feature construction process to learn from incremental changes, reducing the difficulty of time-series regression. Experiments on 100 datasets from the M4 forecasting benchmark demonstrate that micro-step regression significantly enhances prediction accuracy compared to traditional methods using raw time steps. Further analysis reveals that features trained on micro-step data evolve into simpler structures, promoting both generalization and interpretability.

Keywords: Time-Series Forecasting · Evolutionary Feature Construction · Genetic Programming · Dynamic Systems

1 Introduction

Most real-world phenomena develop over time [4]. To effectively model these dynamics, time-series regression techniques are widely used. Formally, given a time-series $\{y_t\}_{t=1}^{T}$, where y_t represents the observed value at time t, the objective of time-series regression is to predict future values based on past observations. Specifically, using evolutionary feature construction [13,25] for time-series regression, the goal is to search for an optimal set of features Φ, constructed via genetic programming (GP), along with a machine learning model f, such that:

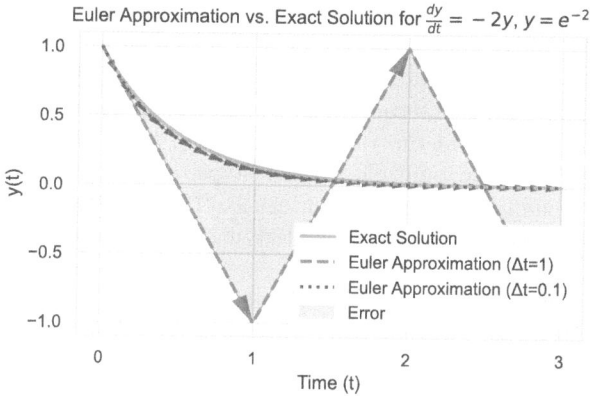

Fig. 1. Euler Approximation vs. Exact Solution. The arrow lines represent the first Euler approximation based on the ground truth, with each arrow indicating a prediction step based on the Euler Approximation.

$$\hat{y}_{t+1} = f(\Phi(y_t, y_{t-1}, \ldots, y_{t-p+1})) \tag{1}$$

where $\hat{y}_{t+1}$ denotes the predicted value at time $t+1$, and $y_t, y_{t-1}, \ldots, y_{t-p+1}$ are the past p observations. Equation (1) can be straightforwardly extended to multi-step prediction: for a horizon of h future values, $\hat{y}_{t+1}, \hat{y}_{t+2}, \ldots, \hat{y}_{t+h}$, predictions can be made iteratively, using prior predictions as inputs in a recursive manner.

In time-series forecasting, non-stationarity often arises, meaning that statistical properties, such as $\mu_t = \mathbb{E}[y_t]$, change over time. A widely used approach to handle non-stationarity is the autoregressive integrated moving average (ARIMA) model [1], which addresses this issue through differencing. Rather than directly estimating $\hat{y}_{t+1}$, ARIMA with first-order differencing estimates the change Δy over a single time unit at each step t, with the next value calculated as $\hat{y}_{t+1} = y_t + \Delta y$. The unit interval-whether a day, hour, or minute-varies depending on the application. From the perspective of system identification, this approach represents a first-order approximation of the system's dynamics, akin to the Euler method [8]. However, real-world systems often evolve continuously rather than discretely, meaning that a single time step may only approximate these gradual changes. This limitation can lead to substantial errors when predicting over longer intervals, as the model is forced to extrapolate over time spans where finer-scale dynamics are active but not fully captured.

In this paper, we observe that in time-series forecasting, even when the ground truth ordinary differential equation (ODE) $\frac{dy}{dt}$ is known, directly predicting the value at the next time step $t+1$ using the approximation $y_t + \frac{dy}{dt}$ can lead to significant errors. This issue is demonstrated in Fig. 1, where the ground truth ODE is $\frac{dy}{dt} = -2y$. Suppose a machine learning model successfully identifies the ground truth ODE. In this case, predicting the next value y_{t+1} based on the current value y_t becomes $y_{t+1} = y_t - 2y_t$. However, as shown in the figure, this approach yields inaccurate results due to the large step size, which prevents the approximation from capturing the system's gradual, continuous changes over the unit time interval. This limitation highlights the inadequacy of conventional time-series forecasting methods in accurately modeling the continuous evolution of dynamic systems.

Motivated by this challenge, we propose a micro-step time-series regression method based on a linear interpolation technique. Rather than predicting the value at $t+1$ directly from the value at time step t, we decompose the time-series regression task into multiple smaller steps, allowing the model to make autoregressive predictions over micro intervals. This approach more effectively captures the continuous nature of the underlying dynamics, enhancing both prediction accuracy and robustness[1].

The key contributions of this paper are as follows:

- We analyze and demonstrate the impact of inappropriate step sizes in dynamic systems prediction. Specifically, we observe that even when the ground truth values are known, large step sizes can lead to inaccurate predictions, highlighting a fundamental limitation in standard time-series forecasting models.
- We propose a novel data-augmentation technique based on linear interpolation, designed to enrich the training data for fitness evaluation in GP. By augmenting the training set, evolutionary feature construction can capture fine-grained temporal dependencies, thereby allowing the time-series forecasting model to predict small, continuous changes instead of large, discrete jumps, achieving better prediction accuracy.

2 Background

2.1 Genetic Programming for Time-Series Regression

Genetic programming (GP) has been widely applied to time-series regression tasks, including applications like quality of service forecasting [6] and streamflow prediction [17]. GP is valued for its ability to evolve interpretable models that capture nonlinear and complex temporal dependencies [18]. Early work demonstrated that GP could compete with traditional methods such as Auto-Regressive Integrated Moving Average (ARIMA) [1] and Exponential Smoothing (ETS) [2] in time-series forecasting. Recent advancements have further enhanced GP by incorporating unit compatibility into the objective function [18] and integrating it into neuroevolution frameworks [19]. Additionally, GP has been applied as a routing model to direct different segments of time-series data to specialized base models [11]. Despite these innovations, most approaches rely on fixed time steps predefined by the problem. The impact of time-step granularity on prediction accuracy remains an open area for research.

2.2 Dynamic Systems and Time-Series Forecasting

A dynamic system describes how the values of state variables evolve over time, typically modeled by ordinary differential equations (ODEs). Formally, an ODE system with n state variables $y^{(1)}(t), y^{(2)}(t), \ldots, y^{(n)}(t)$ can be defined as:

---

[1] Source code: https://anonymous.4open.science/r/MicroStepSR/experiment/methods/MicroSR.py.

$$\begin{cases} \frac{dy^{(1)}}{dt} = -0.28 \cdot y^{(1)} \cdot y^{(3)} \\ \frac{dy^{(2)}}{dt} = 0.28 \cdot y^{(1)} \cdot y^{(3)} - 0.47 \cdot y^{(2)} \\ \frac{dy^{(3)}}{dt} = 0.47 \cdot y^{(2)} - 0.3 \cdot y^{(3)} \\ \frac{dy^{(4)}}{dt} = 0.3 \cdot y^{(3)} \end{cases}$$

$y^{(1)}(t_0) = 0.6, \; y^{(2)}(t_0) = 0.3,$
$y^{(3)}(t_0) = 0.09, \; y^{(4)}(t_0) = 0.01$

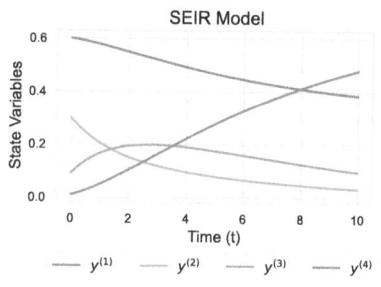

Fig. 2. Example of a 4-variable ODE system, and the corresponding trajectory of each state variable for the given initial conditions. The system in the example is the Susceptible-Exposed-Infectious-Recovered (SEIR) model used in epidemiology, set with parameters and initial conditions from the ODEBench system identification benchmark suite [5].

$$\frac{dy^{(i)}}{dt} = f_i(t, y^{(1)}, y^{(2)}, \ldots, y^{(n)}), \quad i = 1, \ldots, n, \tag{2}$$

where $\frac{dy^{(i)}}{dt}$ represents the rate of change of the i-th variable with respect to time t, and f_i defines the relationship of each state variable's derivative to time and the other state variables.

An ODE system can be solved given initial conditions $y^{(i)}(t_0)$ for $i = 1, \ldots, n$, a time step Δt, and a maximum time T, producing a trajectory that represents the values of each state variable over the interval $[t_0, T]$. An example of a 4-variable ODE system and its trajectory is shown in Fig. 2.

In practice, the first-order Euler method [8] can be used to approximate values over time. The value at the next time step t_{n+1} can be approximated as:

$$y_{n+1} = y_n + h \cdot f(t_n, y_n), \tag{3}$$

where h is the step size, y_n is the value at step n, and $f(t_n, y_n)$ is the derivative at step n. However, the accuracy of the first-order Euler method can be insufficient for systems with complex dynamics, motivating the exploration of finer-grained prediction techniques.

2.3 Genetic Programming for System Identification

System Identification (SI) involves the task of reconstructing a predictive model of a dynamic system [9], typically represented by a set of Ordinary Differential Equations (ODEs), starting from data points obtained from temporal trajectories. Symbolic Regression (SR), a technique capable of reconstructing free-form equations from data [22], is a natural choice for SI. Due to its flexible representation and gradient-free search mechanism, GP has been widely adopted as a symbolic regression method to solve the SI problem [10,24]. Nevertheless, several other techniques have also been applied to SI, ranging from general black-box neural networks [3] to physics-informed neural networks for specific applications [21], and transformer-based approaches that deliver human-readable ODEs [5].

While relatively small ODE benchmarks for SI have been available for a while, a recent publication [5] introduced a thorough system identification benchmark suite, ODEBench, with 63 different systems, along with experimental results for various SI techniques. The introduction of ODEBench now makes it possible to study system identification techniques in a more comprehensive manner.

3 Motivations and Modeling

The motivation for micro-step time-series regression stems from the field of system identification, particularly the analysis of how the interval between data points, Δ_t, influences forecasting accuracy on ODEBench benchmarks with known ground truth ODEs.

3.1 The ODEBench Benchmark Suite

ODEBench [5] is a Python-based benchmark suite designed for system identification of ODE systems. It includes 63 distinct combinations of systems and parameters, covering a variety of ODE configurations: 23 single-variable systems, 28 two-variable systems, 10 three-variable systems, and 2 four-variable systems. For each system, ODEBench provides both a training trajectory and a test trajectory, each generated from different initial conditions over the time interval $[0, 10]$, with uniformly sampled points spaced by Δ_t.

3.2 Influence of Δ_t When Ground Truth is Known

Similar to time-series regression, system identification can be transformed into a standard tabular learning format. A common approach is based on the Forward Euler method [8]. Specifically, given a state variable y, the goal of the Forward Euler method is to learn a symbolic model $F_y(\Delta_t, y)$ that predicts the incremental change in state over Δ_t in a single step, specifically $y_{t+\Delta_t} - y_t$. For a known ground truth ODE $y'(t)$, the model F_y can be expressed as $F_y(\Delta_t, y) = y'(t) \times \Delta_t$, derived from the mathematical relationship: $\frac{\partial F_y(\Delta_t, y)}{\partial \Delta_t}\bigg|_{\Delta_t = 0} = y'(t)$.

Since ODEBench provides the ground truth equations for each system, the ground truth incremental changes $y_{t+1} - y_t$ for each state variable can also be computed. Analyzing these ground truth values and the forecasts made by $F_y(\Delta_t, y)$ using the ground truth $y'(t)$ reveals that some cases yield considerably lower performance, as illustrated in Table 1. However, detailed examination shows that performance improves significantly when the default interval $\Delta_t \approx 0.06$ in ODEBench is reduced to $\Delta_t \approx 0.03$ or $\Delta_t \approx 0.015$. This improvement arises because the system evolves continuously over the interval between

Table 1. Impact of Reduced Discretization Interval (Δ_t) on R^2 Performance Metrics Across Various Systems and State Variables.

System Id	State Variable	Trajectory	R^2 for $\Delta_t \approx 0.06$	R^2 for $\Delta_t \approx 0.03$	R^2 for $\Delta_t \approx 0.015$
11	x_0	1	0.002478	0.769415	0.948090
		2	0.957637	0.990291	0.997697
26	x_0	1	0.531413	0.896984	0.976729
		2	0.838427	0.964003	0.991650
	x_1	1	0.670316	0.926977	0.983374
		2	0.884917	0.974147	0.993958
41	x_0	1	0.835624	0.961387	0.991028
		2	0.849260	0.924593	0.980166
	x_1	1	0.993183	0.998353	0.999599
		2	0.988084	0.997064	0.999260
48	x_0	1	0.897235	0.975271	0.994358
		2	0.976794	0.994000	0.998487
	x_1	1	0.973322	0.993513	0.998514
		2	0.996567	0.999107	0.999775
49	x_0	1	0.946908	0.987823	0.997125
		2	0.854351	0.962177	0.990543
	x_1	1	0.976213	0.994499	0.998698
		2	0.883269	0.968993	0.992211

y_t and $y_t + \Delta_t$. Predicting over a large interval Δ_t may fail to precisely approximate the integral: $\int_t^{t+\Delta_t} y'(t)\,dt$, leading to significant errors. By reducing Δ_t, predicting the next state with $y_t + F_y(\Delta_t, y)$ becomes better aligned with the continuous dynamics of the system, thus resulting in more accurate forecasts.

Based on the experimental results in Table 1, a key conclusion is that a large discretization interval can lead to poor predictive accuracy, even when the ground truth function is known. In such cases, a machine learning model may attempt to fit a complex function to capture all points. However, this complex function often could not represent the true underlying dynamics, leading to poor generalization. More concretely, in the context of time-series regression, if data is sampled at a weekly interval while the ground truth model represents a daily-evolving system, it may be difficult to learn an accurate model based on the weekly samples alone. Even if a model is found that fits all points in the training data, it may generalize poorly to unseen data.

4 Proposed Approach

4.1 Base Learner

In this paper, we aim to enhance traditional ARIMA-based time-series regression by integrating GP constructed features. Each GP individual consists of m GP trees, $\phi_1, \ldots, \phi_m$, where each tree represents a distinct constructed feature derived from the time-series data. These constructed features are applied to an ARIMA model (p, d, q) to make predictions, where p is the autoregressive order, d is the differencing degree for stationarity, and q is the moving average order.

Let $\{y_t\}$ denote the observed time-series at time t. In this work, we employ first-order differencing ($d = 1$) to address non-stationarity, defined as:

$$y'_t = y_t - y_{t-1}. \tag{4}$$

The model is designed to predict the incremental change y'_t rather than the raw next value y_t, allowing it to capture short-term variations more effectively. With GP-constructed features, the ARIMA-based model for predicting y'_t is defined as:

$$y'_t = c + \sum_{i=1}^{m} \beta_i \phi_i(\{y_{t-j} \mid j = 1, \ldots, p\}, \{\epsilon_{t-k} \mid k = 1, \ldots, q\}), \tag{5}$$

where c is a constant, β_i are feature weights, and $\phi_i(\{y_{t-j} \mid j = 1, \ldots, p\}, \{\epsilon_{t-k} \mid k = 1, \ldots, q\})$ represents the constructed features derived from past values y_{t-j} (for $j = 1, \ldots, p$) and past errors ϵ_{t-k} (for $k = 1, \ldots, q$). When predicting multiple steps ahead on unseen data, as in a long-horizon prediction $t + k$, the past errors used in predictions include those derived from future predicted points, such as ϵ_{t+k-1}. Since calculating errors for unseen data is challenging due to the absence of future labels y_{t+k-1}, we set $q = 0$ to omit the moving average term, focusing on autoregressive and differencing terms. To alleviate overfitting, we apply L2 regularization to the coefficients β_i, with a regularization strength of $\lambda = 1$.

4.2 Evolutionary Framework

The proposed algorithm is built upon a conventional multi-tree GP based evolutionary feature construction framework. The framework consists of the following steps:

- Population Initialization: A population of GP individuals is initialized, where each individual comprises multiple GP trees [14,26]. Each tree is initialized using the ramped-half-and-half method. The functional set includes mathematical operators (as detailed in Sect. 5.4), and the terminal set consists of lag features $y_{t-1}, \ldots, y_{t-p}$.
- Fitness Evaluation: For each individual Φ, m features are constructed, and an ARIMA model is trained on these features to forecast the time-series values. Specifically, the data are augmented using a linear interpolation technique, expanding the time-series from T to $\hat{T} = T \times (k + 1)$, where k is the augmentation parameter. The details of this augmentation process are provided in Sect. 4.3. The training loss for each individual Φ on a specific training case t_j, representing a specific time step, is denoted by $L_{i,j}$. Here, $L_{i,j}$ measures the forecasting error of individual Φ_i on t_j in terms of prediction accuracy for that time step. Collectively, these errors form the loss matrix $\mathbf{L} \in \mathbb{R}^{n \times \hat{T}}$, where n is the number of individuals in the population and $\hat{T}$ is the number of training cases, providing a semantic information for selection.

- Parent Selection: Automatic ϵ-Lexicase selection [12] is applied to select individuals based on their diverse performance across training cases, leveraging the loss matrix $\mathbf{L} \in \mathbb{R}^{n \times \tilde{T}}$ from a population with n individuals. Rather than aggregating the time-series forecasting errors, ϵ-Lexicase selection filters individuals sequentially by evaluating each training case in a random order. For a given training case t_j, only individuals meeting the condition $L_{i,j} \leq \min_k L_{k,j} + \epsilon_j$ are retained, where $\min_k L_{k,j}$ is the minimum loss on t_j across all individuals, and ϵ_j is set as the median absolute deviation. This filtering process continues through the sequence of training cases until only one individual remains, which is selected as the parent.
- Offspring Generation: Random subtree crossover and mutation are used to crossover and mutate the selected parents to generate offspring.
- Archive Maintenance: The best individual with the highest fitness is stored in an archive, which will be used for final predictions on unseen data. The archive is used to avoid the lose of the best individual during the evolutionary process.

The steps of fitness evaluations, parent selection, offspring generation, and archive maintenance are repeated iteratively until a stopping criterion is met.

4.3 Training for Micro-step Regression

To implement micro-step time-series regression, we augment the training data to enable the model to learn predict small incremental steps rather than predicting the value at step $t+1$ directly from step t. Specifically, we achieve this by linearly interpolating data points between step t and step $t+1$. Formally, given an augmentation parameter k, the interpolated values are generated as:

$$y_{t+\frac{i}{k+1}} = y_t + \frac{i}{k+1} \cdot (y_{t+1} - y_t), \quad i = 1, 2, \ldots, k. \tag{6}$$

This augmentation results in each time step t producing k interpolated points. For a time-series training set with T steps, the augmented time-series $\tilde{y}$ is thus represented as:

$$\tilde{y} = \{y_1, y_{1+\frac{1}{k+1}}, y_{1+\frac{2}{k+1}}, \ldots, y_2, \ldots, y_T\}. \tag{7}$$

The augmented series $\tilde{y}$ is then utilized in the fitness evaluation stage to compute fitness values, as described in Sect. 4.2.

4.4 Prediction Based on Micro-Step Regression

In micro-step time-series regression, the prediction is made for fractional steps rather than a full step. To forecast a future time step $\hat{y}_{T+h}$, the model first predicts $h \times (k+1)$ fractional increments and then aggregates these increments to obtain the final forecast:

$$\hat{y}_{T+h} = y_T + \sum_{i=1}^{h} \sum_{j=1}^{k+1} \hat{\Delta \tilde{y}}_{T+i,j}, \tag{8}$$

where y_T is the last observed actual value, and $\hat{\Delta y}_{T+i,j}$ represents the predicted incremental value at fractional step j within time step $T+i$.

5 Experimental Settings

5.1 Datasets

We conduct experiments on the M4 dataset [15], a widely used benchmark for time-series forecasting. The M4 dataset includes 414 hourly time-series from diverse domains such as finance, industry, and others. For simplicity, we use the first 100 time-series from these 414 in this paper.

5.2 Baseline Methods

The baseline method is training a GP method on raw features without decomposing the regression task into smaller steps. In addition to GP, the proposed method is also compared against several widely used machine learning algorithms: ElasticNet, Decision Trees (DT), Random Forests (RF), XGBoost, and Support Vector Regression (SVR). We also include ODEFormer [5] for comparison, as it represents a state-of-the-art deep learning approach for system identification. To ensure a fair comparison, all baseline algorithms, except ODEFormer, use lag features and a differencing mechanism consistent with those employed in the GP method. In this setup, ElasticNet can be viewed as a baseline ARIMA model augmented with L1 and L2 regularization. In contrast, ODEFormer is specifically designed to model system dynamics directly without relying on lagged features, avoiding the need to transform data into a tabular format with time-lagged features. Consequently, training R^2 scores are not reported for ODEFormer, and only the test MAPE is provided.

5.3 Evaluation Protocol

For evaluation, the training and test splits provided by M4 [15] are used. The samples are not randomly shuffled, as it is essential to maintain temporal order to avoid using future data to predict past values. To simulate a scenario where the underlying system changing interval is more fine-grained than the sampled interval, data are subsampled by selecting the first data point at each 6-hour interval from the hourly-level data. The six-hour interval is a commonly used setting in both medical [16] and physics domains [7]. In the standard M4 benchmark, the task is to predict the next 48 h. However, after subsampling, this prediction horizon becomes the next 8 data points. To ensure robustness, all experiments are repeated 30 times with different random seeds. A Wilcoxon signed-rank test [23] with a significance level of 0.05 is employed to compare the performance of the proposed method against the baseline.

Table 2. Parameter settings for GP.

Parameter	Value
Maximal Population Size	200
Number of Generations	100
Crossover and Mutation Rates	0.9 and 0.1
Initial Tree Depth	0–3
Maximum Tree Depth	10
Number of Trees	5
Elitism (Number of Individuals)	1
Functions	+, −, *, AQ, Square, Log, Sqrt, Max, Min, Sin, Cos, Abs, Negative

Table 3. Statistical comparison of training R^2 scores across 100 datasets. (Symbols +/∼/- indicate the method in the row performs significantly better, similarly, or worse than the method in the column).

	GP	ElasticNet	DT	RF	XGB	SVR
Micro-GP	73(+)/25(∼)/2(-)	100(+)/0(∼)/0(-)	0(+)/0(∼)/100(-)	0(+)/0(∼)/100(-)	0(+)/0(∼)/100(-)	100(+)/0(∼)/0(-)
GP		100(+)/0(∼)/0(-)	0(+)/0(∼)/100(-)	0(+)/0(∼)/100(-)	0(+)/0(∼)/100(-)	100(+)/0(∼)/0(-)
ElasticNet			0(+)/0(∼)/100(-)	0(+)/0(∼)/100(-)	0(+)/0(∼)/100(-)	100(+)/0(∼)/0(-)
DT				100(+)/0(∼)/0(-)	100(+)/0(∼)/0(-)	100(+)/0(∼)/0(-)
RF					0(+)/0(∼)/100(-)	100(+)/0(∼)/0(-)
XGB						100(+)/0(∼)/0(-)

5.4 Parameter Settings

The parameter settings, shown in Table 2, are common for evolutionary feature construction [27]. To handle division safely, we use the analytical quotient (AQ) function [20], defined as $AQ(x,y) = \frac{x}{\sqrt{y^2+1}}$, which avoids division by zero. Similarly, the logarithmic function is defined as $\log(1 + |x|)$ to prevent undefined values from negative inputs. To reduce runtime, we implement early stopping: if the fitness does not improve over 10 generations, the evolutionary process is terminated. For the ARIMA model, the lag parameter is set to 4, corresponding to a 24-h cycle given a 6-h sampling interval. The augmentation parameter k is set to 1, meaning that each pair of time steps is augmented with one interpolated point.

6 Experimental Results

6.1 Comparison on Training R^2

To evaluate performance on the training data, we use the R^2 metric to assess the model's ability to predict the value at the next time step $t+1$ based on the current time step t, with all time steps belong to the training set. The R^2 metric

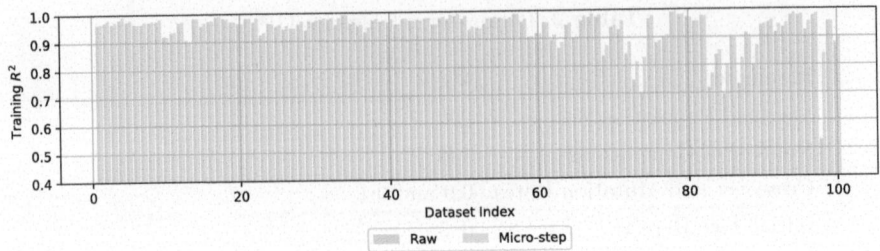

Fig. 3. Training R^2 scores across 100 time-series. Higher values indicate better performance.

is defined as $R^2 = 1 - \frac{\sum_{i=1}^{n}(y_i - \hat{y}_i)^2}{\sum_{i=1}^{n}(y_i - \bar{y})^2}$, where n is the total number of data points, y_i is the actual value at time step i, $\hat{y}_i$ is the predicted value at time step i, and $\bar{y}$ is the mean of the actual values. The median training R^2 scores over the 30 runs across 100 time-series, using both micro-step regression and raw data, are shown in Fig. 3. A statistical comparison between GP and machine learning methods over the 100 time-series is presented in Table 3. The experimental results demonstrate that micro-step regression significantly improves the training R^2 of the evolutionary feature construction method compared with the baseline approach that directly uses the original series. This improvement indicates that data augmentation effectively decomposes the challenging, originally hard-to-fit dataset into easier-to-fit datasets, thus improving training accuracy. When comparing Micro-GP's training R^2 with other machine learning algorithms, results indicate that Micro-GP significantly outperforms ElasticNet across all datasets, underscoring the effectiveness of feature construction over a purely linear model. In contrast, Micro-GP yields lower training R^2 compared to tree-based models, including DT, RF and XGB. This discrepancy arises because tree-based models can precisely memorize the training points by recursively partitioning the feature space until only one point exists in each region. Consequently, these tree-based models achieve superior training performance.

6.2 Comparison on Test MAPEs

For the test data, the objective is not to predict only the next time step $t + 1$ based on the previous time step t. Instead, the goal is to forecast over a horizon that extends beyond the training period. To evaluate model performance over this interval, we use the Mean Absolute Percentage Error (MAPE), which is defined as $\text{MAPE} = \frac{1}{n}\sum_{i=1}^{n}\left|\frac{y_i - \hat{y}_i}{y_i}\right| \times 100\%$, where n is the total number of predictions. The median MAPE over 30 runs across 100 time-series is presented in Fig. 4, along with a statistical comparison in Table 4. These results demonstrate that micro-step regression significantly improves the generalization performance of the evolutionary feature construction method compared to the standard approach applied to the original series. This improvement suggests that micro-step prediction is an effective strategy for enhancing time-series regression

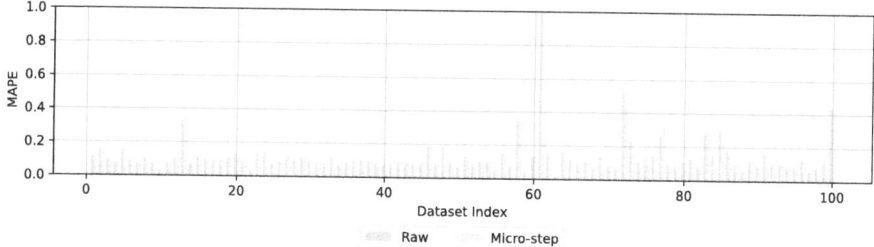

Fig. 4. Test MAPE across 100 series. MAPE values are capped at 1, with lower values indicating better performance. Accumulation of errors in time-series forecasting results in high MAPE for certain datasets.

Table 4. Statistical comparison of test MAPE across 100 datasets. (Symbols +/∼/- indicate the method in the row performs significantly better, similarly, or worse than the method in the column).

	GP	ElasticNet	DT	RF	XGB	SVR	ODEFormer
Micro-GP	21(+)/74(∼)/5(-)	32(+)/44(∼)/24(-)	51(+)/22(∼)/27(-)	35(+)/27(∼)/38(-)	38(+)/25(∼)/37(-)	96(+)/3(∼)/1(-)	98(+)/1(∼)/1(-)
GP		17(+)/48(∼)/35(-)	30(+)/48(∼)/22(-)	20(+)/43(∼)/37(-)	25(+)/34(∼)/41(-)	74(+)/23(∼)/3(-)	76(+)/22(∼)/2(-)
ElasticNet			52(+)/18(∼)/30(-)	48(+)/6(∼)/46(-)	52(+)/0(∼)/48(-)	99(+)/0(∼)/1(-)	100(+)/0(∼)/0(-)
DT				22(+)/16(∼)/62(-)	31(+)/12(∼)/57(-)	93(+)/2(∼)/5(-)	93(+)/2(∼)/5(-)
RF					48(+)/8(∼)/44(-)	96(+)/1(∼)/3(-)	98(+)/1(∼)/1(-)
XGB						94(+)/0(∼)/6(-)	96(+)/0(∼)/4(-)
SVR							53(+)/1(∼)/46(-)

accuracy, particularly when the underlying system exhibits fine-grained dynamics. Comparing the test MAPE results of Micro-GP with other machine learning algorithms, the performance varies across datasets. Specifically, Micro-GP significantly outperforms ElasticNet on 32 datasets but performs worse on 24 datasets. This outcome highlights that while evolutionary feature construction can enhance generalization, its effectiveness depends on the specific dataset and context. A similar pattern emerges in comparisons between Micro-GP and other algorithms: results are mixed across the 100 datasets, indicating that both GP-based and non-GP methods are suited to different types of datasets, and there is no universal algorithm that performs optimally across all datasets. When comparing the proposed method to pretraining-based modeling approaches, the proposed method demonstrates significantly better performance than ODEFormer. ODEFormer utilizes a Transformer-based architecture designed to capture relationships between data and dynamic equations derived from synthetic dynamic systems. However, this pretraining on synthetic data presents challenges when applied to real-world data, such as that in the M4 benchmark, which may exhibit patterns and complexities not present in the synthetic systems. As a result, ODEFormer struggles to generalize effectively to these real-world scenarios, leading to poor performance in the M4 benchmark.

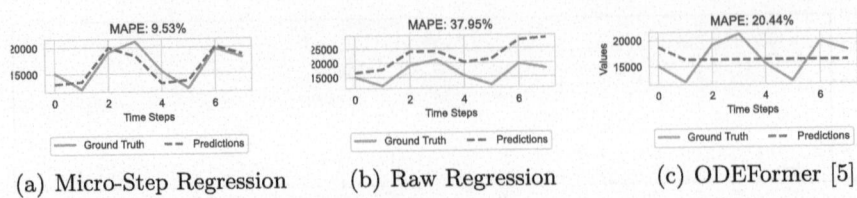

(a) Micro-Step Regression (b) Raw Regression (c) ODEFormer [5]

Fig. 5. Prediction Results for Time-Series 21 in M4 Using Different Regression Techniques.

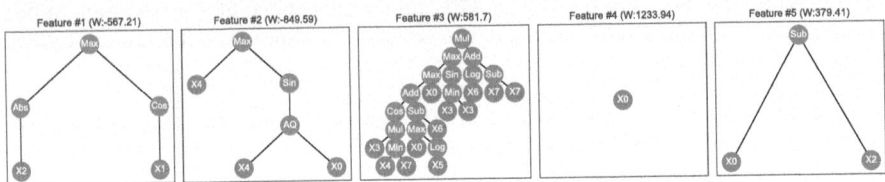

Fig. 6. Examples of Features constructed by GP based on Micro-Step Time-Series Regression.

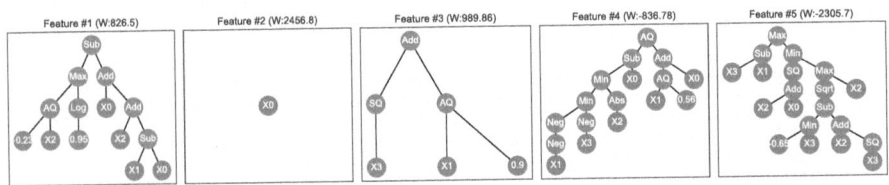

Fig. 7. Examples of Features constructed by GP based on Raw Time-Series Regression.

6.3 Further Analysis

Prediction Plot. To investigate the advantages of Micro-GP for time-series forecasting, we compare prediction results on time-series 21 in M4, shown in Fig. 5(a) and Fig. 5(b). These figures represent the prediction outcomes for models trained with Micro-Step Time-Series Regression and Raw Time-Series Regression, respectively. The model trained on micro-steps demonstrates strong generalization, achieving accurate multi-step predictions on unseen data. In contrast, while the model trained on raw features performs reasonably at the initial prediction step, its accuracy deteriorates significantly from the second step onward, indicating poorer generalization. When compared with ODEFormer, as shown in Fig. 5(c), ODEFormer fails to capture the underlying patterns in the data, highlighting the advantage of the GP-based method for modeling the dynamics of real-world systems.

Example Evolved Programs. To further examine why micro-step learning outperforms raw time-series regression, we analyze the coefficients of the constructed features in Fig. 6 and Fig. 7. Notably, there are no excessively large linear coefficients, suggesting that the observed poor generalization is not due to

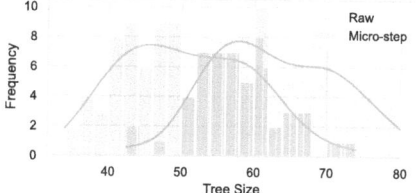
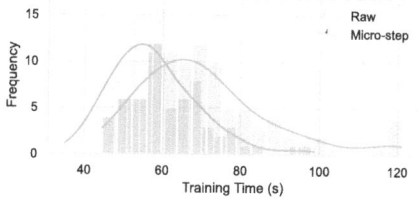

Fig. 8. Distribution of median model sizes across 30 runs on 100 time-series.

Fig. 9. Distribution of median training time across 30 runs on 100 time-series.

ill-conditioned coefficients. The primary distinction between models lies in the complexity of the constructed features. Specifically, features constructed from raw time-series data tend to be more complex than those constructed from micro-step data. This suggests that, to capture dynamics over larger time steps, GP have evolved more intricate features, which may lead to overfitting on the training data. In contrast, when learning on smaller steps, changes are more gradual, enabling GP to evolve simpler, more interpretable features that fit the data effectively. Consequently, these simpler features are less prone to overfitting, enhancing the model's generalization on unseen data.

6.4 Comparisons on Model Size and Training Time

Model Size (Number of Nodes). As analyzed, decomposing the problem into smaller steps can simplify the time-series regression task, leading to better training and generalization performance. A comparison of model sizes between these two strategies is presented in Fig. 8. As shown, the evolved trees on micro-step data are smaller than those on raw data, indicating that learning GP features on micro-step data is more efficient and results in more compact models. By contrast, learning directly from raw data requires more complex GP trees to capture changes over larger time steps, which significantly increases the search space and complicates the learning process. Therefore, from a search efficiency perspective, the micro-step time-series regression technique offers more advantageous.

Training Time. Regarding training time, a comparison using micro-step data versus raw data is presented in Fig. 9. The results indicate that employing micro-step data does not significantly increase training time, even with the doubling of training data due to augmentation. This is likely due to the reduction in model complexity achieved through micro-step regression. While the amount of training data increases, the less complex GP trees resulting from micro-step regression offset the additional computational complexity, leading to only a modest increase in training time.

7 Conclusions

In this paper, we identified that large step sizes adversely affect performance in dynamic system identification tasks, even when the ground truth ODEs are known. To mitigate this issue, we proposed a micro-step time-series regression technique that augments training data with interpolated points between observed time steps. This approach allows the model to predict smaller, incremental changes rather than large, discrete jumps, effectively decomposing the regression task into simpler sub-problems. This decomposition allows GP to evolve simpler yet effective features that capture meaningful patterns in the data, leading to improved accuracy in time-series regression and the evolution of more compact models. Experimental results demonstrate that the proposed method significantly enhances generalization performance compared to models trained directly on raw time-series data, achieving superior results in over 21 of the 100 evaluated series. Moreover, the method outperformed pre-trained dynamic system modeling approaches, such as ODEFormer, highlighting the advantages of evolutionary feature construction for time-series regression. For future work, a promising direction is to develop a method for automatically determining the optimal step size in micro-step time-series regression, which could enhance the adaptability and applicability of the proposed method to real-world problems.

References

1. Box, G.E., Jenkins, G.M., Reinsel, G.C., Ljung, G.M.: Time Series Analysis: Forecasting and Control. Wiley (2015)
2. Brown, R.: Statistical Forecasting for Inventory Control, vol. 2, pp. 443–473. McGraw-Hill (1959)
3. Chen, T.Q., Rubanova, Y., Bettencourt, J., Duvenaud, D.: Neural ordinary differential equations. In: Bengio, S., Wallach, H.M., Larochelle, H., Grauman, K., Cesa-Bianchi, N., Garnett, R. (eds.) Advances in Neural Information Processing Systems 31: Annual Conference on Neural Information Processing Systems 2018, NeurIPS 2018, 3–8 December 2018, Montréal, Canada, pp. 6572–6583 (2018)
4. Cramer, S., Kampouridis, M., Freitas, A.A., Alexandridis, A.: Stochastic model genetic programming: deriving pricing equations for rainfall weather derivatives. Swarm Evol. Comput. **46**, 184–200 (2019)
5. d'Ascoli, S., Becker, S., Schwaller, P., Mathis, A., Kilbertus, N.: ODEFormer: symbolic regression of dynamical systems with transformers. In: The Twelfth International Conference on Learning Representations, ICLR 2024, Vienna, Austria, 7–11 May 2024. OpenReview.net (2024)
6. Fanjiang, Y.Y., Syu, Y., Huang, W.L.: Time series QoS forecasting for web services using multi-predictor-based genetic programming. IEEE Trans. Serv. Comput. **15**(3), 1423–1435 (2020)

7. Galtieri, J., Reno, M.J.: Intelligent sampling of periods for reduced computational time of time series analysis of PV impacts on the distribution system. In: 2017 IEEE 44th Photovoltaic Specialist Conference (PVSC), pp. 2975–2980. IEEE (2017)
8. Gaucel, S., Keijzer, M., Lutton, E., Tonda, A.: Learning dynamical systems using standard symbolic regression. In: Lecture Notes in Computer Science, pp. 25–36. Springer, Heidelberg (2014)
9. Iba, H., deGaris, H., Sato, T.: A numerical approach to genetic programming for system identification. Evol. Comput. **3**(4), 417–452 (1995)
10. Kronberger, G., Kammerer, L., Kommenda, M.: Identification of dynamical systems using symbolic regression. In: Computer Aided Systems Theory–EUROCAST 2019: 17th International Conference, Las Palmas de Gran Canaria, Spain, 17–22 February 2019, Revised Selected Papers, Part I 17, pp. 370–377. Springer (2020)
11. Kuranga, C., Pillay, N.: Genetic programming-based regression for temporal data. Genet. Program Evolvable Mach. **22**(3), 297–324 (2021). https://doi.org/10.1007/s10710-021-09404-w
12. La Cava, W., Helmuth, T., Spector, L., Moore, J.H.: A probabilistic and multi-objective analysis of lexicase selection and ε-lexicase selection. Evol. Comput. **27**(3), 377–402 (2019)
13. La Cava, W., Moore, J.H.: Learning feature spaces for regression with genetic programming. Genet. Program Evolvable Mach. **21**(3), 433–467 (2020). https://doi.org/10.1007/s10710-020-09383-4
14. La Cava, W., Silva, S., Danai, K., Spector, L., Vanneschi, L., Moore, J.H.: Multidimensional genetic programming for multiclass classification. Swarm Evol. Comput. **44**, 260–272 (2019)
15. Makridakis, S., Spiliotis, E., Assimakopoulos, V.: The M4 competition: results, findings, conclusion and way forward. Int. J. Forecast. **34**(4), 802–808 (2018)
16. Matowe, L.K., Leister, C.A., Crivera, C., Korth-Bradley, J.M.: Interrupted time series analysis in clinical research. Ann. Pharmacother. **37**(7–8), 1110–1116 (2003)
17. Mehr, A.D., Gandomi, A.H.: MSGP-LASSO: an improved multi-stage genetic programming model for streamflow prediction. Inf. Sci. **561**, 181–195 (2021)
18. Murari, A., Peluso, E., Spolladore, L., Rossi, R., Gelfusa, M.: Upgrades of genetic programming for data-driven modeling of time series. Evol. Comput. **31**(4), 401–432 (2023)
19. Murphy, J., Desell, T.: Minimizing the EXA-GP graph-based genetic programming algorithm for interpretable time series forecasting. In: Proceedings of the Genetic and Evolutionary Computation Conference Companion, pp. 1686–1690 (2024)
20. Ni, J., Drieberg, R.H., Rockett, P.I.: The use of an analytic quotient operator in genetic programming. IEEE Trans. Evol. Comput. **17**(1), 146–152 (2012)
21. Raissi, M., Perdikaris, P., Karniadakis, G.: Physics-informed neural networks: a deep learning framework for solving forward and inverse problems involving nonlinear partial differential equations. J. Comput. Phys. **378**, 686–707 (2019)
22. Schmidt, M., Lipson, H.: Distilling free-form natural laws from experimental data. Science **324**(5923), 81–85 (2009). https://doi.org/10.1126/science.1165893
23. Virgolin, M., Alderliesten, T., Witteveen, C., Bosman, P.A.: Improving model-based genetic programming for symbolic regression of small expressions. Evol. Comput. **29**(2), 211–237 (2021)
24. Winkler, S., Affenzeller, M., Wagner, S., Kronberger, G., Kommenda, M.: Using genetic programming in nonlinear model identification. In: Identification for Automotive Systems, pp. 89–109 (2012)

25. Zhang, H., Chen, Q., Xue, B., Banzhaf, W., Zhang, M.: Modular multi-tree genetic programming for evolutionary feature construction for regression. IEEE Trans. Evol. Comput. (2023)
26. Zhang, H., Chen, Q., Xue, B., Banzhaf, W., Zhang, M.: A semantic-based hoist mutation operator for evolutionary feature construction in regression. IEEE Trans. Evol. Comput. (2023)
27. Zhang, H., Zhou, A., Chen, Q., Xue, B., Zhang, M.: SR-forest: a genetic programming based heterogeneous ensemble learning method. IEEE Trans. Evol. Comput. (2023)

Correction to: Introducing Crossover in SLIM-GSGP

Gloria Pietropolli, Davide Farinati, Luca Manzoni, Mauro Castelli, Sara Silva, and Leonardo Vanneschi

Correction to:
Chapter 7 in: B. Xue et al. (Eds.): *Genetic Programming*, **LNCS15609, https://doi.org/10.1007/978-3-031-89991-1_7**

The book was published with a typo in the author affiliation of Chapter 7 in this book.

The updated version of this chapter can be found at
 https://doi.org/10.1007/978-3-031-89991-1_7

© The Author(s), under exclusive license to Springer Nature Switzerland AG 2025
B. Xue et al. (Eds.): EuroGP 2025, LNCS 15609, p. C1, 2025.
https://doi.org/10.1007/978-3-031-89991-1_16

Author Index

A
Abdallah, Zahraa S.   1
Alharthi, Khulud   1

B
Bakurov, Illya   68
Banzhaf, Wolfgang   68
Bonin, Lorenzo   120
Briesch, Martin   207

C
Carlet, Claude   18
Castelli, Mauro   103
Chen, Qi   240
Clark, David   85
Cui, Henning   139
Cussat-Blanc, Sylvain   173

D
De La Torre, Camilo   173
De Lorenzo, Andrea   120
Doquet, Guillaume   190
Đumić, Mateja   156
Đurasević, Marko   18, 156

F
Farinati, Davide   35, 103, 120
Franchet, Camille   173

G
Geiger, Alina   207
Gil-Gala, Francisco Javier   156

H
Hähner, Jörg   139
Hauert, Sabine   1
Heider, Michael   139
Hurta, Martin   52

J
Jakobović, Domagoj   18, 156

K
Kianinejad, Marzieh   68
Kocherovsky, Mark   68

L
Langdon, William B.   85
Lavinas, Yuri   173
Luga, Hervé   173
Lutton, Evelyne   240

M
Manzoni, Luca   103, 120
Mariot, Luca   18
Medvet, Eric   224
Mrazek, Vojtech   52

N
Nadizar, Giorgia   173, 224

O
Ochoa, Gabriela   224
Ovesna, Anna   52

P
Picek, Stjepan   18
Pietropolli, Gloria   35, 103, 120

R
Rothlauf, Franz   207
Rovito, Luigi   120
Rusin, Francesco   224

S
Schwob, Robin   173
Sekanina, Lukas   52

© The Editor(s) (if applicable) and The Author(s), under exclusive license to Springer Nature Switzerland AG 2025
B. Xue et al. (Eds.): EuroGP 2025, LNCS 15609, pp. 257–258, 2025.
https://doi.org/10.1007/978-3-031-89991-1

Silva, Sara 103
Sobania, Dominik 207

T
Tonda, Alberto 240
Trautwein, Julian 139

V
Vanneschi, Leonardo 35, 103, 120

W
Wilson, Dennis 173

X
Xue, Bing 240

Z
Zhang, Hengzhe 240
Zhang, Mengjie 240

The manufacturer's authorised representative in the EU is Springer Nature Customer Service Centre GmbH, Europaplatz 3, 69115 Heidelberg, Germany. If you have any concerns regarding our products, please contact ProductSafety@springernature.com

Printed and bound by CPI Group (UK) Ltd, Croydon, CR0 4YY

26/03/2026

02078935-0010